Coastal Research Library

Volume 16

The aim of this book series is to disseminate information to the coastal research community. The Series covers all aspects of coastal research including but not limited to relevant aspects of geological sciences, biology (incl. ecology and coastal marine ecosystems), geomorphology (physical geography), climate, littoral oceanography, coastal hydraulics, environmental (resource) management, engineering, and remote sensing. Policy, coastal law, and relevant issues such as conflict resolution and risk management would also be covered by the Series. The scope of the Series is broad and with a unique cross-disciplinary nature. The Series would tend to focus on topics that are of current interest and which carry some import as opposed to traditional titles that are esoteric and non-controversial. Monographs as well as contributed volumes are welcomed.

More information about this series at http://www.springer.com/series/8795

Anja M. Scheffers • Dieter H. Kelletat

Lakes of the World with Google Earth

Understanding our Environment

Springer

Anja M. Scheffers
Department of Geoscience
Southern Cross University
Lismore, Australia

Dieter H. Kelletat
Department of Geography and Education
University of Cologne
Köln, Germany

Every effort has been made to contact the copyright holders of the figures and tables which have been reproduced from other sources. Anyone who has not been properly credited is requested to contact the publishers, so that due acknowledgment may be made in subsequent editions.

ISSN 2211-0577 ISSN 2211-0585 (electronic)
Coastal Research Library
ISBN 978-3-319-29615-9 ISBN 978-3-319-29617-3 (eBook)
DOI 10.1007/978-3-319-29617-3

Library of Congress Control Number: 2016934460

Printed on acid-free paper

This Springer imprint is published by Springer Nature
The registered company is Springer International Publishing AG Switzerland

Preface

Before we discuss lakes, we will briefly touch upon the general principles of water.

Water exists on Earth in different forms: as a liquid, frozen in snow, hail or ice, and as vapour. Water can be found in oceans and lakes, flowing in rivers, as ground water, liquid or frozen in permafrost areas, as glaciers and as sea ice near the surface or far below (age-old), as moisture in the soil, in the biosphere, or as one of the elements in different kinds of rock. The volume of water stored in these various reservoirs, however, is very different (Fig. 1.1).

Water is the most valuable resource to our planet. As important as the vast oceans covering approximately 360 Mio km^2 are the more than 300 Mio lakes that occur on all continents, a number by far too large to imagine (Downing et al. 2006). The terrestrial environment covers an area around 150 Mio km^2 whereby the average statistical size to one of these many lakes (altogether with max. 0.6 % of all water for the continents) is calculated to approximately 0.5 km^2, an area of around 700 × 700 m. As 16 of the largest lakes on Earth reach a cumulative size of 1 Mio km^2, it becomes clear that more than 99 % of terrestrial lakes are much smaller than their statistical average of 0.5 km^2.

To form a lake (pond or swamp), a closed depression/basin to store water is needed. These formations can be created by endogenic processes—triggered by forces from Earth's interior like tectonics or volcanic forces—or from active processes at the surface that in many cases occur close to ocean levels along coastlines of the world. The filling of a basin/depression can be from freshwater (being groundwater, precipitation, snow and ice melting, and river run off), or by salt water (becoming salty by higher rates of evaporation than discharge into the basin), and some salt lakes transform into saltpans (and vice versa). Lakes maybe very old and persist over a wide range of geological epochs and climates, or short lived or seasonal or just appearing for geological moments as days or weeks. Additionally, shallow lakes may develop into swamps, bogs, or other kinds of wetlands. For all kinds of lakes, ponds, or wetlands, however, water is needed.

In landscapes enriched with many lakes (e.g. Finland, Alaska, Canada, northern Russia), their distribution and in particular their forms and morphologic patterns are almost impossible to identify from the surface. For this reason we will again (after dealing with coastlines of the world and the forms on continents in two earlier books[1]) take the aspect through space via satellite imagery (from Google Earth programs) to get impressions on the variety of Earth's lakes.

Lakes as bodies of water (fresh or saline) in natural depressions may be extremely large (like the Caspian Sea covering over 420,000 km^2), but also very small and miniscule. These bodies of water, in particular left from former glaciations or occurring in permafrost landscapes of the Arctic, maybe so small (and often shallow) that the term lake may be regarded as an euphemism and "pond" maybe the more correct name for them. However, there is no scientific definition for ponds. Because of their small size, they often are overlooked in studies, and as they number much higher than of lakes and are zones of high ecological significance for

[1]Scheffers AM, Scheffers SR, Kelletat DH (2012) The coastlines of the world with Google Earth: understanding our environment. Coastal Research Library, Dordrecht, Springer.

Scheffers AM, May SM, Kelletat, DH (2015) Landforms of the world with Google Earth: understanding our environment, Dordrecht, Springer.

many organisms (insects, amphibians, migrating birds), ponds derive our interest. Their actual numbers have never been counted and they tend to appear or disappear within a very short time. These bodies of water often show a transition into wetlands as swamps or bogs and/or may dry out periodically or episodically depending on the conditions of climate and rainfall in the catchment, the availability of groundwater, alternatively the pond may freeze to the ground in the cold period of each year. Potentially, as they will occur again as lakes, these features are also included in this book.

Special chapters are devoted to saline (or salt) lakes, and—in their evaporated forms—to saltpans.

From the many aspects of lake science including water budgets, temperature regimes, mixing types, biology/ecology, and chemistry (to mention a few), we shall concentrate on the genesis of lakes and other closed forms containing water, moisture/swamps, or minerals. The organisation of the book follows different forms of lake origin (often connected to the question of age) such as extra-terrestrial meteor impacts, structural depressions by tectonic activity, patterning of joints or faults, volcanic origin e.g. in craters, or the forming influence of glacier ice, subterranean permafrost, littoral processes, running water (rivers), wind (aeolian dune landscapes or by deflation), and solution of rocks (karst forms).

An important aspect to the character of depressions as sedimentary archives identifies former conditions of landscapes including climate change. From this we learn that in former times under different climates, lakes may have existed where today there is an extreme shortage of water (as in deserts) or as giant lakes once dammed by Ice Age glaciers. These and other aspects demonstrate how vast the science of lakes is and a book with only several hundred figures can never adequately cover all of these subjects. More information on the genesis of the world's coastlines and its terrestrial forms based on satellite images from the Google Earth program are in two previous books.

Although on a scientific background, we will use language clear to everyone and in particular use a high number of figures (satellite images and other figures) whereby visual impressions can convey far more information than words. Indeed this is more a picture book than a scientific novel! The text is simply to give framework to the different aspects of lakes concerning their distribution, forms, and forming processes. Only a few from the many thousands of publications are given in a list of references (for further reading not necessarily discussed in the text) for those who are inquiring on more aspects of lakes. The captions present the geographical co-ordinates and an indication to its scale, plus more data if available. As service to our readers, we incorporated a large amount of numerical data and in particular the captions (latitudes, longitudes, length and area of lakes, volume of water, average/max depth and more). The given data is often approximate, as many regions are not investigated in any detail and data is taken from the technical instruments of Google Earth program (or older) and more recent sources and references may offer differing data, explained by errors due to the rapid change of lakes, swamps, and pans.

Although it may seem peculiar to write a book on only 0.4 % of the world's water, there are over 300 million lakes and as the eyes of the continents are extremely diversified: from majestic to miniscule, brilliant clear to muddy, deep to shallow, found high in mountains or near the coast, fashioned by rivers and glaciers, formed by solution of ground rock, be it permafrost decay or even by cosmic impacts. We believe it worthwhile to present this high diversity in the larger number of images, taken mostly from a point high above the Earth's surface, as from satellites in the Google Earth program.

The authors have numerous decades of experience through their fieldwork in all continents on the landscapes and climates of the globe, with a strong connection to the forming processes of water, e.g. along the coastlines of the world. Their expertise is geomorphology and landscape evolution, climate change, the period of the Ice Ages (the Pleistocene epoch of the last 2.4 Mio years), as well as the chronology of evolutionary steps via the analyses of sediments and different techniques of numerical dating. The authors hope this short overview on the lakes of the world will encourage the reader to develop their knowledge on continental waters, the most valuable resource for the future of humankind.

Mullumbimby, NSW, Australia — Anja M. Scheffers
Mülheim an der Ruhr, Germany — Dieter H. Kelletat

About Google Earth

Virtual, web-based globes such as Google Earth, NASA World Wind, or Microsoft Virtual Earth allow all of us to become travellers visiting the most remote places and tour our planet or even outer space at speeds faster than a rocket. Any computer user can easily, at no charge, download and use Google Earth (for both PC and Mac computers).

If you have not done so already, download Google Earth, (the new version) or Google Earth Pro (higher resolution and more features), from earth.google.com, install it on your computer, and prepare yourself to fly around the globe on your own research expedition. You can travel to millions of locations and look for the context of all landscape features of interest to you (geography, geology, vegetation, man-made structures, and more). You can also see these objects from different altitudes (i.e. in different scales), perspectives, and directions; you can view a chosen area around 360° from an imaginary point in the air; and you can fly deep into canyons and craters. You can look straight down in a traditional 2D perspective or enable an oblique view in 3D, you can hover above one location, circle around, or fly like a bird over countries, continents and oceans. In this book we focus on geologic and geographic features, but that is only a snapshot of what Google Earth is providing with their virtual globe. There is no room here for a complete tutorial, but you will find that the program is so easy to use and understand that you will be an expert after working with it for a few minutes. Please visit the Google Earth web page for a complete free Google Earth tutorial that is constantly updated to reflect the improvements in different versions of Google Earth (http://earth.google.com/support/bin/answer.py?hl=enandanswer=176576).

We hope that the diversity of the landforms of the world will come alive for you and stimulate your curiosity to become an explorer of these fascinating places either as a hobby or profession.

Acknowledgements

This book would not have been possible without the support provided by our editors at Springer Publishing, Petra van Steenbergen and Hermine Vloemans. We would personally like to thank them for their consistent cooperation and full support.

Dr. Charlie Finkl – thank you for considering this volume for the Coastal Research Library.

Very kindest regards to Anne Hager (editing, graphics), Frank Schmidt-Kelletat (graphics) and Stan Kinis (language) for their time, assistance and guidance. Their support was vital to this book.

To friends along the way..., thank you for listening and learning together.

And finally, to our extended, ever supportive, loving families. We love you all.

Mullumbimby, NSW, Australia — Anja M. Scheffers
Mülheim an der Ruhr, Germany — Dieter H. Kelletat
October 2015

Contents

Introduction: General Approach, and a Short Look at the Diversity of Lakes

1

Abstract

Although continental water represents only a very small percentage of all our global water, it occurs in a multitude of different facets and whether hidden under the surface as groundwater, in the lithosphere, the biosphere or the atmosphere, a larger part of all continental water is transformed into snow and glacier ice, and only a small part can be found in rivers and lastly in lakes, which is the focus of this book. The number of lakes exceeding 300 Mio is beyond our imagination. After presenting quantitative data on the largest lakes, the simple shape of lakes is our first objective. Frequently it offers a first insight into the origin of a lake or at least toward the origin of the basin in which it lies. Clusters of lakes comprise more information on the environment and the processes under which large groups of lakes (often with the same size or contours) are formed. The colour appearance of a lake helps in identification of surface conditions, the water depth, content and concentration of suspension materials, or solutions such as salt. In addition, tiny organisms may give a lake its characteristic colour, particularly if the water is saline or if only a salty pan remains. The latter can often be found in lakes that have no through-flow but are endorheic (ending in a basin without outflow). Following, we present succinct chapters on groups of formations with special characteristics such as former Ice Age lakes that stored a huge amount of meltwater for a short period of geological time, and sub-glacial lakes as they occur in Antarctica and ephemeral meltwater ponds on glacier ice. The main focus will then follow the genetic types of lakes, combined with the origin of the lake basins or topographic depressions.

The science of water and its distribution on the Earth's surface is *hydrology*, separated into *oceanography* and *limnology*, the latter concerning the water on land. These water bodies generally occur in two different manners: flowing/moving as in rivers (previously summarized under the term *potamology*), and not moving as in lakes (and related features). However movement is also possible in large lakes, by waves and wind-induced currents or flows powered by temperature changes and differences in density (triggered by temperature variances and/or those of salinity). Extensive tests and long-time registrations have shown that a wind event of Bft. 4 (36 km/h) is able to mix a lake's water to a depth of 6–12 m and after many hours set the lake in motion with a velocity of 630 m/h or 16 km/day (Schwoerbel and Brendelberger 2005).

Although we focus on the different morphologic and genetic types of lakes, ponds, swamps and pans, there are many other aspects that make each lake a specific ecosystem (i.e. defining conditions to the development of life). Just to mention the most important ones, this model is

- the relation of size to depth, determining the kind and intensity of mixture
- the temperature for adaption of atmospheric gases including oxygen for the organisms, dissolved inorganic and organic matter, often determined by sediment/rock and water input from the wider environment
- the type of organisms and their population, lastly biomass production

A.M. Scheffers, D.H. Kelletat, *Lakes of the World with Google Earth*, Coastal Research Library 16,
DOI 10.1007/978-3-319-29617-3_1

- the food chain, the relation of plankton to nekton
- the benthic organisms and sedimentation

A lake is any body of slowly moving or standing water that occupies an inland basin of appreciable size. Definitions that accurately classify lakes, ponds, swamps, and even rivers and other bodies of non-oceanic water are yet not well established. Nonetheless, it can be said, that rivers and streams are relatively fast moving, marshes and swamps contain notably large quantities of grasses, trees or shrubs, and ponds are quite small in comparison to lakes. In terms of geological time spans, lakes are temporary bodies of water and like glaciers, are short living or adept to strong changes.

The total area of all lakes covers 1.8 % of the entire landmass on Earth (about 2.5 Mio km^2) and the total water volume of all lakes is approximately 275,000 km^3 of which at least 40 % is saline. Ultimately however, lakes barely contain 0.26 % of all continental fresh water (Fig. 1.1), an extremely small amount compared to the vast amount of continental ice stored in Antarctica, Greenland and in high mountain regions of the World.

According to calculations made by an international team of scientists (Downing et al. 2006), the world has more than 300 million lakes and includes all the natural lakes but excludes man-made reservoirs such as dams. This extraordinary total measures lakes a minimum of 0.1 ha (1000 m^2) in area (Fig. 1.2). Less than 10 % of all lakes (~27 Mio) exceed one hectare with only 16 exceeding 10,000 km^2 (Table 1.1). The U.S. state of Alaska alone accounts for more than three million lakes with surface areas greater than 8 ha (80,000 m^2 or a quadrat of approximately 300 m sides). Of all larger lakes, approximately 60 % are located in Canada covering an impressive 7.6 % of its landscape (compared to all lakes covering only 1.8 % of the world's landmass) and the deepest in Canada is the Great Slave Lake of 614 m depth, even though its water level is only 157 m asl.

As explained above, usually lakes are young geologic features spanning only several thousands of years that occasionally reach 10 times that figure and yet compared to the age of

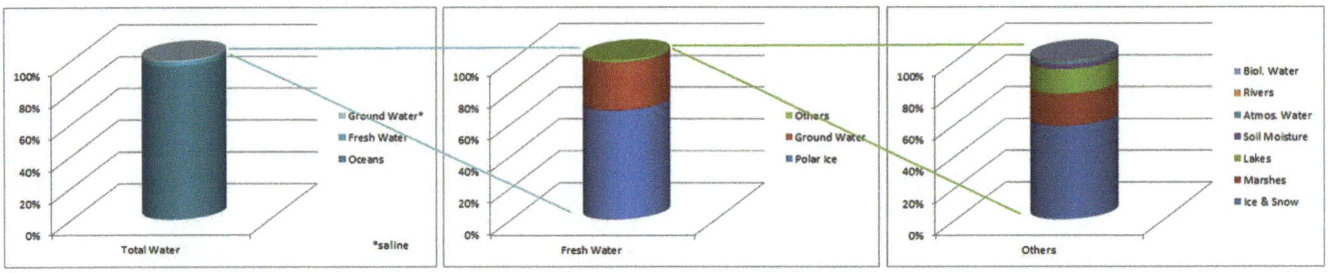

Fig. 1.1 Distribution of Global Water: Total Water (100 %) constituting: 96.5 % in oceans, 2.5 % as fresh water, and 0.93 % as saline groundwater. Fresh water (2.5 % of all water) is composed: 68.6 % polar ice, 30.1 % groundwater, and all other types combined 1.3 %. This 1.3 % of fresh water constitutes: Ice and snow outside the polar regions 0.93 %, marshes 0.33 %, lakes 0.26 %, soil moisture 0.047 %, atmospheric water 0.037 %, rivers 0.006 %, and biological water only 0.003 % (Data from Alberta Environment, Graphics: Frank Schmidt-Kelletat)

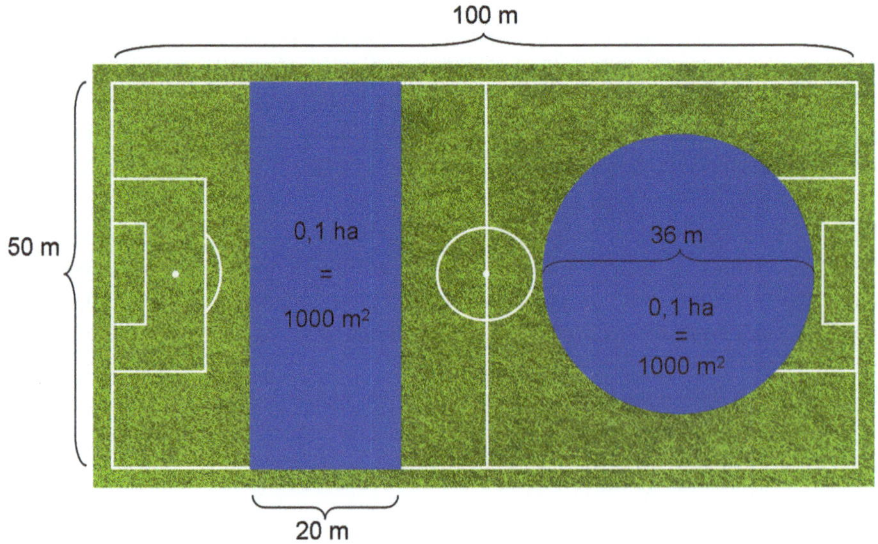

Fig. 1.2 Image to demonstrate the size of a 0.1 ha (1000 m^2) lake compared to a soccer field (100 × 50 m): it covers a rectangle 20 m wide and 50 m long or a circle with a diameter of 36 m (Graphics: Frank Schmidt-Kelletat)

Table 1.1 Number and surface area of lakes at least 500 km^2 large on the continents

Continent	No. of large lakes	Surface area km^2
Asia	61	559,000
Africa	23	197,000
Europe	25	61,000
North America	122	486,000
South America	14	46,000
Oceania/Australia	8	23,000
Total	253*	1,372,000

Source: http://www.endoh7735.com/lakes/english/world_lakes.htm [Accessed 06 May 2015]

other continental geomorphological features, this is a relatively short time. One extreme exception is Lake Baikal in Russia, many millions of years old (some sources document over 20 Mio years) lies in a tectonic rift (graben) where its age can also be detected from the 7 km thick benthic sediments. Again located in rift zones, extreme ages can also be found in the narrow and long lakes of Lake Tanganyika and Lake Malawi/Nyassa in east Africa.

A special group of lakes are hidden from sight, located in caves or under glaciers and in particular under the Antarctic ice masses. Natural lakes (ponds or swamps) mostly have simple contours resembling circles, ovals or long forms, whereas artificial lakes, e.g. valleys closed by a dam, exhibit extremely complex shorelines (Fig. 1.3a, b).

Although there is millions of natural lakes on Earth, each require special investigation and explanation since lakes depend on a special constellation, which is rare to our planet—forming a closed depression needs special parameters as it will only occur if material has been transported against gravity. The distribution of lakes on Earth is also very different and not governed by climates only (less common in dry regions). By far most lakes occur in regions formerly covered by inland glacier ice or modern permafrost and are at higher latitudes or altitudes with normal to higher precipitation and reduced evaporation making water available.

1.1 Examples of Large Continental Lakes

The large number and genetic variety of lakes and their extremely different history and fate is the focus of this book and we will only briefly touch upon the largest lakes on Earth that are reviewed in many books and shorter scientific publications owing to facts on them can be found more easily than for the remaining 99 % of lakes.

The distribution of large lakes on the different continents depends primarily on the climatic zones in which these continents lie. The humid zones of the inner tropics and the temperate to cool latitudes have high precipitation rates and are best suited to fill any depression with water, regardless of the endogenic and exogenic processes which shaped these basins. A concise table (Table 1.1) presents several figures explaining these facts.

The Caspian Sea (422,000 km^2) is by far the world's largest inland sea but partly saline, so much so that the combined surface areas of the Great Lakes of USA and Canada (242,000 km^2) represent the largest fresh water reservoir of the World. By volume, Lake Baikal in Russia reaches a maximum depth of 1741 m and is the largest fresh water mass on earth with ~23,000 km^3 (compared to Caspian Sea ~79,000 km^3; see also Figs. 1.4, 1.5, 1.6 and Table 1.2). The second largest body of freshwater is Lake Tanganyika in eastern Africa (approximately 19,000 km^3 and max. depth 1435 m) and following Lake Superior the largest of the Great Lakes in North America with 12,100 km^3. Although Lake Victoria in eastern Africa is third largest in area (69,485 km^2), it is only tenth in volume (2760 km^3) owing to a maximum depth of only 85 m.

The Caspian Sea (Fig. 1.6) is located in a vast and flat landscape between Asia and Europe. It borders Iran, Kazakhstan, Turkmenistan, Azerbaijan and Russia, being the largest inland body of water and biggest salt-water lake in the world. Stretching 1200 km north to south with an average width of 320 km east to west, the overall shoreline reaches 7000 km. The Caspian Sea (422,000 km^2) represents 14 % of the total area of lakes in the world (2.7 million km^2) and approximately 40–44 % (79,000 km^3) of the total volume of the world's lacustrine waters. Over 130 rivers flow into the Caspian, of which the rivers Volga, Ural and Terek from the north contribute 88 % to the total inflow. This inflow guarantees fresh water to the north where the southern region of the Caspian has less inflow and water is saline (mean salinity ~1 %). During extreme arid climatic periods, the sea evaporated and left a wide plain of salty sediments (halite), which have partly been carried out by winds limiting the agricultural use of the surrounding land.

Although the northern part of the Caspian Sea is very shallow with depths of only 5–6 m, the southern part reaches depths of more than 1000 m, forming a wide crypto-depression (the base is below the ocean level). Located close to the wide inlands of Asia with its very cold winters of the

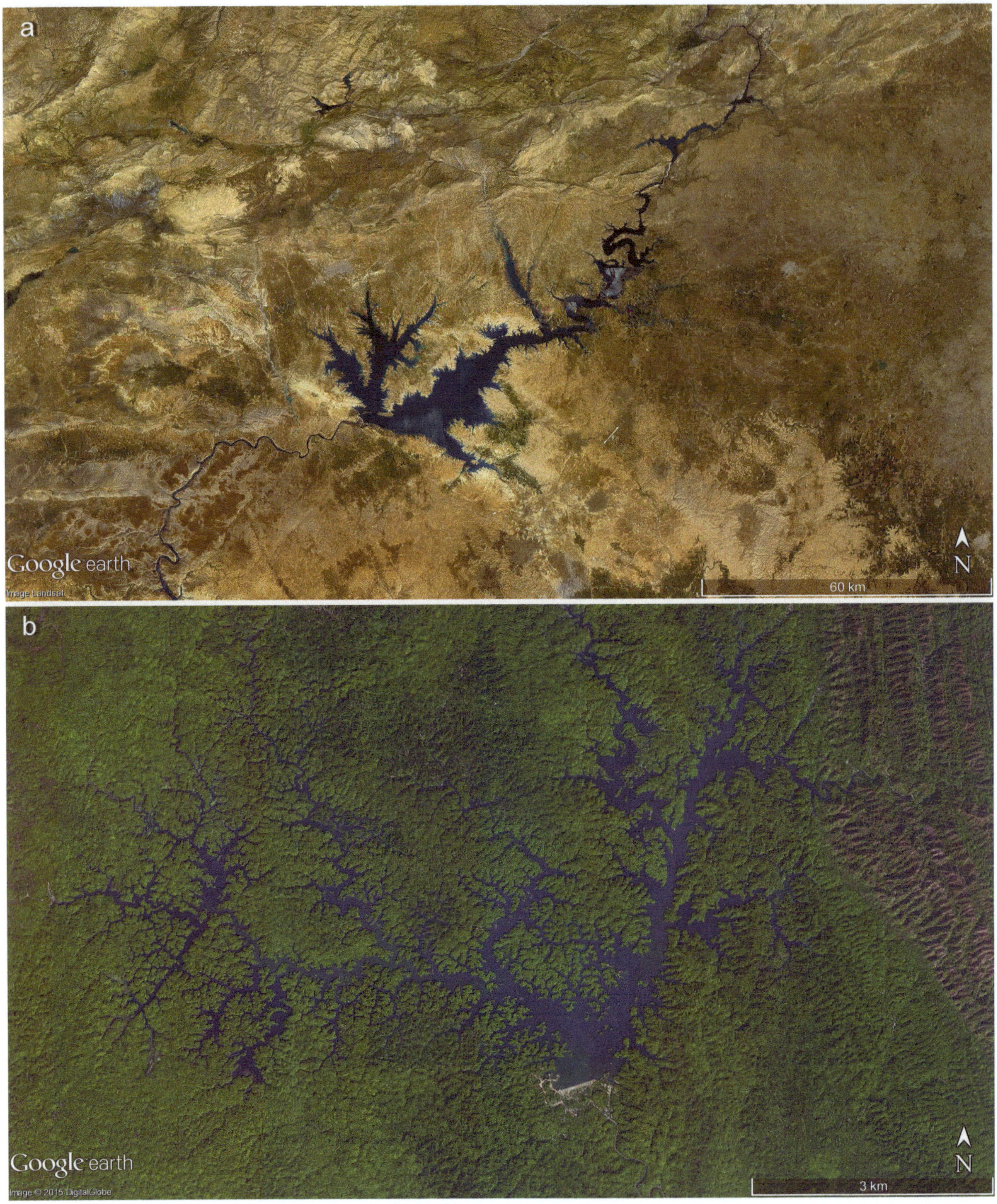

Fig. 1.3 Substantially incised and complex shorelines are typical for rising water and drowning of a fluvial landscape, as is along artificial reservoirs. (**a**): Atatürk Reservoir in Turkey at *~37°07'N, 38°37'E.* Scene is 100 km wide. (**b**) Reservoir in Myanmar, constructed 2011 AD, at. Scene is 12 km wide (Images credits: ©Google earth 2015)

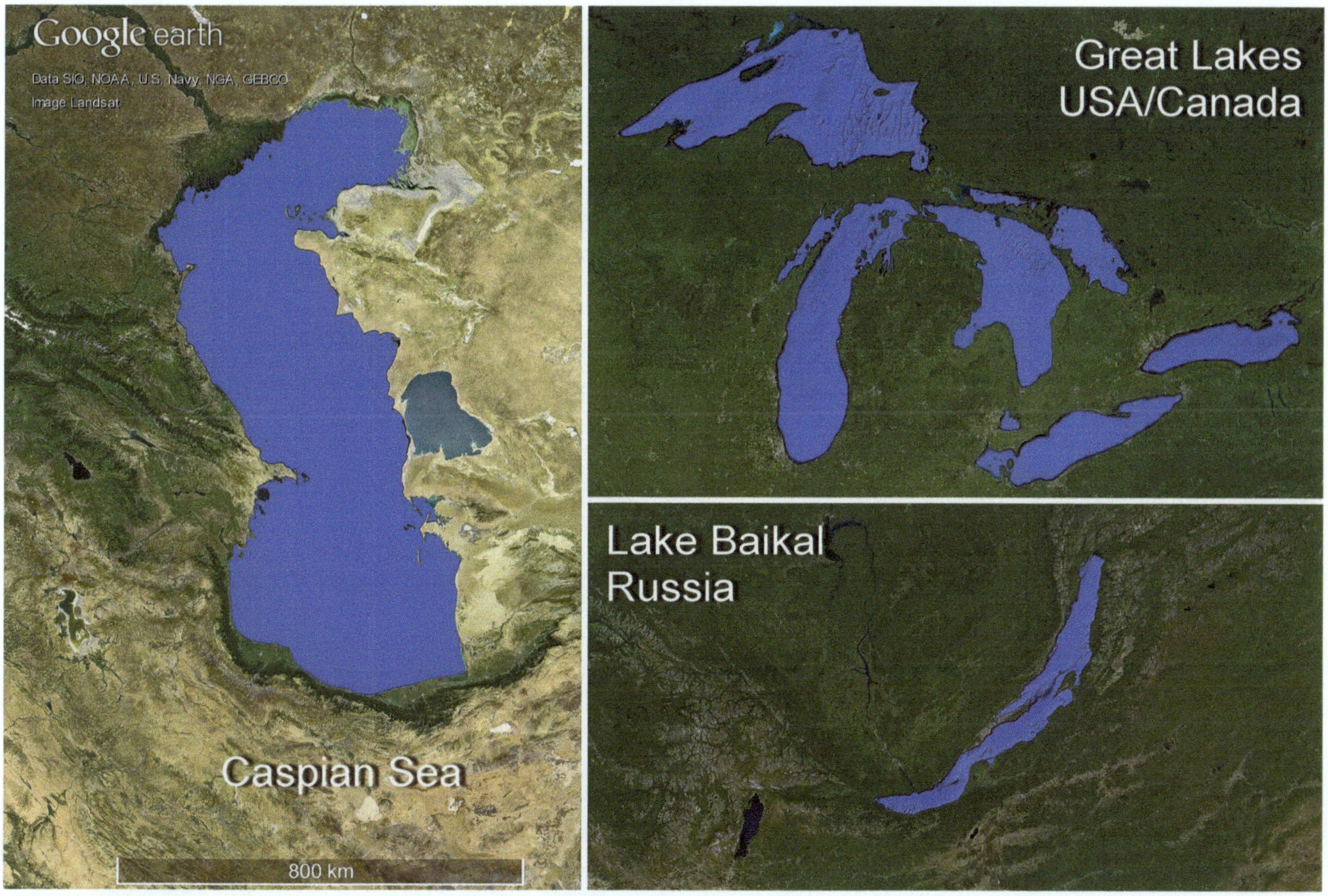

Fig. 1.4 Comparison between the Caspian Sea, the Great Lakes of Canada/USA, and Lake Baikal (Russia) presented in the same scale (Image credit: ©Google earth 2015)

north, the Caspian Sea normally freezes in winter, whereas ice cover is rare in the south because of high salinity and vertical mixing of deep water, storing the warmth of high summer radiation. Because the Volga River is the major contributor of water to the Caspian Sea (~80 %), its discharge determines the lake level that currently sits 28 m below sea level and alternates from year to year (within the last decades) to about 3 m. One cause for low lake levels is the deflection of Wolga waters used in wide irrigation practices of cotton fields in Russia and Kazachstan.

Lake Baikal (Fig. 1.7) in Southern Siberia of Russia (max. depth 1741 m and 1186 m below sea level) approaches 700 km in length and is the largest freshwater body in the world containing more water than all North American lakes combined and represents 20 % of all the world's fresh lake waters (approximately 23,000 km^3). As its shape reveals, Lake Baikal fills an ancient rift valley first formed 25 million years ago, making it the oldest lake in the world. In fact, Lake Baikal is not only gradually expanding by rift processes spreading at a rate near 2 cm/year, but also the depth of the lake is being reduced from over 300 rivers filling it with sediments whereby the long periods of deposition have created a thick 7 km sediment benthos. The broad expansiveness of the lake from west to east allows for strong winds to mix the waters to great depth and create a niche for life in those deep zones that are void in other deep waters as in the Caspian Sea.

On the footings of existing structural depressions, the Great Lakes of the USA have been carved out by Pleistocene glaciers (Figs. 1.8 and 1.9) and collectively have a little less freshwater than Lake Baikal alone. The Great Lakes basins were filled by ice of the last Ice Age until 12,000 years ago, and the form of a tongue-like glacier lobe still can be seen in the shape of Lake Michigan. The level to the Great Lakes—interconnected by east flowing river channels—descends from Lake Superior (183 m asl) to Lake Ontario of (75 m asl), the highest variance between Lake Erie and Lake Ontario (Niagara Falls being 55 m in height). All the Great Lakes send their water to the northern Atlantic Ocean via St. Lawrence River and with the exception of Lake Erie, the depth of all the Great Lakes reaches below sea level.

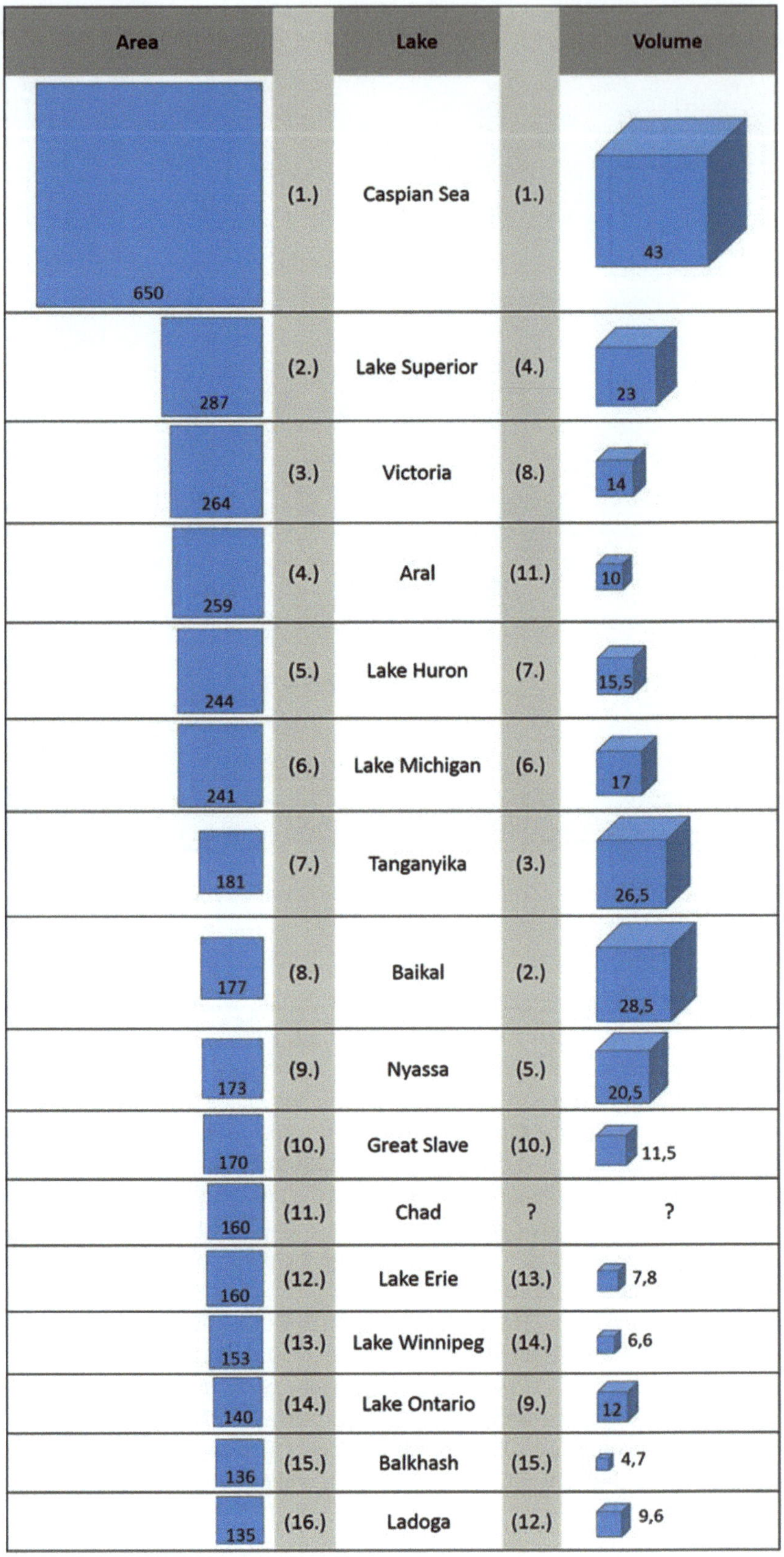

Fig. 1.5 *Left*: The 16 largest lakes in the world, ordered by size (surface area) and transformed into quadrats with side-lines in km (compare Ruttner 1963; in Burgess and Morris 1987). *Right:* Volume of the 16 largest lakes on the continents oriented by surface size and visualized by cubes. The largest cube representing the Caspian Sea has a volume with 43 km of edge lengths (43 × 43 × 43 km), Lake Baikal would have edge lengths of 28.5 km. The smallest (by volume) of the 16 shown Lake Balkhash represents a water-cube of 4.7 km of edge lengths (Graphics: Frank Schmidt-Kelletat)

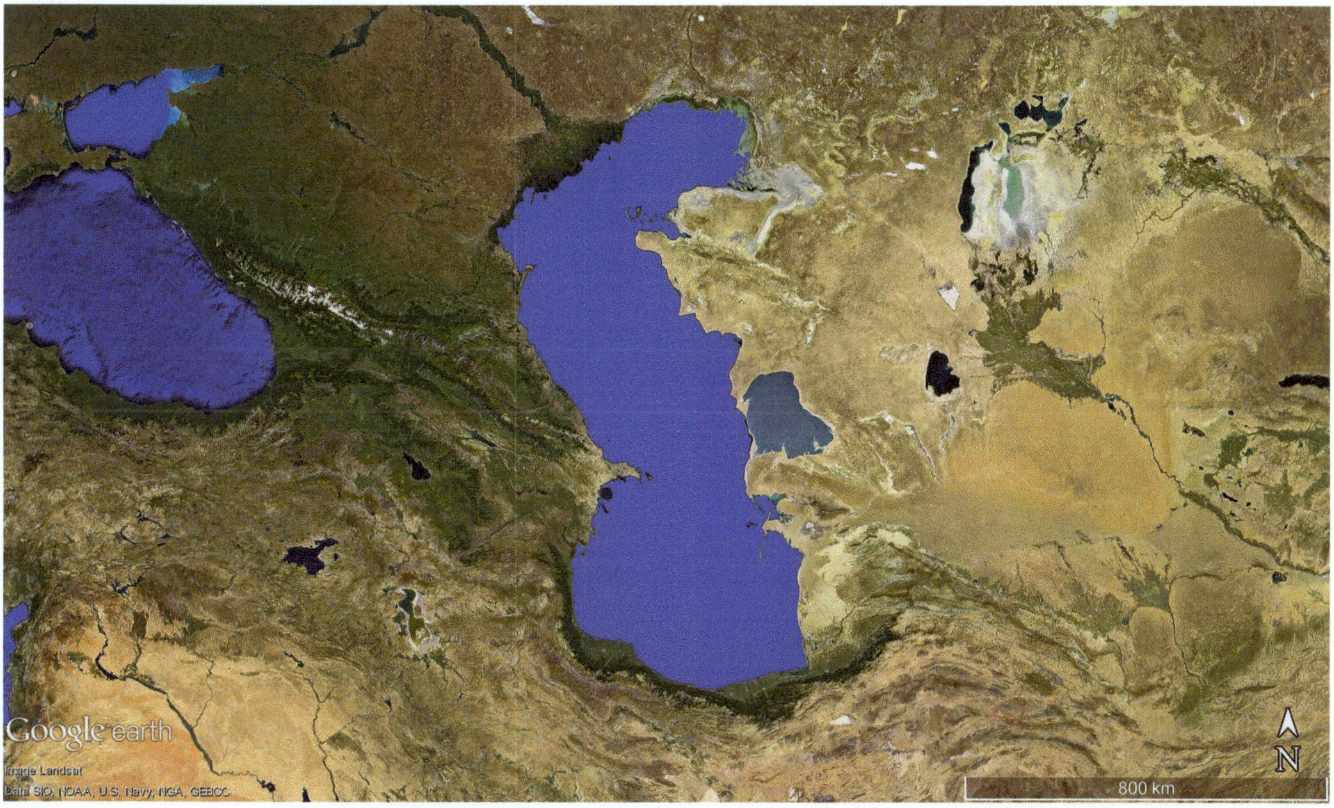

Fig. 1.6 The Caspian Sea is a remnant of an ancient part of the Mediterranean via the Black Sea and until about 11 Mio years ago the Caspian Sea was partly situated on an oceanic crust and connected with other parts of the world's oceans (Image credit: ©Google earth 2013)

Table 1.2 The largest lakes on the continents (exceeding 10,000 km^2)

	Area	Volume	Equivalent to a cube of	Mean depth	Maximum depth	Maximum length
Name	km^2	km^3	km	m	m	km
Caspian Sea	422,000	79,000	43.0	187	1072	1125
Superior	82,414	12,100	28.5	149	406	580
Victoria	69,485	2760	26.5	40	85	400
(Aral)	(66,000)	(1064)	23.0	(16)	(69)	(355)
Huron	59,596	3540	20.5	59	229	370
Michigan	58,016	4918	17.0	85	282	517
Tanganyika	32,893	18,900	15.5	570	1435	640
Baikal	31,500	23,615	14.0	730	1741	607
Nyasa/Malawi	30,044	8400	12.0	292	706	558
Great Slave	28,930	1580	11.5	41	614	477
(Chad)	(25,760)	?	?	?	(7)	230
Erie	25,719	484	10.0	19	64	390
Winnipeg	23,553	294	9.6	12	62	412
Ontario	19,477	1639	7.8	86	244	300
Balkhash	18,428	106	6.6	58	27	470
Ladoga	18,130	908	4.7	51	225	188

Data compiled from different sources

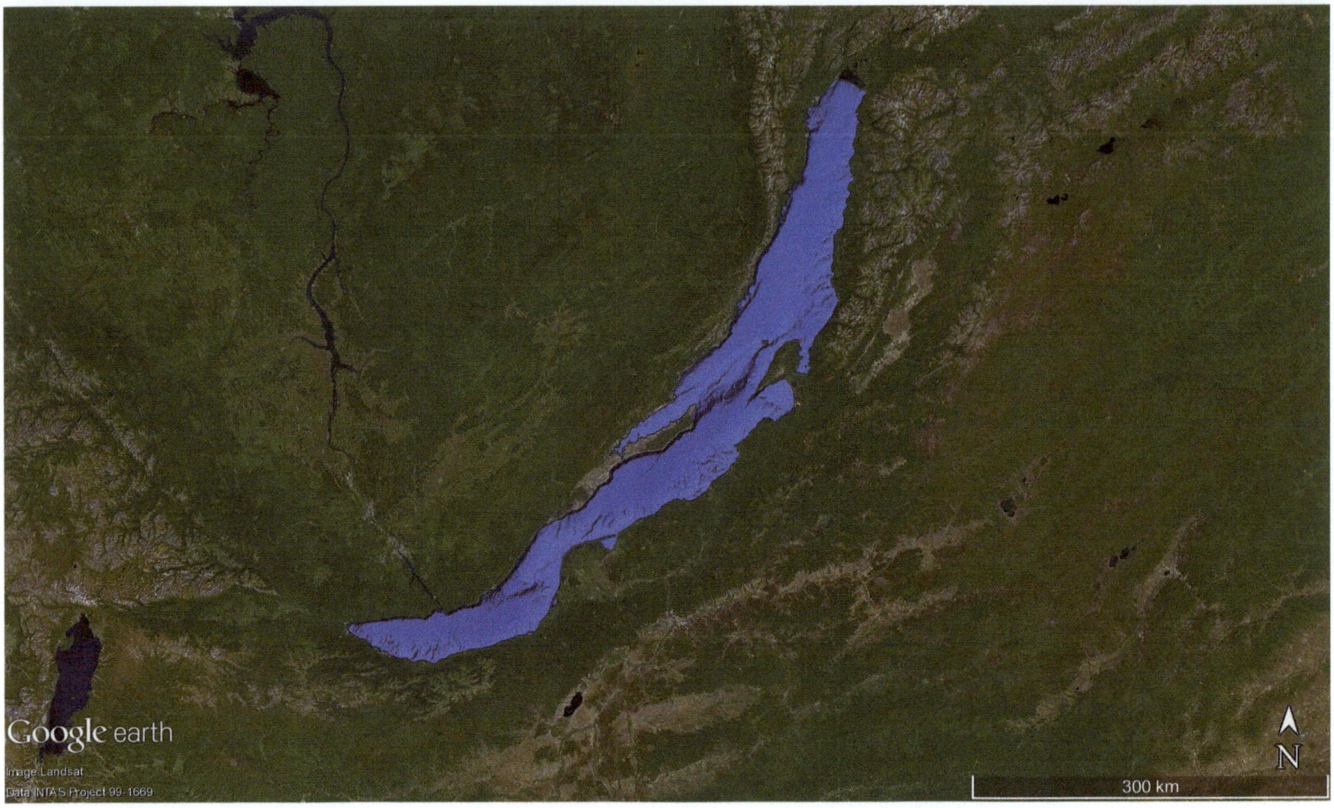

Fig. 1.7 Lake Baikal (Russia) is the deepest and oldest lake in the World with a total length of approximately 700 km, situated in a rift valley (Image credit: ©Google earth 2013)

Some figures show the dissimilarities to the five "Great Lakes":

	Lake Superior	Lake Huron	Lake Michigan	Lake Erie	Lake Ontario
Surface area [km^2]	82,000	60,000	58,000	25,660	19,000
Water volume [km^3]	12,100	3540	4918	484	1639
Mean depth [m]	149	59	85	19	86
Maximum depth [m]	406	229	282	65	244

The total surface area of the five large lakes is 242,000 km^2, and the total volume is about 22,000 km^3. From the five main lakes, Lake Superior (Fig. 1.9) as its name suggests is the largest and is more than 600 km in length running from east to west, and reaches 250 km widths from north to south. With 12,000 km^3 discharge of nearly 200 rivers, Lake Superior stores more than half of the water of all five lakes.

Lake Victoria (Fig. 1.10a, b) is the largest lake in Africa, and the second largest (by surface area) freshwater lake in the world (second only to Lake Superior). Situated in Central Africa, Lake Victoria borders Uganda, Tanzania and Kenya. Formed in a basin flanked by uplifted shoulders of two major east African rift valleys (Fig. 1.10b), Lake Victoria comprises an area of 70,000 km^2 and stretches north to south for 335 km and 240 km from east to west with a shoreline longer than 3220 km. The lake lies at an altitude of 1134 m and because of an average depth of only 40 m, and during former dryer climates (end of the Pleistocene glaciation about 14,500 years ago) the lakebed was dry. Several terraces located at elevations of 3 m, 13 m and 19 m along the lake document different fillings of the basin. Understanding lake level changes can be identified through analysis of its benthic sediments that point to weathering conditions and vegetation cover in the wider environment.

Lake Aral (Figs. 1.11 and 1.12) was initially open water in the 1960s stretching 435 km north to south and 290 km from east to west. Having an average depth of 16 m, Lake Aral covered 64,000 km^2 then making it the fourth largest lake in the world. Decades later and the lake has receded up to 90 % caused by water taken from its main tributaries Amu

Fig. 1.8 Great Lakes at the border of Canada and USA. Scene is 1600 km wide! (Image credit: ©Google earth 2013)

Fig. 1.9 Lake Superior, the largest of the Great Lakes at the border of Canada and USA drains into Lake Huron in the SE corner of the image (Image credit: ©Google earth 2013)

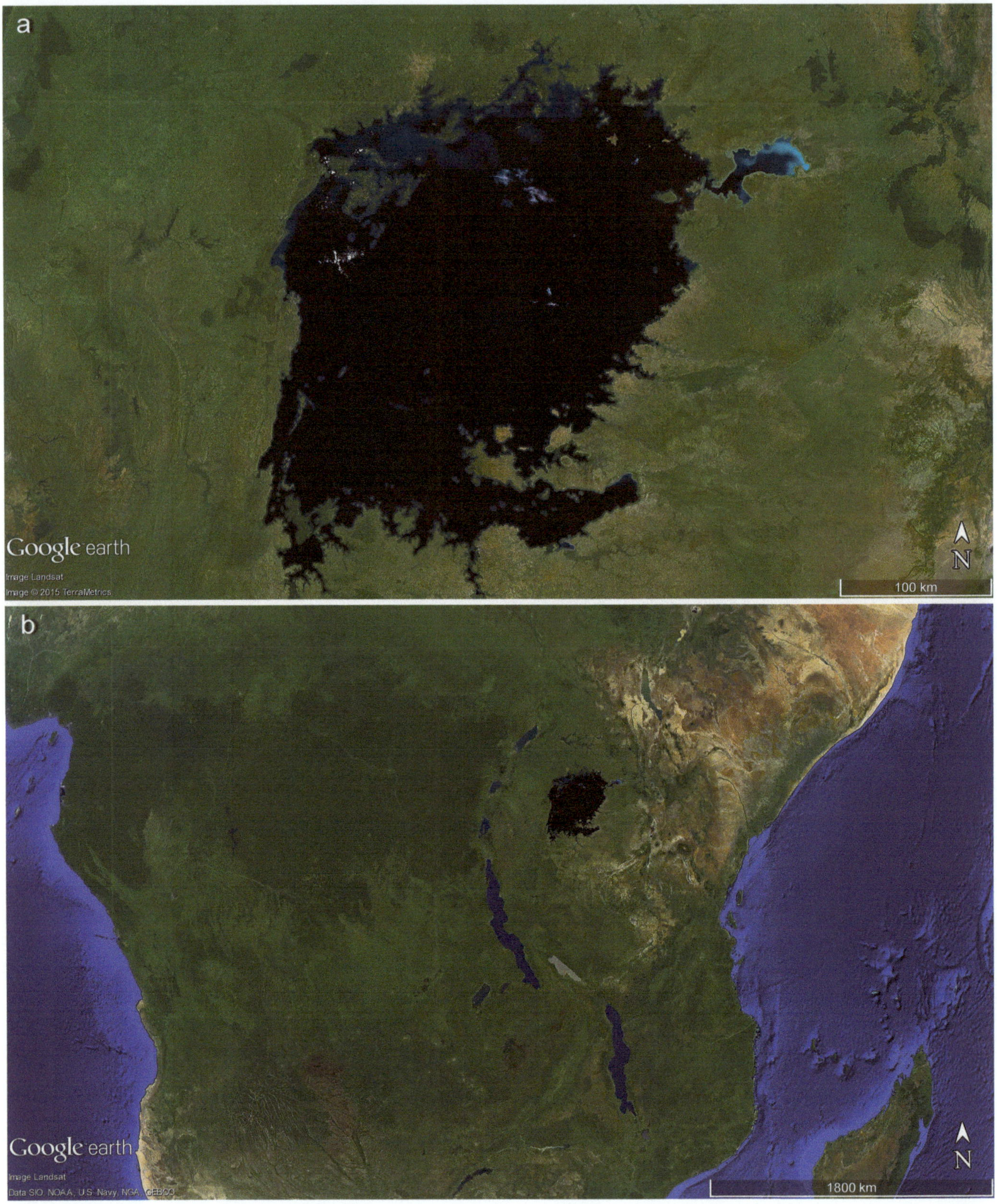

Fig. 1.10 (**a**) Lake Victoria, eastern Africa, with a coastline of 3220 km and a surface area approximately 70,000 km^2 (**b**) Lake Victoria is situated in a shallow depression between the uplifted shoulders of two deep East African graben systems (rift valleys) (Image credit: ©Google earth 2013)

Fig. 1.11 Lake Aral, at the border of Kazakhstan in the north and Usbekistan to the south. Image is late 1973 when water level was high. Width of scene is 620 km (Image credit: ©Google earth 2013)

Darya and Syr Darya for intensive irrigation required for cotton agriculture in the desert.

By 2005 Lake Aral fragmented into the North Aral Sea and the two basins of the South Aral Sea. In 2009 the east basin had become waterless but was reflooded for short periods from winter snow and glacier meltwater. The drying of the seabed created sandy salt flats extending for thousands of km^2, delivering salt by wind in all directions and killing most of the surrounding vegetation. It is expected Lake Aral will fluctuate into the future, mostly regarding surface area and much less its total water content.

Lake Chad (Figs. 1.13 and 1.14) on the southern fringe of Sahara desert is another example of a lake drying out, most probably a result of climatic change rather than anthropogenic impacts.

Lake Chad (Fig. 1.14) remnant of the former inland sea palaeo-lake Mega-Chad had its largest extension around 7000 years ago. In its time, Lake Mega-Chad was the largest of four Saharan palaeo-lakes and is estimated to have covered an area of 400,000 km^2 nearly as large as Caspian Sea today. Until 1960s, Lake Chad still belonged to the four largest lakes of Africa and reached more than 26,000 km^2. A slightly dryer climate during the 1970s coupled with a rising demand for water by local populations has since reduced the lake considerably.

The extreme size changes of Lake Chad can be measured through satellite images. In 1966 the lake covered 22,000 km^2 and just 7 years later barely extended 10,000 km^2 regressing to only 500 km^2 in 1982, down from 2270 km^2 in 1983. In 1994, a satellite measured surface area exceeding 1700 km^2, partly replenished by floods from the southern equatorial rains. With a mean depth of only 1.5 m (and a volume of 72 km^3) presents extreme size influences caused by evaporation processes of the hot sub-Saharan environment.

The shrinkage of Lake Chad is the result of climate change with increased evaporation, intensified by the misuse of water and overgrazing. Still, a large region of the lake remains in a style of wetlands.

The largest lake on the South American continent Lake Titicaca (Fig. 1.15) approaching 8400 km^2 is at the border of Peru and Bolivia and although in a moderately dry climatic environment, the lake contains fresh water. Lake Titicaca is by far the largest lake nestled at the exceptional altitude of 3810 m asl on the Altiplano of the central Andes. Altiplano describes a "high plain", and in fact over 6000 m high Andean mountain chains and volcanoes rise from the Altiplano that sits higher than most summits of the European Alps, or the Rocky Mountains and Sierra Nevada of USA. The next high-elevation lake in size to Lake Titicaca is Qinghai Lake (3198 m asl) in northern China (Tibet, around *36°53'N* and

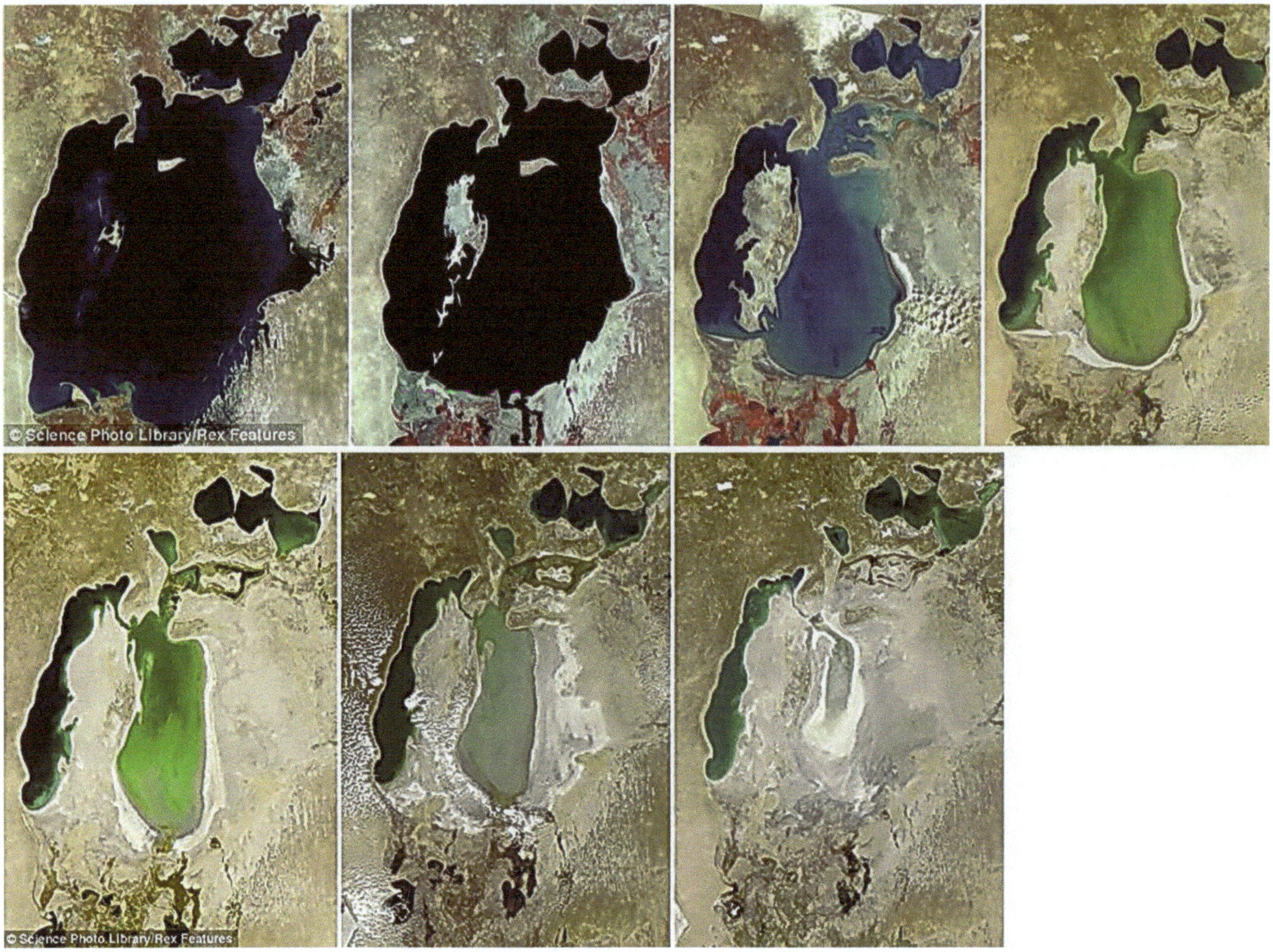

Fig. 1.12 Landsat satellite photographs of the Aral Sea from 1973 to 2009. Since the 1960s the Aral Sea has lost more than half of its volume and these images clearly demonstrate the size decrease in the past 40 years. The first image *(top left)* taken in 1973 is followed from left to right *(top row)*: 1987, 1999 and 2001. The bottom row shows photos from 2004, 2007 and 2009. By 2010, the size has declined to only 10 % of its original area. The shrinkage of Aral Sea is due to the overuse of the feeder rivers Syr Darya and Amu Darya to irrigate cotton and paddy fields. The previous depth of 68 m has been reduced to below 10 m and between 1993 and 2008 the level dropped by 8 m! Several former harbours are now located 150 km from water and lie within a salt desert. The Aral lake disaster remains as one of the largest man-made catastrophes of the world within the last 100 years (Credit: http://www.dailymail.co.uk/sciencetech/article-2,177,202/The-changing-face-Earth-Dramatic-high-resolution-satellite-images-world-changed-decades.html#ixzz2RO6G2M81 [Accessed 07 Aug 2015])

100°11'E). Qinghai Lake (the largest inland water of China) contains salt water and covers more than 5000 km^2 with a length of 108 km (particularly shrinking in recent decades). Due to its limited depth (max. 27 m, mean 21 m) the volume of the lake accounts for only 120 km^3 being only 15 % than that of Lake Titicaca. The largest high-elevation lake in Asia, however, is Yssykköl (translates to Yssyk Lake) north of the Tarim Basin in NE Kirgistan (China, around *42°26'N* and *77°16'E*). Yssykköl is surrounded by the Altai Mountains and lays 1603 m asl covering an area of 6200 km^2 with a capacity of 1738 km^3 (because of depths reaching 668 m). Yssykköl has a salt content of about 0.6 % and although its volume is 2.5 times that of Lake Titicaca, the latter remains the largest freshwater body in extreme altitudes.

1.2 Different Shape of Lakes: From Simple to Complex

The shape of lakes can give first impressions on their hydrological and ecological character: the surface area of a lake compared with the length of its shoreline may explain the influence of sediments and nutrients delivered to the lake, and the extension of ecologic zones (those along the shore and in shallow water with aquatic vegetation and living space for other organisms). The maximum length of a lake gives considerations on the maximum potential fetch a wind may strike over a lake to form waves (and mix the waters). The mean depth values for lakes evaluate light zones and what extent of deep zones lack light for photosynthetic processes.

Fig. 1.13 Lake Chad in 1963 covering an area of approximately 26,000 km² at the border of Chad, Niger, Nigeria and Cameroun. Width of scene is 400 km. In 2005 the area receded to only 1350 km², by 95 %! (Image credit: ©Google earth 2013)

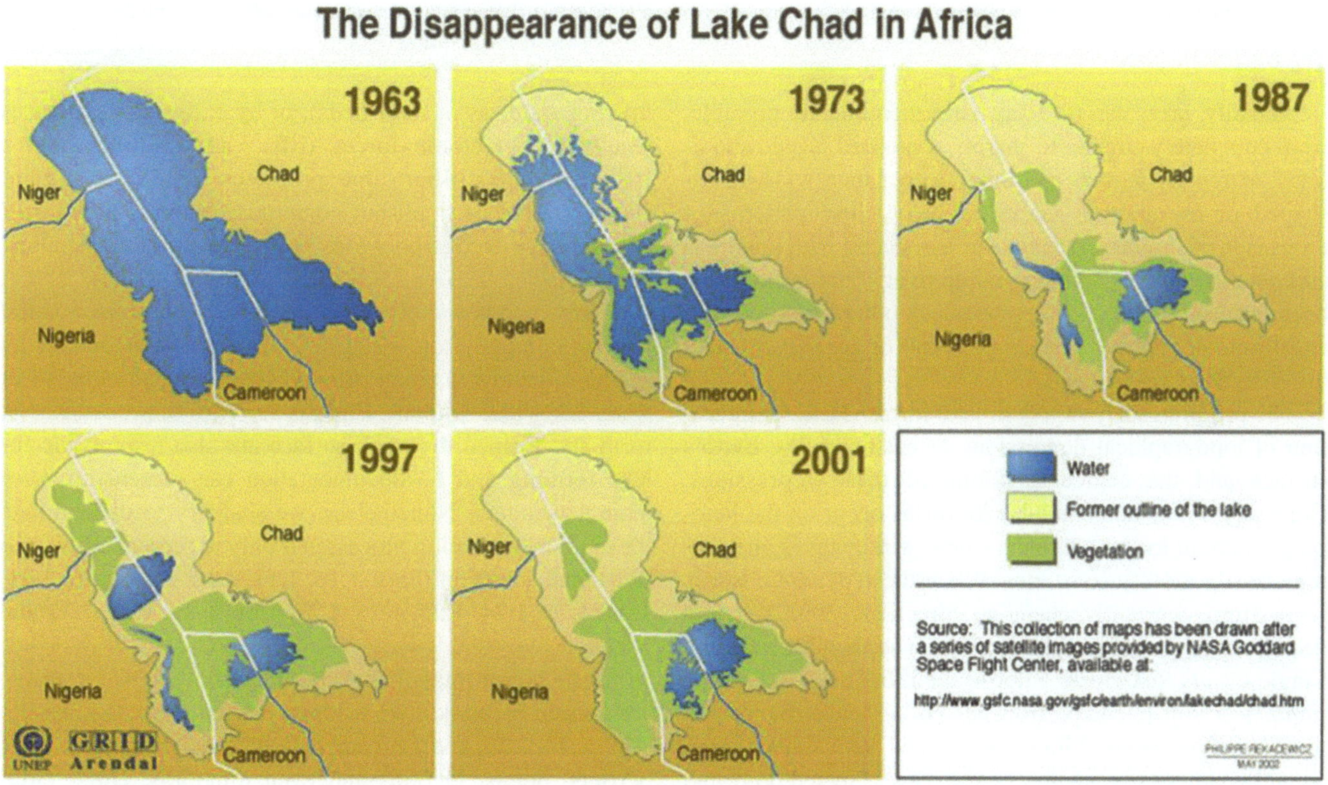

Fig. 1.14 The disappearance of Lake Chad between 1963 and 2001 (Credits: NASA and Philippe Rekacewicz, UNEP GRID Arendal)

Fig. 1.15 Lake Titicaca sitting 3810 m asl *(15°55'S, 69°21'W)* and divided by the Peru-Bolivian Border is the largest lake at this altitude on Earth (190 km long, size 8372 km², max. depth 281 m). With a volume of 893 km³ Lake Titicaca is almost twice the volume of Lake Erie—one of the Great Lakes in North America, and eight times the volume of Lake Nicaragua—the largest lake of Central America. Lakes of this magnitude in high altitudes are rare as a consequence of natural events, whereby over a long time span around 20 Mio years of uplift to this part of the lithosphere was ample time for river back-cutting and the emptying these lake depressions (Image credit: ©Google earth 2013)

Actually, lakes can take any form theoretically possible: from completely circular to sharply contoured jagged shorelines, extremely narrow and long, deeper than wide, heart-shaped or as question marks, in very close patterns of thousands of similar size, single or associated with hundreds and countless more. Some resemble rivers (and in fact have been formed by meandering processes), others are oval or in egg-form, others have quadrangle forms or are polygons, all different sizes.

The origin to such variety in lake forms is the configuration of topographical depressions or basins on the Earth's surface and the degree of filling of these depressions. Generally, round and smooth lake forms occur on flat landscapes and on loose sediments where wind, waves and currents easily modify coastlines. When these dynamic forces work from constantly changing directions, within a few decades a smooth or rounded lake shape may develop. Single events that create a dot in the relief such as cosmic impacts or volcanic explosions also often leave round contours of depressions. More complicated lake shapes often occur by either exogenic forces that leave directed, long and narrow sculptures in the relief (as moving glaciers, continuous wind directions as in the trade wind zones, constant flowing water as in river beds), or by endogenic processes resultant of tectonic forces (rifts valleys, folds, faults, fractures/joints). Short-time processes that form angular shapes (rather as small but numerous forms) is permafrost where its ice wedge polygons push up a frame for shallow ponds.

Lakes, ponds and pans with multiple shapes also exhibit various colours: from black to white, green to brown, all shades of yellow to orange through deep red, particularly salt water lakes and saltpans. Endless varieties of processes that form and formed the earth surface are also responsible for lake forming and no documentation can present all these natural wonders. Nonetheless, we shall try to differentiate the main lake forming processes if only to present a first step of a small selection (Figs. 1.16, 1.17, 1.18, 1.19, 1.20, 1.21, 1.22, 1.23, 1.24, 1.25, 1.26, 1.27, 1.28, 1.29, 1.30, 1.31 and 1.32).

Along larger inland lakes, particularly with wave action, are coastal forms known to open oceans (Figs. 1.25, 1.26, 1.27, 1.28, 1.29, and 1.30) such as elevated shorelines, spits, barriers, deltas, lagoons and even tombolos. Other elongated lake or pan shapes occupy former valleys (Fig. 1.31) and structural influence can be found around a lakes distinc-

Fig. 1.16 Round lakes in different landscapes (**a**) Sub-arctic Canada at *50°59'N, 83°54'W* (Nash Lake, Cochrane, Ontario, 1.4 km diameter). (**b**) A sinkhole 2.8 km across in karst landscape of central Florida at *29°37'N, 81°59'W* (Image credit: ©Google earth 2015)

Fig. 1.17 (**a**) An oval shaped lake (2.3 km wide) in the swamps of the eastern Bolivian lowlands at *13°09'S, 63°03'W*. (**b**) A hidden dome-like structure with a central depression determines this round lake form (10 km across) in eastern Labrador, Canada, at *50°50'N, 59°25'W* (Image credit: ©Google earth 2013)

Fig. 1.18 (**a**) Strange rectangular lake shape in the permafrost landscape of Alaska at *70°09'29.34"N, 147°15'17.11"W*. Scene is 1.2 km wide. (**b**) Rectangular lake 1 km^2 in Alaska at *70°06'N, 147°32'W* (Image credit: ©Google earth 2015)

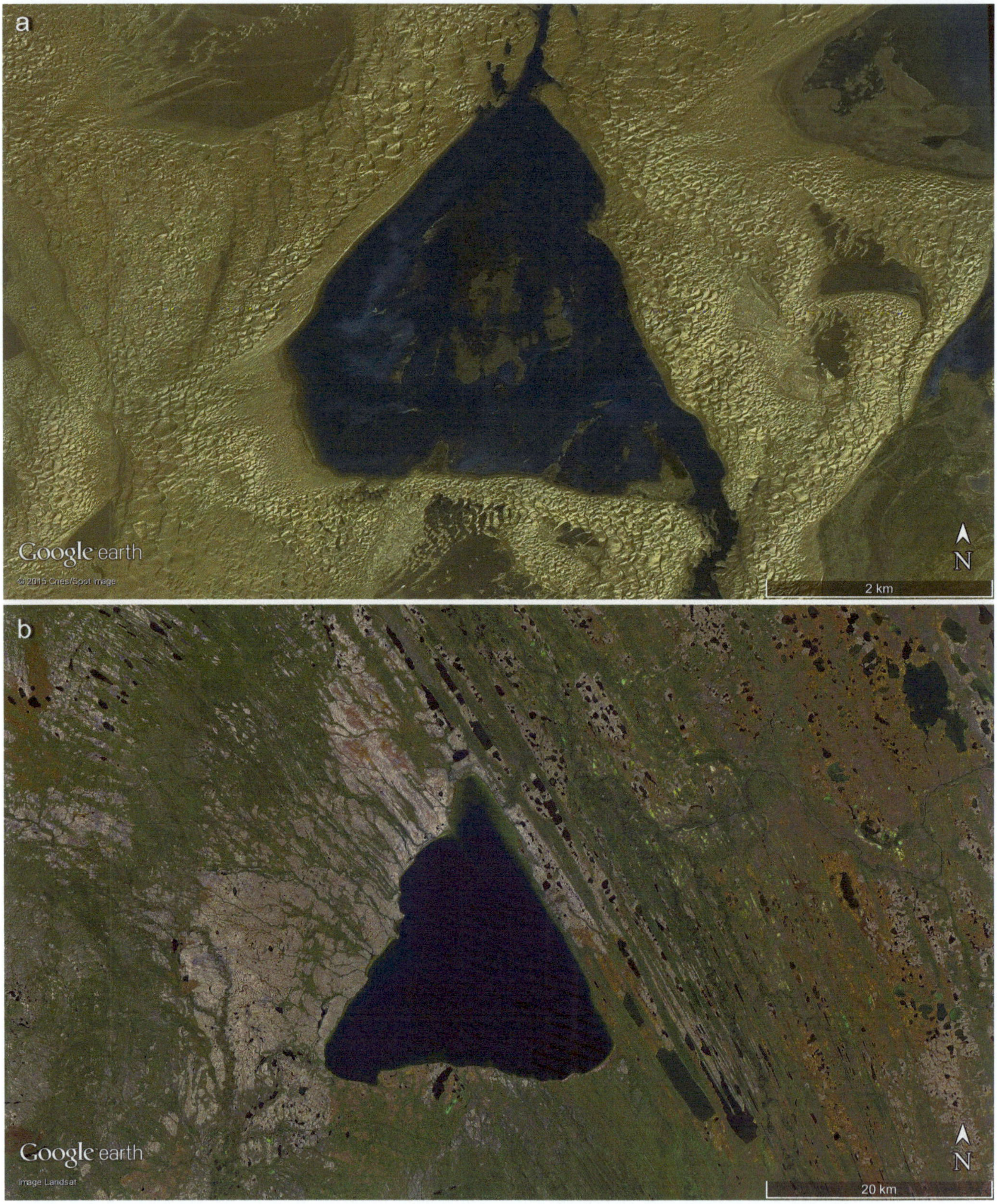

Fig. 1.19 (**a**) A triangular lake in an 11 km wide scene surrounded by 40 m high dunes in the eastern Gobi desert (China) around *39°31'N, 87°56'E*. (**b**) Triangular shaped Keller Lake in the Northwest Territories of Canada at *63°50'N, 121°11'W*. Ice Age glaciers have polished the landscape. Max length of the lake is 28 km and approximately 380 km^2 in area (Image credit: ©Google earth 2015)

Fig. 1.20 Devils Mountain Lake in Alaska resembles a mushroom shape formed by the coalescence of two formerly separated lakes, at *66°23'N, 164°30'W*. Scene is 16 km wide (Image credit: ©Google earth 2015)

tive relief, in particular younger folded mountain chains (Fig. 1.36)

Lakes may change their general form and contour via sedimentary processes from the surrounding slopes, e.g. from rock-glacier debris lobes, fluvial fans or deltas (Fig. 1.33). Extremely shallow basins are periodically or episodically filled by water, whereby the limited depth may allow strong winds to shift the lake's water in a downwind direction subsequently forming complex shore patterns.

1.3 Different Association/Clusters of Lakes

The number, size and clusters of lakes and ponds depend primarily on the topography or type of relief and lastly rely on the forces that formed the specific landscape. In several environments exist very rare basin- or depression like forms and lakes as singular elements, while in other environments lakes may appear in various clusters. Figure 1.34a, b, c demonstrates clusters that only contain two lakes as examples to the wide scope of natural shapes: These two lakes may form a reclamation mark, a pair of drops, or a butterfly form. Figure 1.35, exhibiting rectangular lake shapes, documents that the individual form may or may not depend on dominant landscape patterns. In the incredibly flat landscape of Cape Krusenstern (Alaska), clearly the buttressing of beach ridges opening to the southeast control the elongated rectangular shape, as well as its width and the total pattern of this lake cluster.

Many groups of lakes also may show some regularity in size, shape and distribution in the landscape, or appear to lie chaotically (Fig. 1.38). Mostly however, some features explain what might be the main reason for the distribution of these lake-filled depressions. We see in a 400 km wide scene of Mongolia, W-E and NW-SE-running mountain chains ordered by compressional processes from the S or SW by the continual folding and uplift of the Himalayas, resulting from the collision of India (formerly part of Africa in *Gondwana* times) with Asia. The secondary approximate direction of the mountain chains from S to N with several lakes extending in this direction is a result of ancient glaciers carving valleys from the high mountains in the S to the wide Mongolian basin in the N. Clearly detailed in that landscape shows a structural control by folded mountain chains (Fig. 1.36). The 200 km wide example of Fig. 1.37 identifies tiny lines of cracks and joints in the rock uncovered by glacial erosion during the Ice Age in Canada. The large lake itself has extremely irregular contours.

Fig. 1.21 As a joke of nature, a perfectly apple-shaped lake with a diameter of 1.1 km has developed in the permafrost regions of Alaska at approximately *66°23'N, 164°38'W* (Image credit: ©Google earth 2013)

Although a large cluster of lakes can sometimes have similar contours and size, it is difficult to find order in their overall distribution—with the exception of a slight orientation of the long axes in SSE-NNW direction (Fig. 1.38). Other landscapes of glacial sculpting have unearthed old folding structures and demonstrate a clear orientation of lakes (Fig. 1.39).

The order of lakes in Fig. 1.39 (regular or irregular) is more observable from the accompanying forms and landscape structures: in the western Alaska coastal region lakes are between an old and elevated beach ridge system. From Canada (Fig. 1.40), a broad valley is filled with irregular ground moraine on which numerous lakes have been developed, more or less round in shape due to being in a sedimentary environment affected by wind and freezing over in winter. The round shape to nearly all of the numerous lakes and ponds in Fig. 1.41 can be elucidated by the very flat and windy landscape of a sandy spit at the northern shores of the Black Sea. The slight SW-NE component to the clustering of lakes is due to remnants of beach ridges forming the long Dzharylhats'ka spit, and the slight NW-SE extension of single lakes created by NW-winds dominant to this landscape.

1.4 Colour of Lakes, Ponds and Pans

Specifically working with satellite image interpretation, the colours of the lakes' waters and of its landscape can offer important information on certain qualities. The colour of a lake depends on many factors: it can vary by the angle of sunrays at the moment of satellite imaging, or even by the different angles under which people view the lake that explains why the same object is explained differently by diverse observers (aside subjective feelings).

Another factor to affect the colour of lake is the changing cover of its surface produced by waves (rough surfacing), or even by ice, both affecting the amounts of reflection and absorption of sunlight on and in the water. In the beginning young ice is clear (transparent) similar to glass, but through time fissures occur where snow crystals may accumulate on the ice surface whereby the lake appears lighter (reflecting a higher quantity of sunlight; Figs. 1.42a, b and 1.43).

Very important to satellite image interpretation is depth of water: if lake water is quiet and deep then colours are primarily very dark to black, but in shallow lakes water

Fig. 1.22 Heart-shaped lakes. (**a**) From Canada, near Ompah at *45°01'08.26"N, 76°53'53.23"W* in a 0.75 km wide scene. (**b**) From NW Russia at *70°10'N, 70°49'E* and 8 km across (Image credit: ©Google earth 2013)

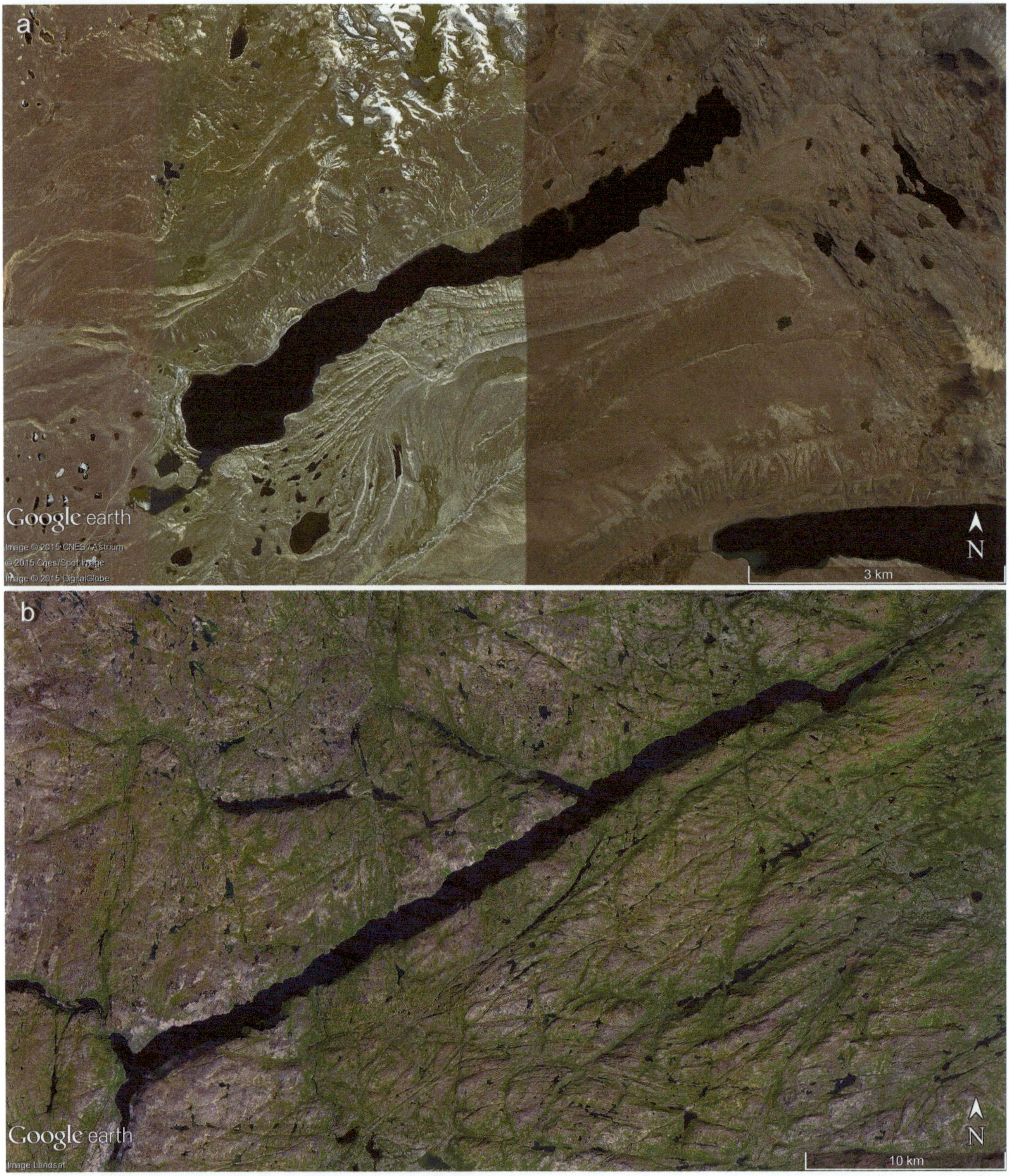

Fig. 1.23 (**a**) A 12 km long narrow lake lies 4548 m asl in an ancient glacial bed within a series of lateral and terminal moraines in western Bolivia at *14°53'S, 69°17'W*. (**b**) A structurally controlled shape of a lake 46 km long but only 1.5 km wide in NE Labrador Canada, at *55°05'N, 61°50'W* (Image credit: ©Google earth 2013)

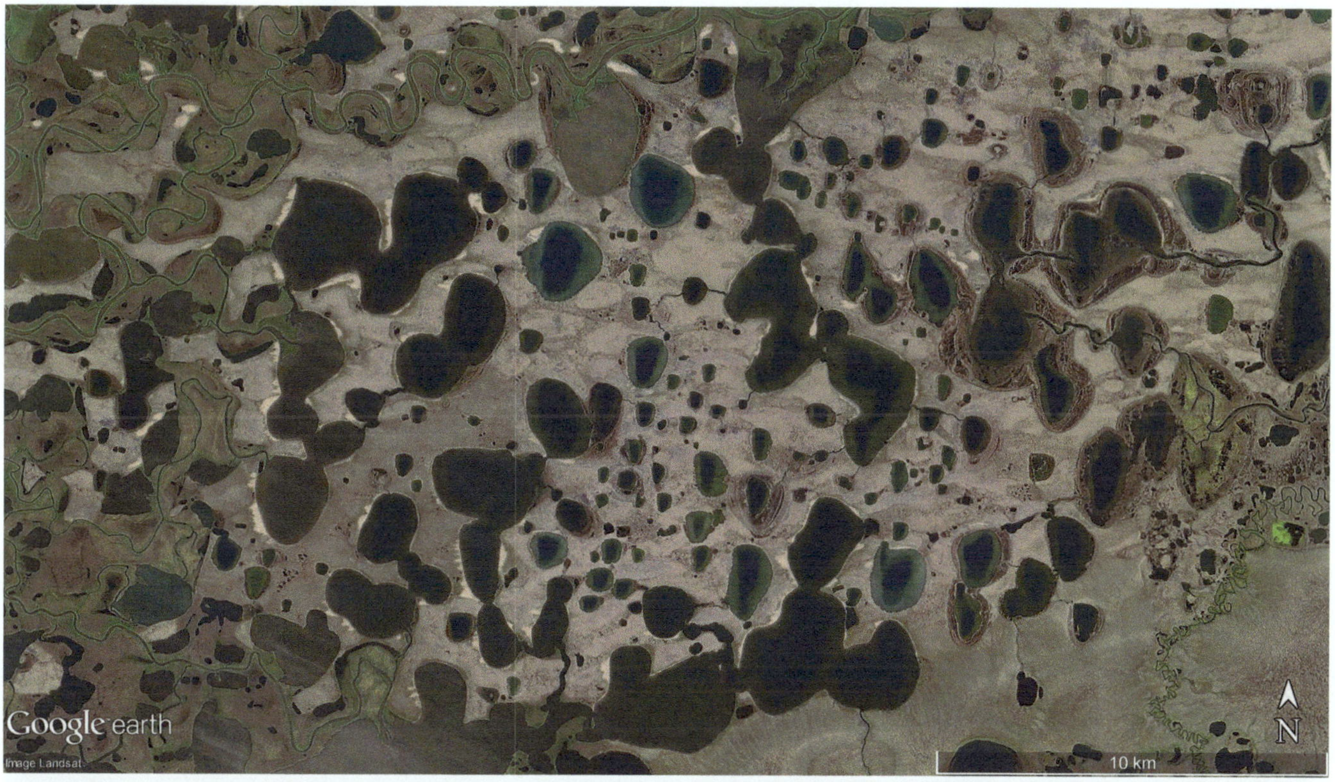

Fig. 1.24 On permafrost ground in a flat landscape of northern Siberia (Russia) lakes tend to coalesce by lateral extension during the thawing period. Global warming during the last century is accelerating this process. The 28 km wide scene is at approximately *70°54'N, 137°40'E* (Image credit: ©Google earth 2013)

Fig. 1.25 The peculiar lake form of Lugmu Tso in NW Tibet *(34°37'N, 80°27'E)* sits at 5010 m asl. Virtually symmetrical to its long axis of 18 km are numerous parallel shorelines and document the long-time shrinking of the lake's extent (Image credit: ©Google earth 2015)

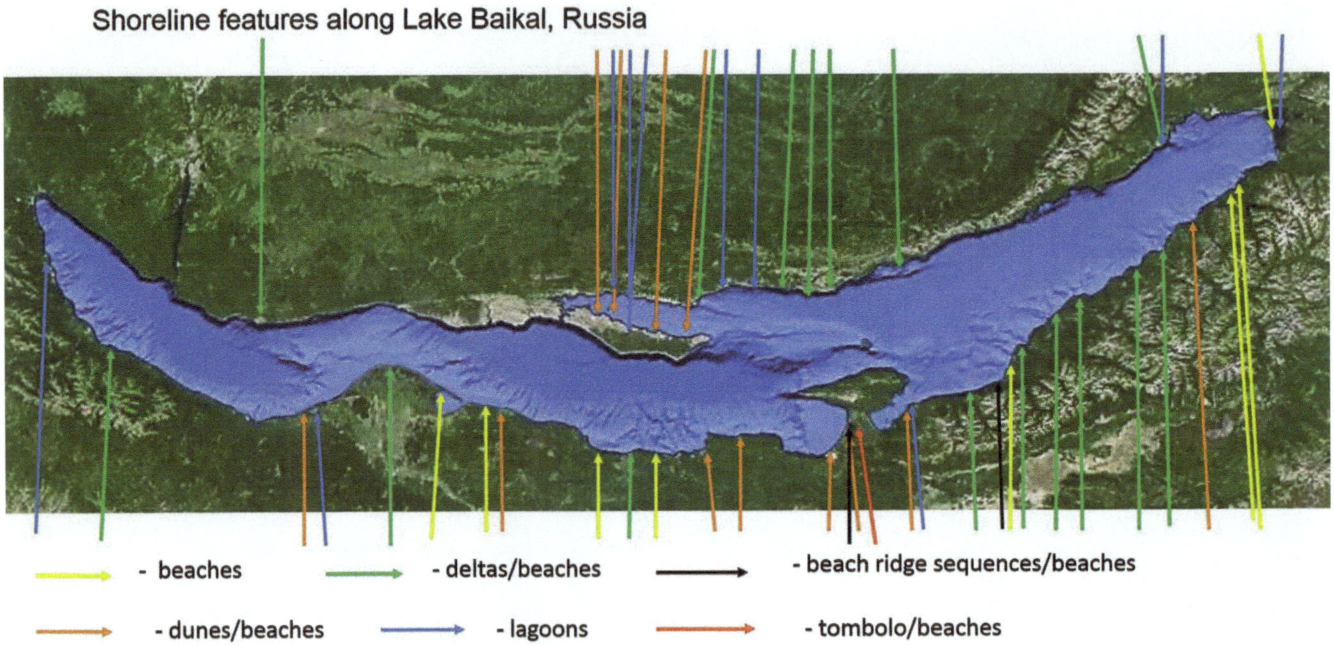

Fig. 1.26 Lake Baikal showing examples of coastal forms typical for oceanic shorelines. Coastal segments not indicated are mostly denudation slopes, sometimes with low cliffs (Image credit: ©Google earth 2013, with own graphics)

Fig. 1.27 The large Lake Hyargas lies 1030 m asl in central Mongolia at *49°09'N, 93°22'E*. Its shoreline is determined by strong sediment inputs from fluvial fans whereby narrow spits have been developed by strong westerly winds causing the southern part of Lake Hyargas to separate into a shallow saline lake. Scene is 190 km wide (Image credit: ©Google earth 2013)

Fig. 1.28 Xijir Ulan Hu in southern Mongolia is a very shallow lake (only 4.7 m deep) resting 4777 m asl. (*35°12'N, 90°23'E*, 35 km wide scene). Spits of different dimensions can be located in its eastern part (Image credit: ©Google earth 2013)

Fig. 1.29 A combination of lakes separated by elegantly shaped narrow barriers with beaches (Bolshoye Eronvoye in the west and Maloye Erovnoye to the east) SE of Lake Baikal *(52°37'N, 111°35'E)* in Russia. Scene is 45 km wide (Image credit: ©Google earth 2013)

Fig. 1.30 A fine tombolo at 4570 m asl. in the SW part of Lake Titicaca in Peru *(16°54'S, 70°01'W)*. Scene is approximately 6 km in width (Image credit: ©Google earth 2015)

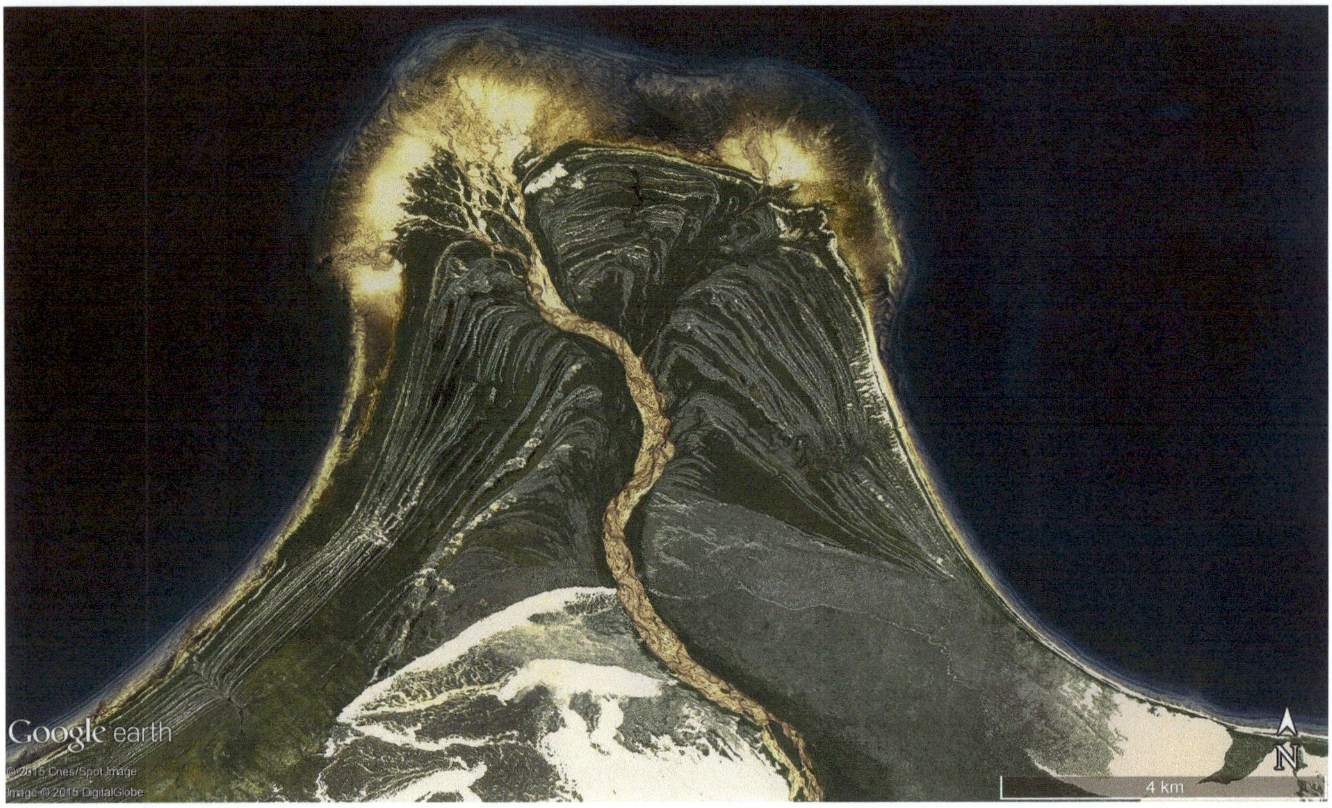

Fig. 1.31 A series of glacio-isostatic uplifted beach ridges along the southern shores of the large Lake Athabasca in N Canada at *59°07'N* and *109°18'W*. Scene is 18 km wide (Image credit: ©Google earth 2013)

Fig. 1.32 Examples of lakes or former lakes (here: saltpans) evidently occupying valley-like depressions in the topography. (**a**) Example is from Kazakhstan at *49°01'N, 48°59'E* as a 3 km scene. (**b**) Image from western Yucatan in Mexico situated around *24°20'N* and *97°49'W* with a scene 2.6 km wide (Images credits: ©Google earth 2015)

Fig. 1.33 A lake in the Bolivian Andes rests 4561 m asl *(16°20'S, 68°10'W)*, formerly 800 m in length this lake is now being filled by alluvial fan processes and looks to disappear in the next few decades. Ridge-like features (terminal moraines) along the southern shores indicate the lake was originally a basin filled by a glacier (Image credit: ©Google earth 2013)

depth can be seen through a sequence of colours ranging from dark to light, radiating from the centre outward to the shores (Figs. 1.44, 1.45, 1.46 and 1.47). In shallow lakes the colour of bottom sediments as well as that of suspension load while mixing of waters or discharge from the surrounding land (or from a glacier), also change the colours of lakes (Figs. 1.48 and 1.49).

Additionally, swimming plants or those growing underwater, algae and micro-organisms, particularly in saline lakes and saltpans, and minerals in solution or at the bottom (Figs. 1.50, 1.51, 1.52 and 1.53a, b), often give off a wide spectrum of vivid and brilliant colours discriminating lakes into three groups of blue, green and brown lakes. Blue lakes often have small catchment areas with a low nutrient supply resulting in clear water and reflect light as the colour blue. Green lakes are green algae resulting from high nutrient inputs and can often change their colour within days or weeks depending on the growth conditions and development of the blue-green algae *Halobacteria* and *Cyanophytae* that can give off yellow and reddish colours in saline lakes or saltpans. Brown lakes form from humic acid as dissolved organic matter from forested environments or bogs.

1.5 Examples for Ice Age lakes

During the Quaternary period (around the last 2.4 Mio years) many ice ages on the continents with a wide distribution of glaciers ensued, interrupted periodically every 100,000 years by warm periods and glacier recession known as interglacial periods. Glacier ice, particularly if thick and moving quickly, is an extraordinary effective modeller of rocky ground. Frequently following in older depressions or valleys, the moving glacial ice carves planate basins leaving behind closed depressions—as a consequence when glaciers recede at the end of an ice age, melt water fills these depressions and may create a high number of lakes, including very large lakes. The wide expansion of large lakes in the western part of the USA at the end of the last ice age resulted from an old structural undulation of basins and ranges, in combination with the supply of melt water from high mountain glaciers after regional deglaciation.

Colder temperatures also sustained the existence of lakes due to reduced evaporation. The development of these large lakes in the western part of the USA started around 14,500 BP, however by around 12,500 BP evaporation rates were rising and large amounts of debris weathered during the ice age had washed down into the basins and filled many basins

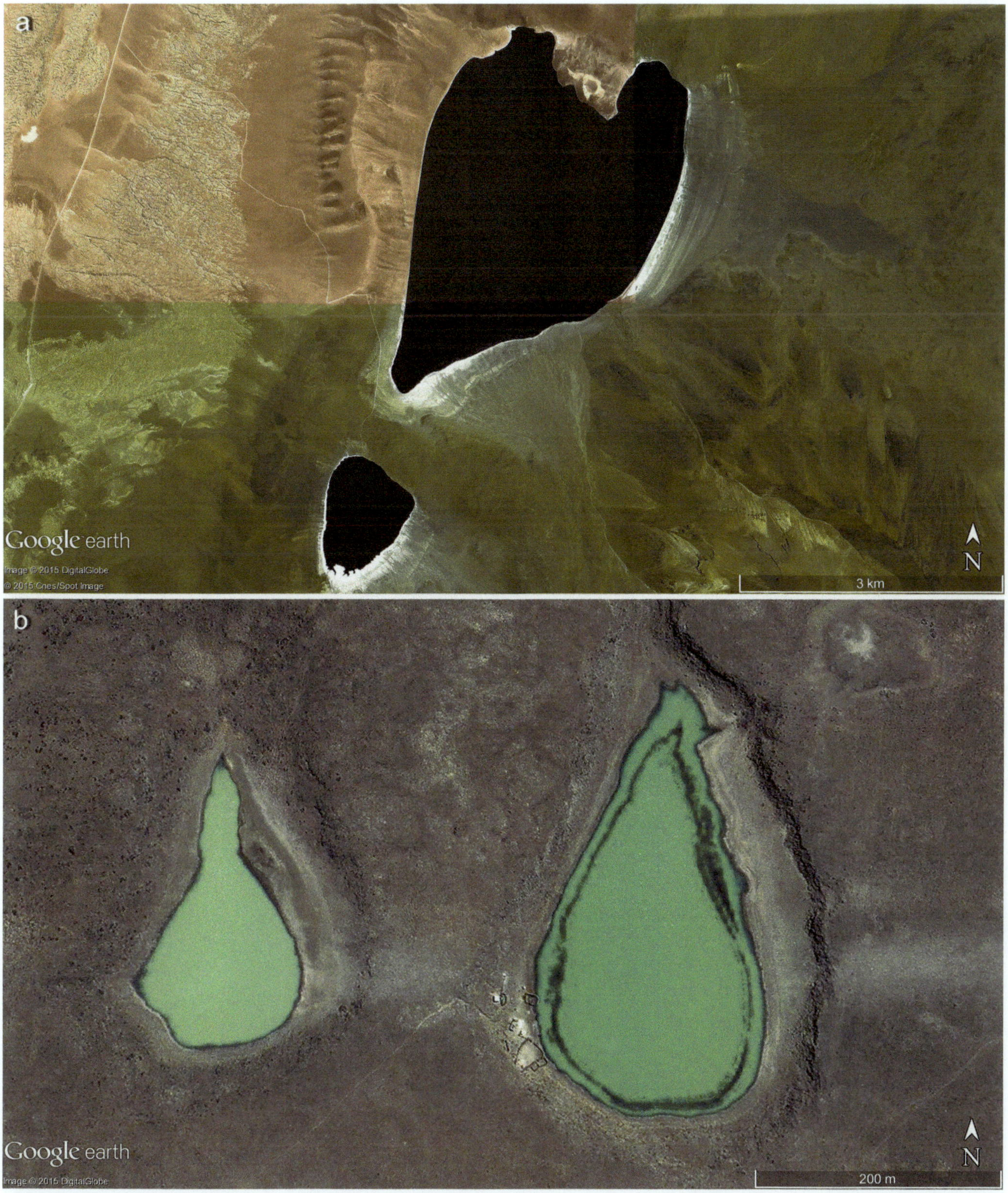

Fig. 1.34 (**a**) Two lakes together resemble an exclamation mark in the northern Chilean Andes at *23°44'S, 67°47'W* and 4,140m asl. The longer lake measures 5.5 km in N-S-extension. (**b**) Two tear shaped maar-lakes in southern Patagonia (Argentina) at *41°22'S, 67°31'W*, the larger lake reaches 410 m in length. (**c**) Twin lakes in a 1.3 km wide scene in western Australia at *21°51'S, 114°42'E* are fashioned into a butterfly shape (Images credits: ©Google earth 2015)

Fig. 1.34 (continued)

Fig. 1.35 Elongated rectangular lakes are structured by sub-parallel beach ridges at Cape Krusenstern at *67°07'N* and *163°37'W* in western Alaska (USA). Scene is 2.2 km wide (Image credit: ©Google earth 2015)

Fig. 1.36 Lake depressions in the high mountains of western China situated approximately *35°16'N* and *92°01'E* are encased by W-E-striking folded mountain belts as structural elements. The large 30 km long lake to the east is Lake Cuorendejia at 4693 m asl (Image credit: ©Google earth 2013)

Fig. 1.37 A very large complex shaped lake (Caniapiscau, now a reservoir) in Labrador, Canada, at approximately *54°24'N* and *69° 47'W*. Scene is 220 km wide (Image credit: ©Google earth 2013)

Fig. 1.38 A 80 km wide scene in northern Labrador, Canada, at approximately *58°06'N, 69°32'W* presents numerous parallel lakes controlled by folding structures and unearthed by several glaciations (Image credit: ©Google earth 2013)

Fig. 1.39 A 30 km wide scene of Western Alaska at *62°16'N* and *165°18'W* shows hundreds of lakes and ponds controlled by a wide beach ridge system more or less parallel to the shoreline, and by sediments deposited along water ways of extremely slight inclines from inland (E) to the sea (W) (Image credit: ©Google earth 2015)

Fig. 1.40 Within a long depression (here shown for 5 km) in central northern Canada *(59°45'N, 113°21'W)* lies a group of lakes consisting of different shapes and sizes. These lakes have been structured on ground moraine since late-glacial times (Image credit: ©Google earth 2013)

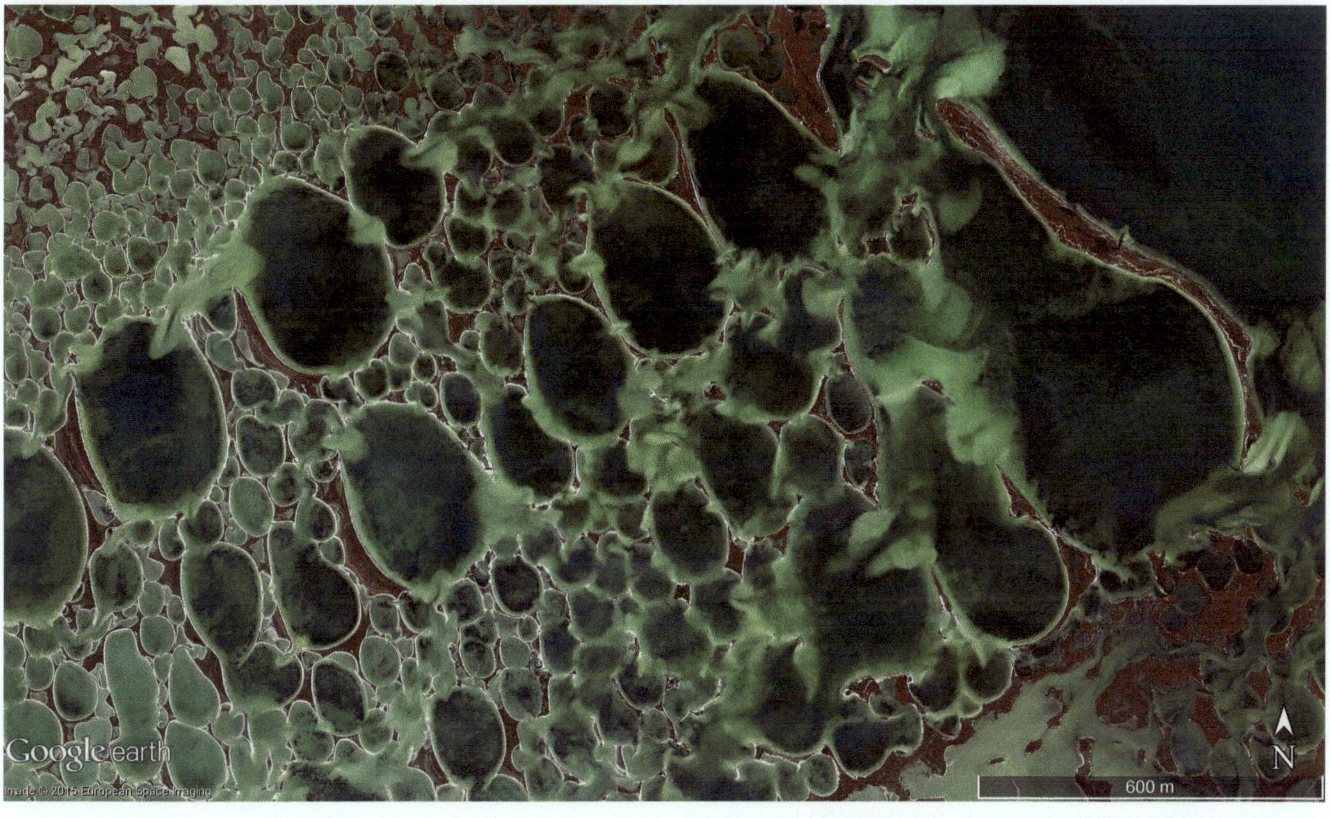

Fig. 1.41 Situated on Dzharylhats'ka spit along the north coast of the Black Sea in Ukraine *(46°02'N, 32°51'E)* are lakes with similar shapes controlled by northwesterly winds. Scene is 2.4 km width (Image credit: ©Google earth 2013)

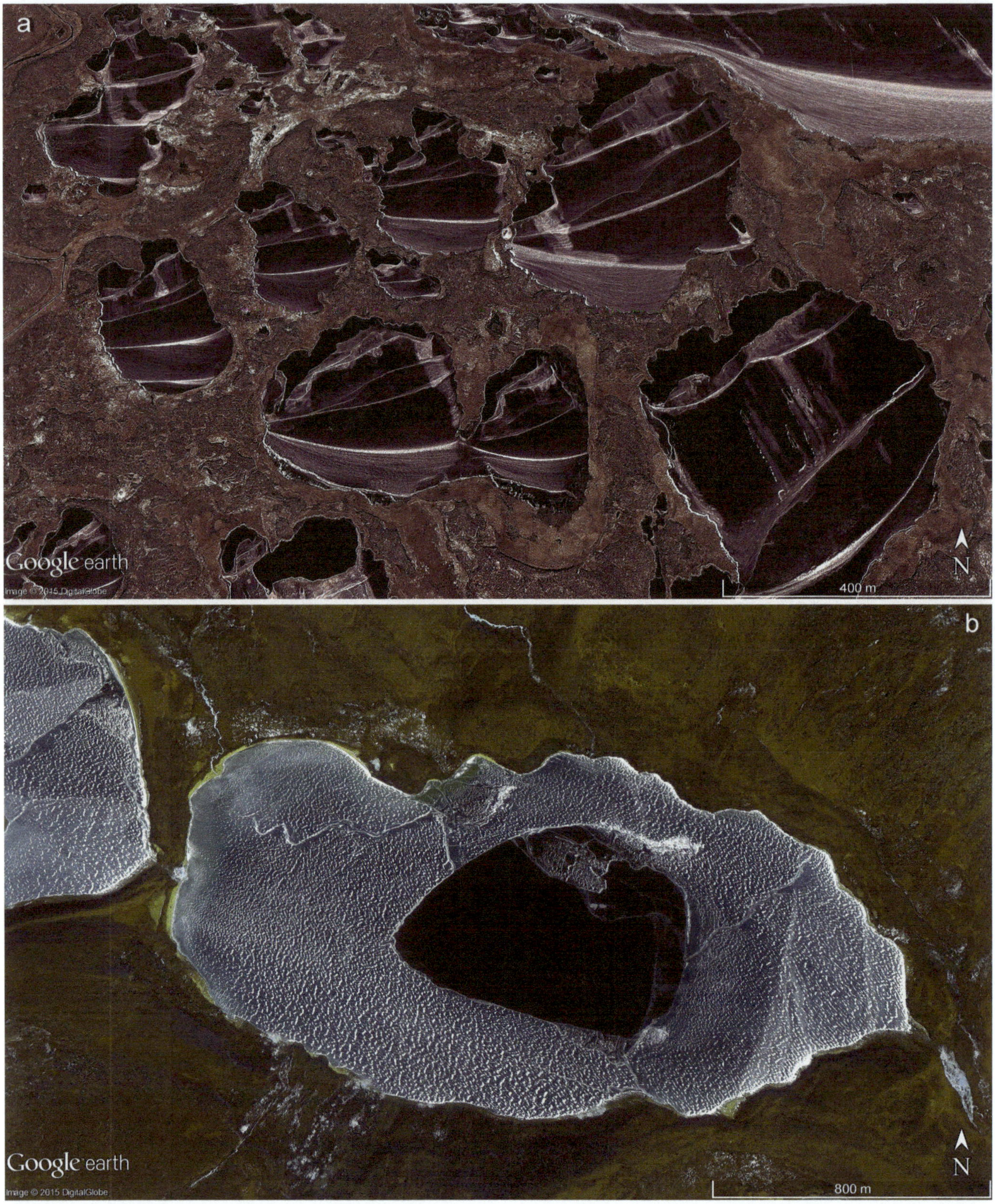

Fig. 1.42 (**a**) Winter ice cover on lakes fragmented by wind stress in SW Alaska at *60°07'N, 164°11'W*. 1.5 km wide scene. (**b**) Lakes over 4000 m asl in the southern part of Sichuan Province in China, frozen over in winter. This example at *29°14'N, 100°18'E* is 2.3 km long and positioned 4494 m asl. The ice-free centre of the lake is extremely dark from the lack of light whereby ice cover and minute snow ripples reflect most of the incoming radiation (Images credit: ©Google earth 2015)

Fig. 1.43 Resulting from multiple pressures, the winter ice cover along the eastern part of Lake Puma Yumco at 5019 m asl in Sichuan Province of China *(28°34'N, 90°23'E)*, has been fractured producing a mixture of old ice and freshly frozen water along polygonal fissures. Additionally, wind drifts of snow and ice crystals have formed more patterns affecting the colour of the ice surface. The scene is 1.5 km wide (Image credit: ©Google earth 2015)

Fig. 1.44 In the high Andes of western Bolivia several ancient glacier beds present several lakes often with different colours (here at *14°55'S, 69°13'W* in a 12.5 km wide scene). The largest lake to the north is Laguna Nueve at 4637 m asl. and in the centre lie the blue lakes of Laguna Kellu (*light blue*) and Laguna Quello (*medium blue*) at a level of 4529 m asl (together with the next two lakes) (Image credit: ©Google earth 2013)

Fig. 1.45 The 35 km long Hala Lake (4081 m asl) in China at *38°17'N, 97°55'E* is centered by deep water surrounded by a shallow shelf to the shoreline. Light coloured sediments in the very shallow water reflect light and emit the *light-blue* aura seen (Image credit: ©Google earth 2014)

Fig. 1.46 A 1.6 km long lake in the permafrost latitudes of NW Canada at *69°58'N, 129°43'W* has been extended in size radiating outward from a deeper centre. Water depths near the shorelines reach levels below 1 m (Image credit: ©Google earth 2015)

Fig. 1.47 Southern shore of a lagoon in the Florida Panhandle with sediment patterning visible in its shallow water *(30°24'27.44"N* und *87°00'24.27"W,* 1.2 km wide scene) (Image credit: ©Google earth 2013)

Fig. 1.48 Suspension load in the southernmost part of Lago Musters, central Argentina, at around *45°S* and *68'W*. Scene is 15 km wide (Image credit: ©Google earth 2015)

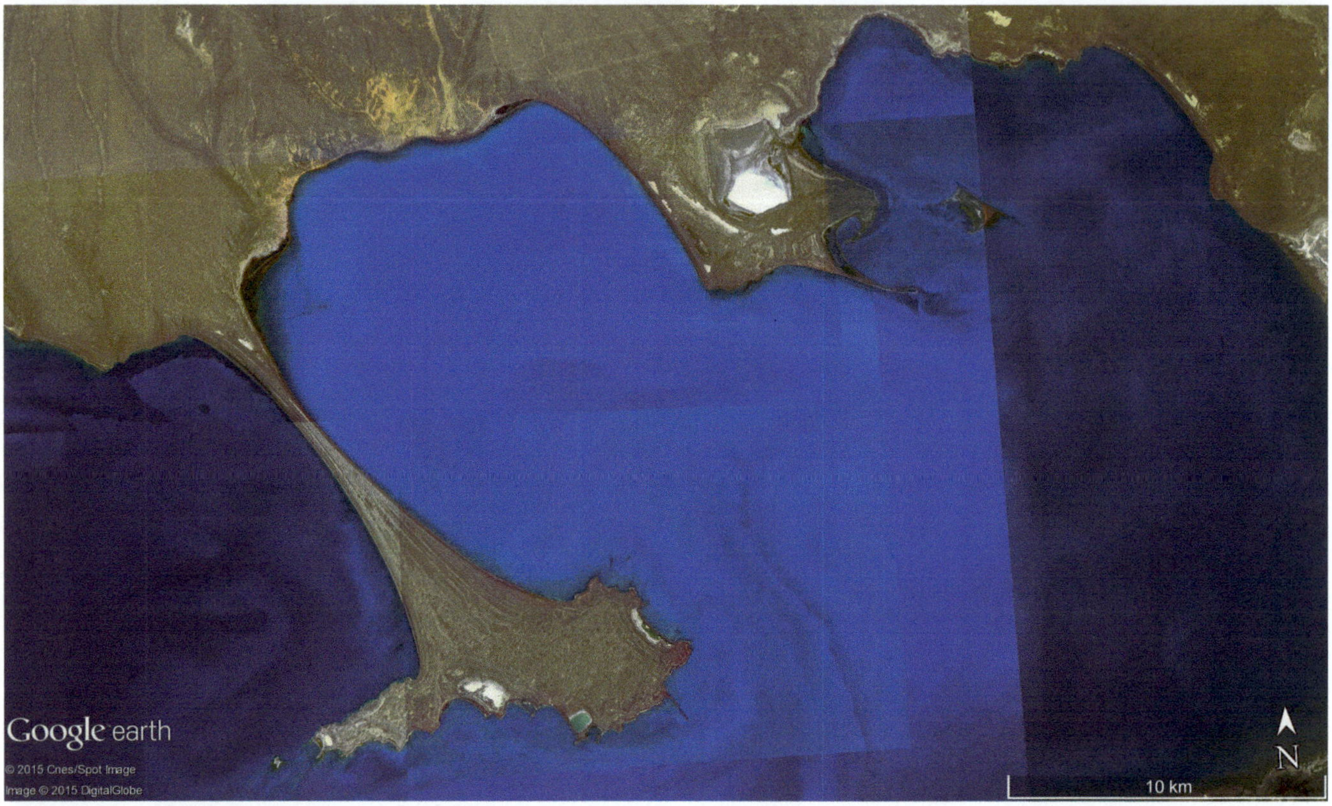

Fig. 1.49 In the west-east-trending Lake Balkhash in Kazachstan at around *46°41'N* and *78°52'W*—one of the largest lakes on Earth (over 18,000 km^2 in area and exceeds 500 km in length)—westerly winds have mixed the waters and brought into suspension fine silt particles. The amount of silt in suspension is reflected in the different colours of the water: the clearer the water the darker the colour blue. A perfectly shaped tombolo forms a natural connection from the main shoreline to a small former island. Scene is 48 km wide (Image credit: ©Google earth 2015)

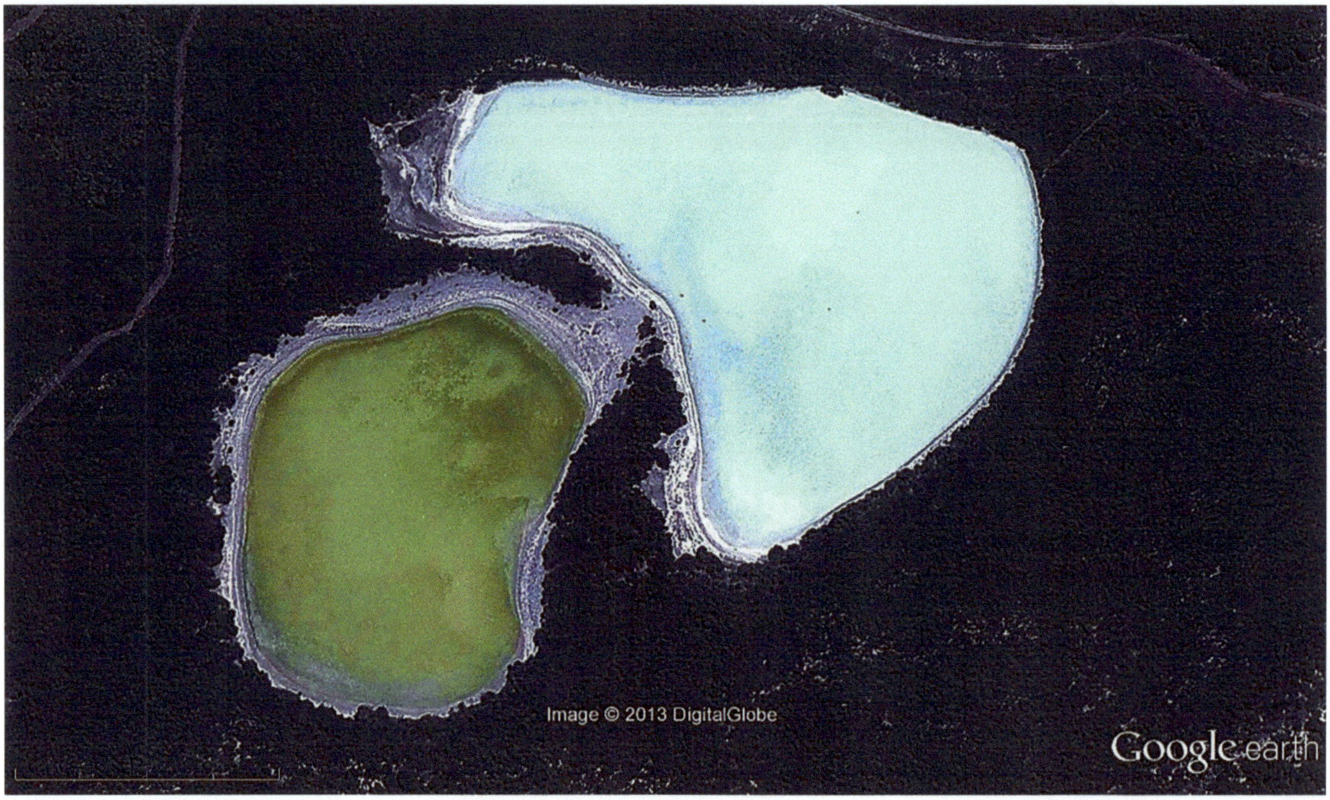

Fig. 1.50 Two lakes differ in colour dependent on their salt content. The highest salt content (*white colour*) is found immediately along the shorelines. A 720 m wide scene in southern West Australia *(31°52'S, 121°E)* (Image credit: ©Google earth 2013)

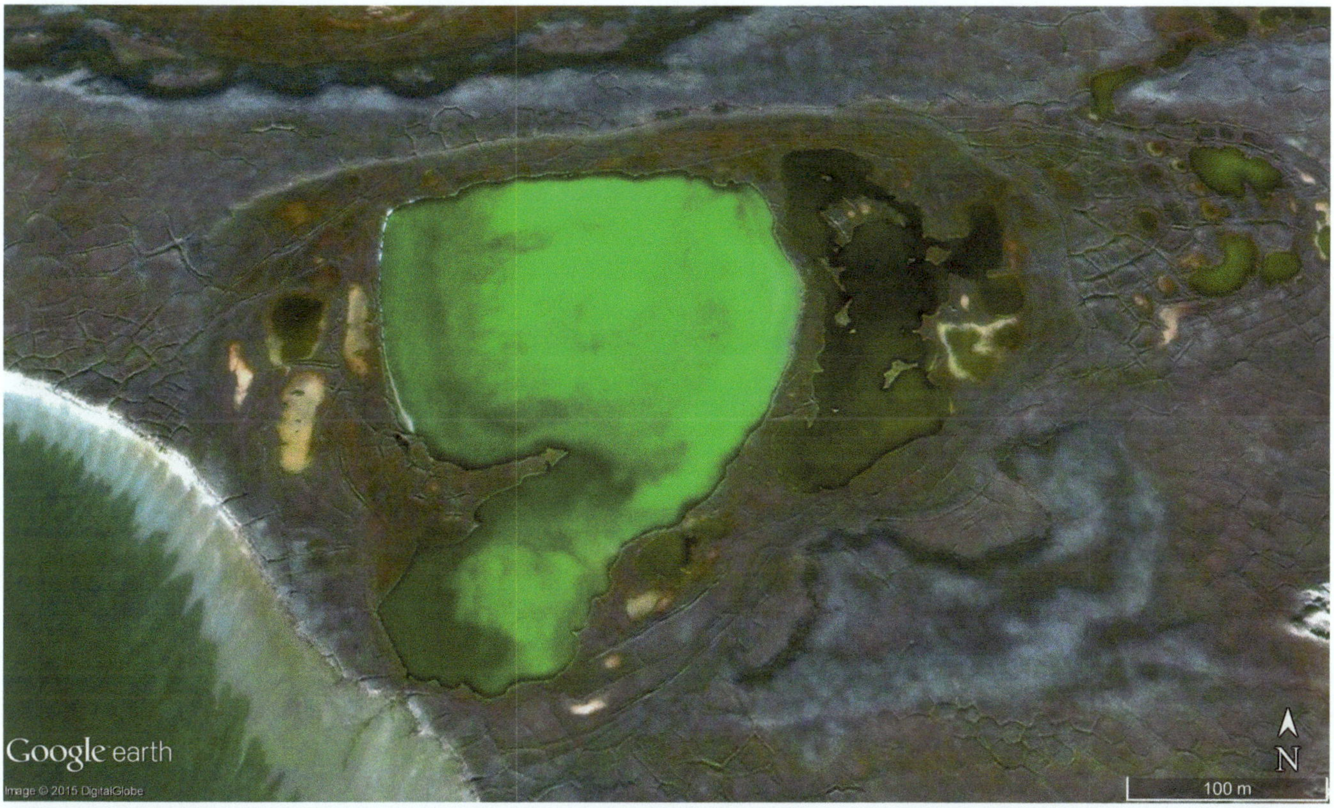

Fig. 1.51 Vivid green colours can occur in high latitude shallow lakes during the long days of spring and summer when algal blooms occur. This 800 m wide scene is on permafrost in NW Canada at *70°11'N* and *129°45'W* (Image credit: ©Google earth 2013)

Fig. 1.52 A small saline lake (only 500 m long) in Argentina at approximately *41°21'S and 68°50'W*. The peculiar colour to the lake points to a certain group of bacteria (Image credit: ©Google earth 2015)

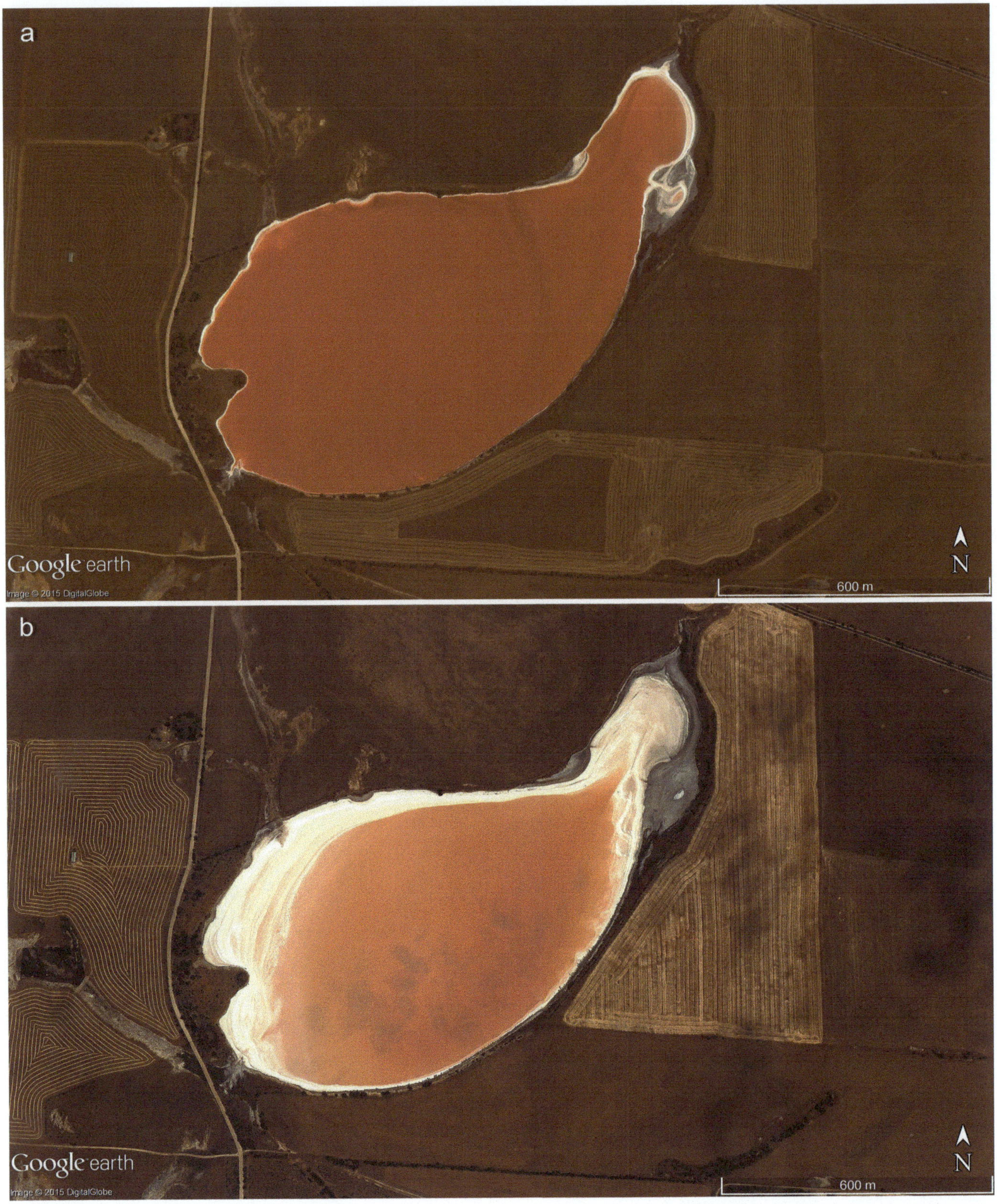

Fig. 1.53 (**a**) Green algae *Dunaliella salina* often mix with the salt bacterium *Halobacteria cutirubrum* and the tiny brine prawn *Artemia* to produce beta carotene (fat-soluble pigments) when temperatures are high. The combination gave this 2 km long saline lake *(34°47'S, 137°46'E)* in South Australia a vivid colour at the beginning of the year 2011 (high summer). (**b**) The same lake toward the end of November (early summer) in 2012 demonstrates an alternative intensity of colours of the bacteria and less water by strong evaporative processes (Image credit: ©Google earth 2015)

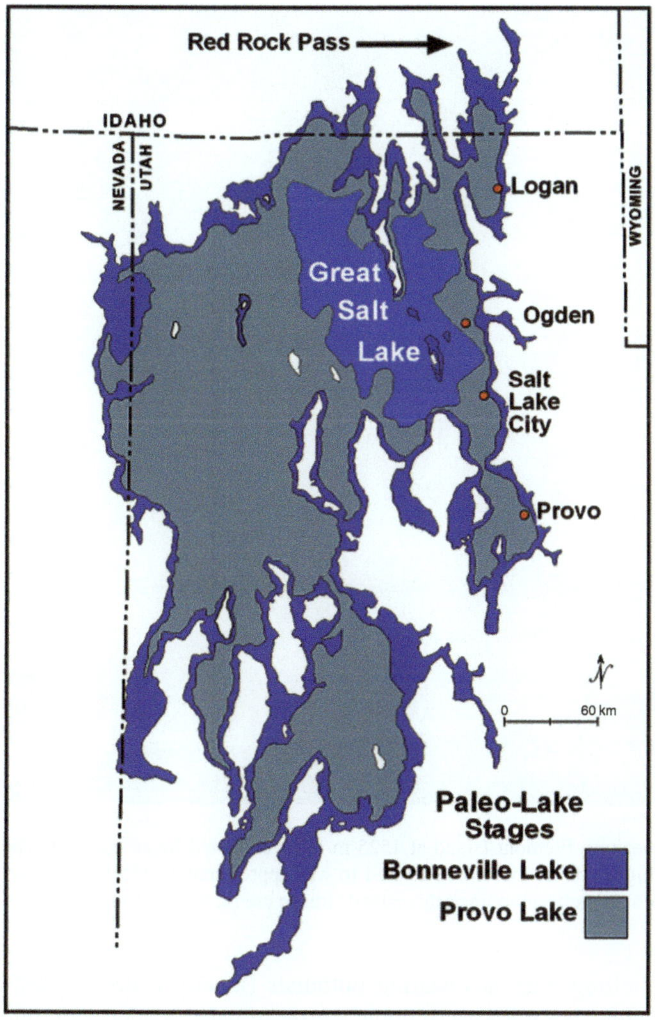

Fig. 1.54 Extension of the former Ice Age Lake Bonneville (16,000–14,500 BP before the overflow of Red Rock Pass) and reduced size following the short Provo Lake phase before the lake dropped to its present level (called Gilbert Phase) (Image credit: Link et al. 1999)

partly or even totally. The outcome was the disappearance of many of these lakes, although in many cases their old and high shorelines (in the form of terraces or ridges) and former maximum extension to the lake can be reconstructed. One of the best examples is former Lake Bonneville, now known as Great Salt Lake in Utah, a small and shallow remnant (Figs. 1.54 and 1.55).

The outflowing of Lake Bonneville occurred at Red Rock Pass. Sediment layers have been rapidly eroded and lowered the lake's level for 107 m (to a level called Provo level), by a rapid overflow of water caused from the second largest flood in the geologic history of the world (Link et al. 1999). An extreme discharge for several weeks was able to transport basaltic boulders up to 3.5 m in diameter far away in a flat landscape. During the lake outburst, alluvial deposits at Red Rock Pass had been undercut, proceeded by a landslide from the Brannock Mountains. Cutting through the sediment layers stopped (for about 1000 years) after reaching resistant bedrock at the base of the Red Rock Pass alluvium. At about 1445 m asl this long lake level (Provo stade) left a prominent strandline often forming an alluvial lake terrace.

Emptying of the large Lake Bonneville resulted in an isostatic rebound of the lithosphere whereby the earths' crust rose to compensate for the diminished load so that the forces dislocated the old shorelines at different levels. Around 12,000 years ago, the end of the Ice Ages reduced the water flowing into the basin resulting in the dominant process of evaporation. The lake increased in saltiness and its elevation was lowered around 11,000–10,000 years ago to what remains todays level (Gilbert stadium). Interestingly, the lake (now Great Salt Lake) still fluctuates for several metres (between 1277 and 1285 m asl) due to small changes in precipitation/inflow and evaporation by higher temperatures occasionally flooding roads and other infrastructure.

Further rationale to the sudden disappearance of formerly large meltwater lakes in higher latitudes of all the continents is their damming by rather unstable glacier ice (Fig. 1.56). Inasmuch as water rises in the temporal reservoir behind these ice dams, outburst of the large lakes may occur with thousands of km^3 of water inundating wide landscapes by very fast moving floods that transport even large boulders for tens of kilometres and sculpted the surface. The Channelled Scablands of the states of Washington, Montana, and northern Oregon scoured by escaping glacial lake floodwaters, lasted just days or a maximum few weeks and belong to the most severe and extended geomorphologic events on the continental surface of the Earth.

One of the large meltwater lakes formed and dammed by glacier ice during the last ice age is Lake Agassiz. Around 12,500 BP Lake Agassiz was fed by melt waters of the receding continental ice sheets of North America becoming the largest freshwater lake on Earth (Fig. 1.56). In any event, the lake could hold no more water from the thawing ice and when it burst its icy banks around 30 % more water than is contained in all of the world's lakes of today flushed into the North Atlantic within a very short period (several weeks maximum). The surge of fresh water was so dramatic it caused a 400-year dip in world temperatures plunging the planet into another brief ice age. Subsequently, rapid global warming at the beginning Holocene epoch led to a phenomenal rise in sea levels—more than 30 m.

1.6 Subglacial Lakes (of Antarctica)

In recent times scientists have discovered over 400 lakes on the Antarctic continent (Fig. 1.57) lying below the coldest ice on Earth. In some regions up to 4000 km of glacier ice conceal lakes hidden between the ice and the rocky or morainic (unstratified glacial drift) ground. The existences of

Fig. 1.55 Old strandlines of Lake Bonneville on Fremont Island (viewed from the east slope, looking north). Modern level of Great Salt Lake rests at 1283 m asl (from 19th Oct., 2013). Castle Rock, the highest peak on Fremont Island at 1525 m asl is encircled by an old cliff. The highest level the lake extended to was approximately 1550 m asl. and completely submerged this island (Image credit: ©Google earth 2014)

water are mostly the result of high pressure (which increases melting temperature) and not meltwater seeping through the ice itself (which can be observed on the margins of inland ice in Greenland). The largest of these subglacial lakes is Lake Vostok and while the lake's basin maybe millions of years old isolated for a long span of time whereby Antarctic ice has never disappeared during the last several millions years, the waters of Lake Vostok itself are relatively young and no older than 13,000 years whereby the water of the lake is dislocated at the base of the glacier ice which moves to Antarctica's boundaries. Under approximately 4 km of ice lies the surface of Lake Vostok that places it at approximately 500 m below sea level. With dimensions of roughly 250 × 50 km and an area of approaching 12,500 km^2 with a mean depth of >400 m (with a maximum of >800 m), the water volume in Lake Vostok calculated by these figures is approximately 5000 km^3 (larger than the Great Salt Lake in Utah, USA by area and a great deal larger than Lake Victoria in eastern Africa!). Iceland is another well-known region with subglacial lake development, however the melting process here is triggered by active volcanism. The consequence is extremely rapid melting with devastating outbursts (locally called *"jökull-haup"*) of millions of cubic metres of water loaded with fine ash ranging up to large boulders.

1.7 Meltwater Ponds on Glaciers

Whereas the vast Antarctic ice dome is normally too cold to produce surface meltwater, glaciers in less extreme latitudes frequently present meltwater ponds on their surface—increasing in numbers and expansiveness in modern times as a result of global warming—in particular the Greenland ice body. Meltwater ponds can even be detected by Google earth satellite images (Figs. 1.58 and 1.59).

Smaller in extension but found in larger groups with deeper water, lakes or ponds occur on valley or piedmont glaciers where ice movement ceases and becomes stagnant. Through melting processes sinkhole-like round depressions appear for many years often the only hints that ice is still present under thick covers of debris and vegetation (Figs. 1.60a, b).

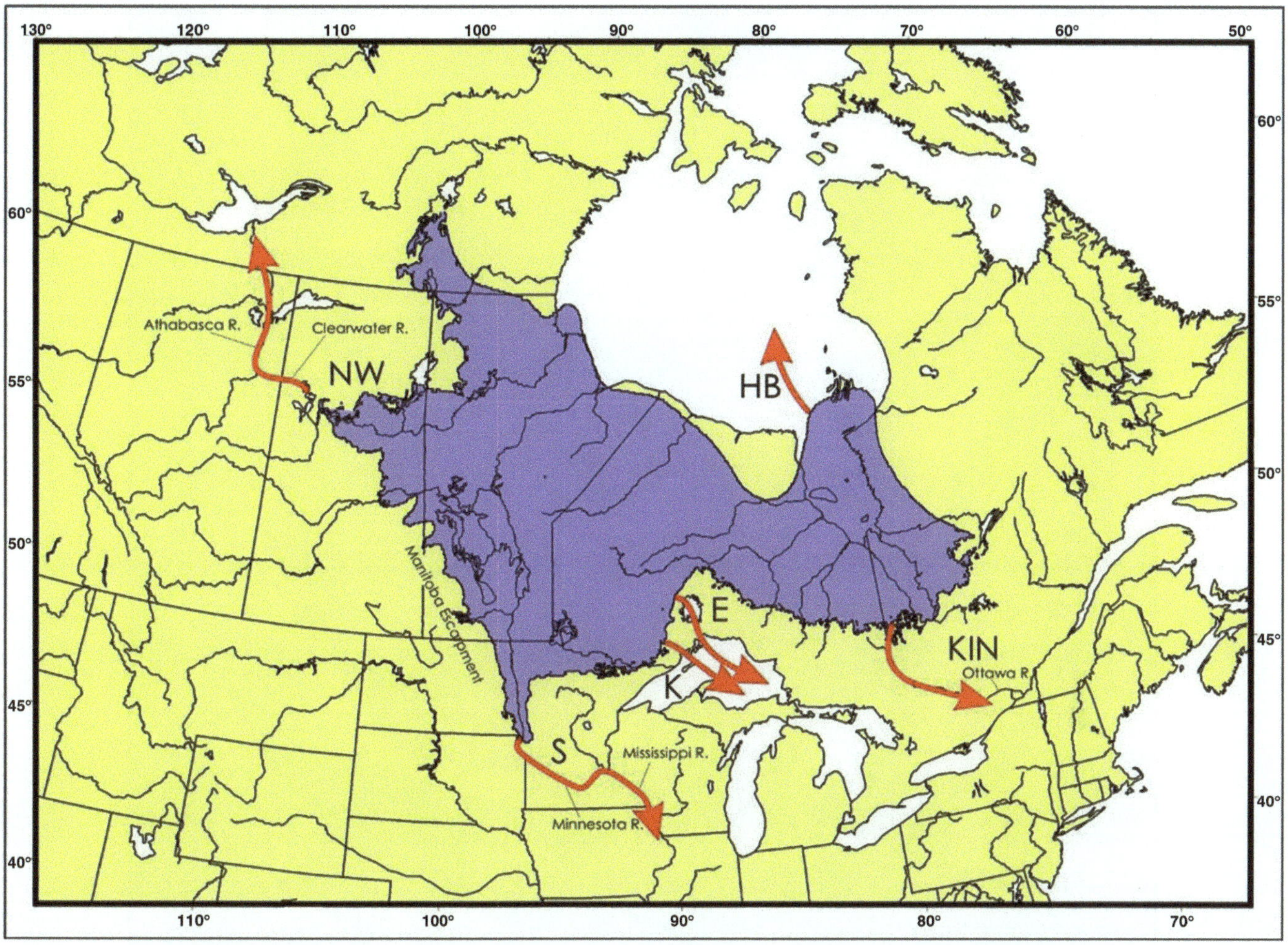

Fig. 1.56 Lake Agassiz is the largest of the ice-margin lakes that once covered what are now parts of Manitoba, Ontario, and Saskatchewan in Canada with North Dakota and Minnesota in the United States. Its largest extension reached 1200 km W-E, and extended 700 km N-S being smaller than the outlined area signifying all the regions once covered by the lakes' different phases between 14,000 and 11,700 years BP (Late-Pleistocene epoch). The Laurentide inland ice body in the north blocked Lake Agassiz so that the waters of the Great Plains could not flow into the present Hudson Bay. As a result, many rivers backed up creating lakes of different sizes along the ice margin. Depending on their water levels, many rivers emptied out into the NW, N, SE and S, (e.g. via the St. Lawrence River, the Mississippi River system, or the Athabaska River). The last outbreak—following the ice margin retreating to the north—was into Hudson Bay (Image credit: Leverington and Teller 2003)

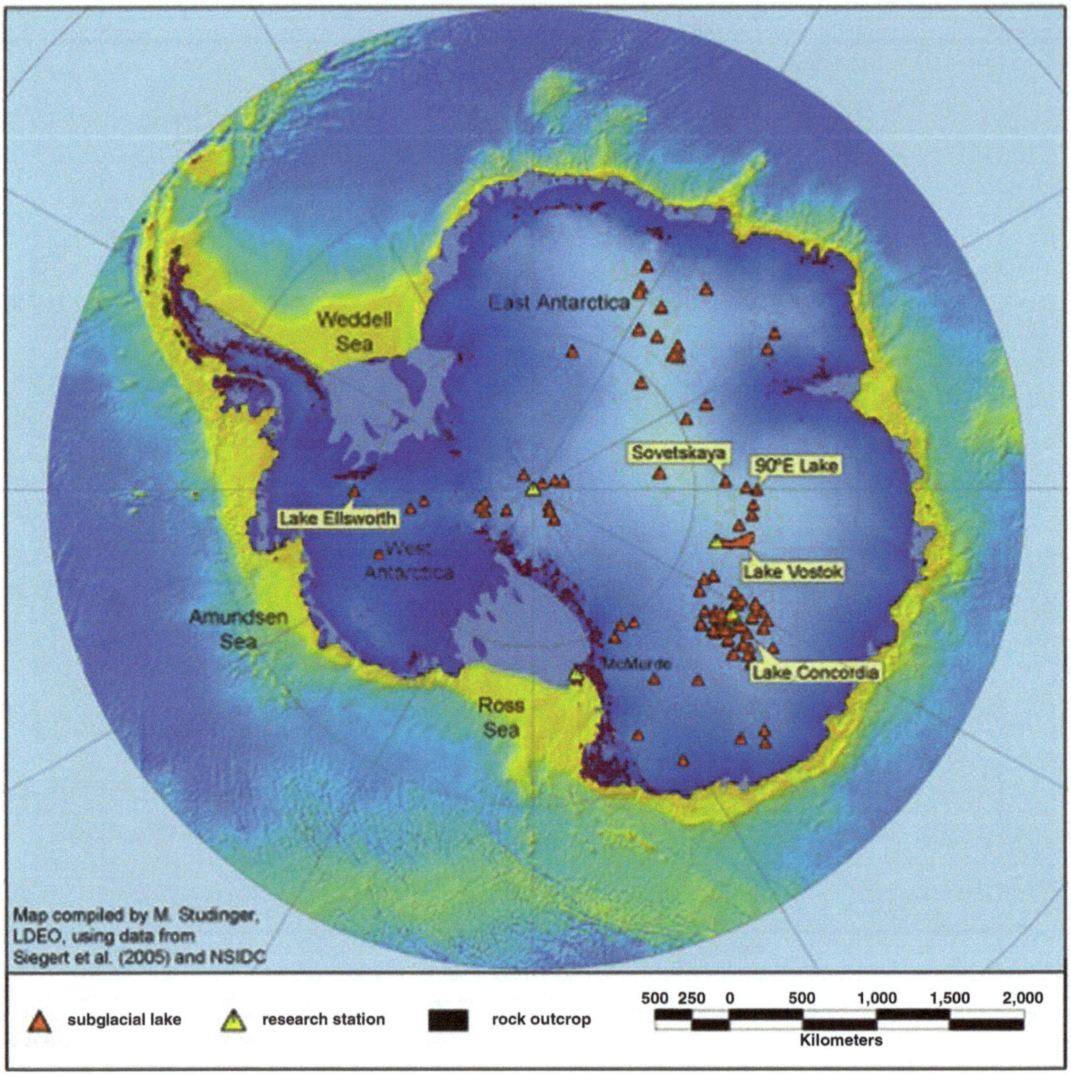

Fig. 1.57 Distribution of lakes under Antarctic ice (2010), the largest is Lake Vostok (Image credit: Michael Studinger, Lamont-Doherty Earth Observatory, Columbia University, New York, USA)

Fig. 1.58 Meltwater lake (700 m across) on southern Greenland inland ice, at *64°01'04.36"N, 48°38'52.49"W* and 1600 m asl (Image credit: ©Google earth 2015)

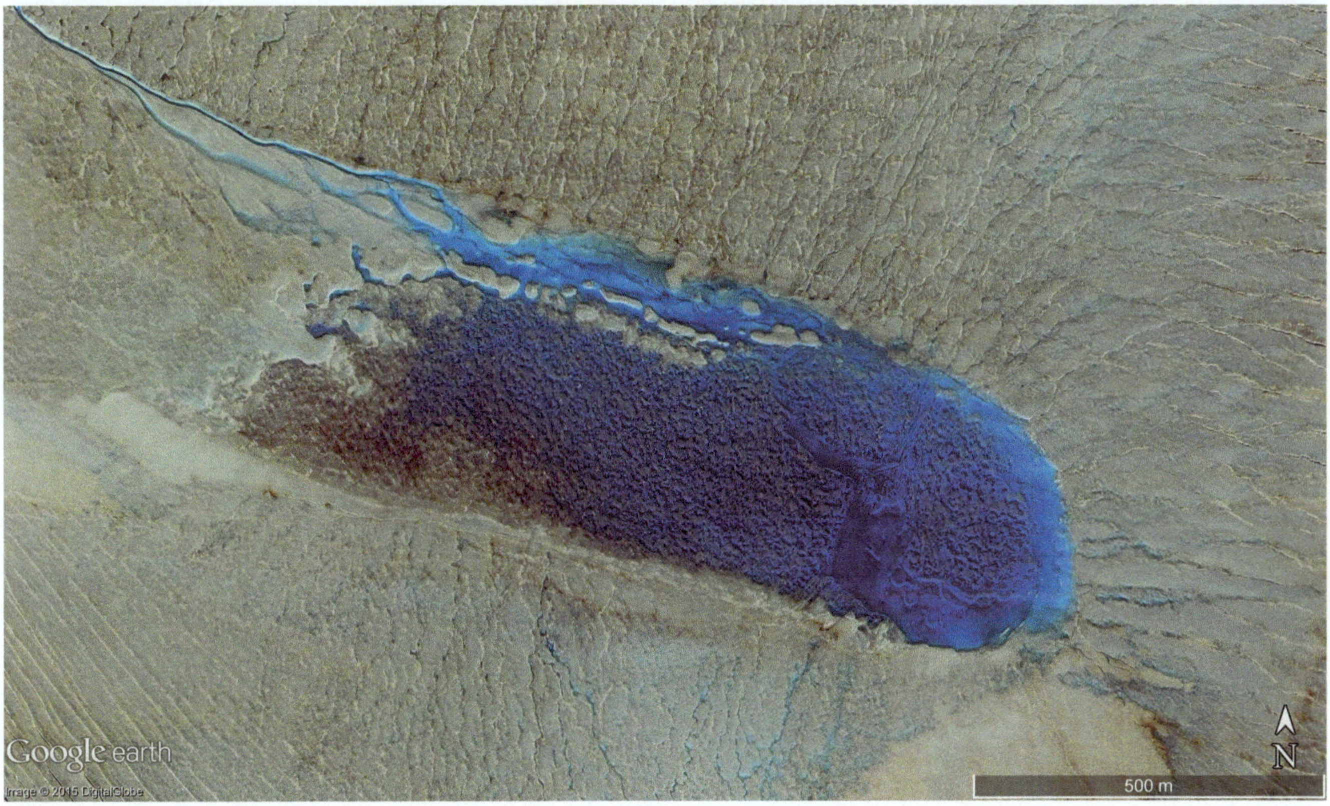

Fig. 1.59 Fed by a meltwater river this shallow lake 1.2 km in length formed in summer 2014 on Greenland inland ice at *64°01'33.29"N* and *48°35'42.63"W* and above 1700 m asl (Image credit: ©Google earth 2015)

Fig. 1.60 (**a**) The 5000 km^2 enormous tongue of the piedmont glacier Malaspina in the Alaska Panhandle *(59°53'N, 140°04'W)* exhibits a thick cover of morainic debris allowing plants to grow on this nearly stagnant glacier ice close to terminus. This 4 km wide scene presents many circular ponds filled with meltwater surrounded by glacier ice. Indeed, the right (*greenish*) part of the image shows well-developed plants such as Willow bushes (*Salix* sp.). (**b**) Detail of glacier ice exposed around the lakes with several small icebergs. Largest lake is 160 m long. Willow bushes grow on the morainic cover (Images credits: ©Google earth 2014)

References

Barnett PJ (1985) Glacial retreat and lake levels, north-central Lake Erie Lake Basin, Ontario. Quaternary Evolution of the Great Lakes. Geol Assoc Canada Spec Paper 30:185–194

Beadle LC (1981) The inland waters of tropical Africa. An introduction to tropical limnology, 2nd edn. Longman, London

Beltran R, Botts L et al (1995) The Great Lakes: an environmental Atlas and resource book, 3rd edn. United States Environmental Protection Agency and Government of Canada, Downsview/Chicago

Bolsenga SJ, Herdendorf CE (1993) Lake Erie and Lake St. Clair handbook. Wayne State University Press, Detroit

Brønmark C, Hansson LH (2004) Biology of lakes and ponds, 2nd edn. Oxford University press, New York

Burgis MJ, Morris P (2007) The world of lakes: lakes of the world. The Freshwater Biological Association, Ambleside

Carpenter SR, Kutchell JE (1993) The trophical cascade in lakes. Cambridge University Press, Cambridge

Chrétien J (2006) The Great Lakes of Africa: two thousand years of history, translated by Scott Straus. Zone Books, Cambridge

Coakley JP (1992) Holocene transgression and coastal-landform evolution in northeastern Lake Erie, Canada. In: Fletcher CH, Wehmiller JF (eds) Quaternary coasts of the United States: marine and Lacustrine systems, Society for Sedimentary Geology, Special Publication 48. Cambridge University Press, Cambridge, pp 415–426

Coe MT, Foley JA (2001) Human and natural impacts on the water resources of the Lake Chad basin. J Geophys Res 106(D4):3349–3356

Cohen A, Soreghan M, Scholz CA (1993) Estimating the age of formation of lakes: an example from Lake Tanganyika, East African rift system. Geology 21:511–514

Delvaux D, Moeys R, Stapel G et al (1997) Paleostress reconstructions and geodynamics of the Baikal region, Central Asia, part 2, Cenozoic rifting. Tectonophysics 282:1–38

Dempsey D (2004) On the brink: the Great Lakes in the 21st century. Michigan State University Press, East Lansing

Downing JA, Prairie YT, Cole JJ et al (2006) The global abundance and size distribution of lakes, ponds, and impoundments. Limnol Oceanogr 51(5):2388–2397

Drake N, Bristow C (2006) Shorelines in the Sahara: geomorphological evidence for an enhanced monsoon from palaeolake Megachad. The Holocene 16(6):901–911

Ellis WS (1990) The aral: a soviet sea lies dying. Natl Geogr 177(2):73–93

Firoozfar A, Bromhead EN, Dykes AP (2012) Caspian sea level change impacts regional seismicity. J Great Lakes Res 38(4):667–672

Fritz SC, Baker PA, Seltzer GO et al (2007) Quaternary glaciation and hydrologic variation in the South American tropics as reconstructed from the Lake Titicaca drilling project. Quat Res 68(3):410–420

Fritz SC, Baker PA, Tapia P et al (2012) Evolution of the Lake Titicaca basin and its diatom flora over the last ~370,000 years. Palaeogeo, Palaeoclim. Palaeoecology 317(318):93–103

George G (ed) (2010) The impact of climate change on European Lakes. Aquatic ecology series 4. Springer, Dordrecht

Gibson JJ, Prowse TD, Peters DL (2006) Partitioning impacts of climate and regulation on water level variability in Great Slave Lake. J Hydrol 329(1):196

Grady W (2007) The Great Lakes: the natural history of a changing region. Greystone Books, Vancouver

Hayworth EY, Lund JWG (1984) Lake sediments and environmental history. Leicester University Press, Leicester

Herdendorf CE (1990) Distribution of the world's largest lakes. In: Serruya C, Tilzer MM (eds) Large lakes: ecological structure and function. Springer, Berlin, pp 3–38

Holcombe TL, Taylor LA, Reid DF et al (2003) Revised Lake Erie post-glacial lake level history based on new detailed bathymetry. J Great Lakes Res 29(4):681–704

Johnson TC, Kelts K, Odada E (2000) The holocene history of lake victoria. Ambio 29(1):2–11

Kalff J (2002) Limnology. Prentice Hall, Upper Saddle River

Kar D (2014) Wetlands and lakes of the world. Springer, Dordrecht

Kasperson J, Kasperson R, Turner BL (1995) The Aral sea basin: a man-made environmental catastrophe. Kluwer Academic Publishers, Boston

Lampert W, Sommer U (1997) Limnoecology: the ecology of lakes and streams. Oxford University Press, New York

Larson G, Schaetzl R (2001) Origin and evolution of the Great Lakes. J Great Lakes Res 27(4):518–546

Lerman A, Imboden D, Gat J (eds) (1995) Physics and chemistry of lakes, 2nd edn. Dordrecht, Springer

Leverington DW, Teller JT (2003) Paleotopographic reconstructions of the eastern outlets of glacial Lake Agassiz. Can J Earth Sci 40:259–127

Likens GE (ed) (2009) Historical estimates of limnicity. Encyclopedia of inland waters. . Elsevier, Amsterdam

Link PK, Kaufman D, Thackray GD (1999) Field Guide to Pleistocene Lakes Thatcher and Bonneville and the Bonneville Flood, Southeastern Idaho. In: Hughes SS, Thackray GD (eds) Guidebook to the geology of Eastern Idaho. Idaho Museum of Natural History, Pocatello, pp 251–266

Meybeck M (1995) Global distribution of lakes. In: Lerman A, Imboden DM, Gat J (eds) Physics and chemistry of lakes. Springer, New York, pp 1–35

Micklin P (2007) The aral sea disaster. Annu Rev Earth Planet Sci 35(4):47–72

Nevers MB, Byappanahalli MN, Edge TA et al (2014) Beach science in the Great Lakes. J Great Lakes Res 40(1):1–14

Pala C (2006) Once a terminal case, the North Aral Sea shows new signs of life. Science 312:183

Pyne AJ (1986) The ecology of tropical lakes and rivers. Wiley, Chichester

Riley JL (2013) The once and future Great Lakes country: an ecological history. McGill-Queen's University Press, Montreal

Roche MA, Bourges J, Cortes J, Mattos R (1992) Climatology and hydrology of the lake Titicaca basin. In: Dejoux C, Iltis A (eds) Lake Titicaca: a synthesis of limnological knowledge. Kluwer Academic Press, Boston, pp 63–88

Sarch MT, Birkett C (2000) Fishing and farming at Lake Chad: responses to lake-level fluctuations. Geogr J 166:156–172

Schwoerbel J, Brendelberger H (2005) Einführung in die Limnologie, 9th edn. Springer-Spektrum, Heidelberg

Scott DL, Etheridge MA, Rosendahl BR (1992) Oblique-slip deformation in extensional terrains: a case study of the lakes Tanganyika and Malawi Rift Zones. Tectonics 11(5):998–1009

Serruya C, Pollinger U (1983) Lakes of the warm belt. Cambridge University Press, Cambridge

Stager C, Johnson TC (2008) The late Pleistocene desiccation of Lake Victoria and the origin of its endemic biota. Hydrobiologia 596(1):5–16

Stumm W (ed) (1985) Chemical processes in Lakes. Wiley, New York

Stumm W, Morgan JJ (1996) Aquatic chemistry, 3rd edn. Wiley-Interscience, New York

Taub FB (ed) (1984) Lakes and rivers. Ecosystems of the world, vol 23. Elsevier, Amsterdam

Tilzer MM, Serruya C (1990) Large lakes. Ecological studies and function. Springer, Berlin
UNESCO (1998) Ecological research and monitoring of the Aral Sea deltas, a basis for restoration, Aral Sea Project 1992–1996, Final science report. UNESCO, Paris
UNESCO (2000) Water-related vision for the Aral Sea Basin for the year 2025. UNESCO, Paris
Van der Leeden F, Troise FL, Todd DK (eds) (1991) The water encyclopedia, 2nd edn. Lewis Publication, Chelsea, pp 198–229
Waples JT (2008) The Laurentian Great Lakes. In: Hales B, Cai WJ, Mitchell BG (eds) North American continental margins. A Synthesis and Planning Workshop, Washington, DC, pp 73–81
Wetzel RG (2001) Limnology. Lake and river ecosystems, 3rd edn. Academic Press, San Diego
Wilcox DA, Thompson TA, Booth RK, Nicholas JR (2007) Lake-level variability and water availability in the Great Lakes. U.S. Geological Survey Circular, 1311
Ylvisaker A (2004) Lake Ontario. Capstone, North Mankato

Important Pre-requisites for the Existence of Lakes: Basin and Depression Forming Processes

2

Abstract

Abstract for Sect. 2.1

All lakes are positioned in basin-like forms on continental surfaces. Many lakes are so large it is difficult to identify these depressions such as the Caspian Sea, Victoria Lake, or the Aral and Chad Lake basins, while other depressions are more significant like those of the North American Great Lakes and Lake Baikal. Depressions can be smooth, shallow and very wide while others are narrow and elongated as observed between folded mountain chains or along rift valleys (Eastern Africa). We can surmise that basins/depressions formed by passive or active crustal movements (tectonics/endogenic processes) are comparatively old (up to several millions years). In contrast, a group of volcanic basins (resemble craters) also of endogenic origin are primarily younger and several simply appear and disappear within a few years. In a geological sense 'young' translates to a million years or less and includes basin-like forms carved into the landscape by exogenic processes from rivers, ice streams and impact craters (1 billion years to just decades old—not necessarily the age of the lakes in these craters and in any case younger than the impact whereas in many cases a great deal younger). All the basins/depressions described so far originally formed in such a manner. If endogenic forces become inactive, the depressions successively infill with sediments and reduce the size, in particular depth. In addition, basins can be breached and emptied if valleys cut backward into higher ground and in geological time scales, this is the most important process of the degradation and emptying of basins. Many lake-containing depressions result from the unearthing of a structural or petrographic weakness uncovered through denudation or erosive processes (wind, water, or glacial ice) and in particular is true for the geometric configurations of lake clusters overlying fracture and joint systems. In this instance, the overall pattern (fractures, joints) may be old and the transformation into open depressions young. The solubility of rock (limestone, anhydrite, gypsum) in water and weak acid solutions over thousands to millions of years leads to the water disappearing into fissures and underground caves. Many closed depressions form in these landscapes, but because of the permeability of the rock, lakes in these depressions are rare. A large group of lakes (and ponds) develop in basins blocked by depositional processes, for instance by damming a former open valley by rock fall, landslide, glacial moraines, or even a living glacier. Many lakes and swamps can be found along all the shorelines of the world such as lagoons separated from the open sea by accumulative processes of waves and storms. Large numbers of depressions also occur in permafrost regions as thermo-karst lakes or small ponds within the polygons of patterned ground.

A.M. Scheffers, D.H. Kelletat, *Lakes of the World with Google Earth*, Coastal Research Library 16,
DOI 10.1007/978-3-319-29617-3_2

Abstract for Sect. 2.2
With the exception of cosmic impacts, the processes that sculpt the surface of the earth (exogenic) are based on weathering processes transforming solid base rock into fragments transportable by moving forces (erosion). Gravity is present everywhere and is the pre-requisite force for most movements. Depending on the mechanical/physical processes of weathering (extension and shrinkage through temperature changes, or joint fracture by frozen water), including chemical processes in the solution of rock, different types and amounts of loose material are produced. The downward dislocation of loose material relies on the manner of moving forces, the size of individual fragments, the combined mass of fragments, and overcoming friction. The erosive forces of wind can only transport fine-grained particles such as clay, silt and sand, in exceptional cases (and for short distances) small pebbles. In contrast to other exogenic processes, wind can also dislocate and transport clasts upward (against gravity), evident in the formation of dunes. Water flows downward requiring consequent (although not uniform) inclination and has the ability to transport clasts the size of boulders when the inclination is steep enough and water supply is strong. Glaciers are the strongest active moving force on Earth (easily move house-sized clasts) and depending on its nourishment a glacier creeps downhill and for short distances even moves uphill. All kinds of weathering and landforms are described and discussed in an earlier book on "Landforms of the World with Google Earth—understanding our environment" (A.M. Scheffers, S.M. May and D.H. Kelletat, Springer 2015). In this chapter we present examples for basins and depressions partly filled by water and formed by the principal exogenic processes that sculpture the Earth surface. Flowing water (as in rivers) typically create depressions in the landscape that are closed along three sides (along both slopes and from the upper origin of the valley) and open to the direction of flow; for this reason this process does not support closed basin formation of lakes and ponds, only if the flow is blocked (by a landslide or rock fall), a basin may then occur. Further, many morphological depressions in dune landscapes occur between the sand hills, but the permeability of sand does not support lakes. Therefore, glacier ice is the best method to actively scour closed and even several hundred metre deep basins. Because of glaciers scouring deep into the crust, basins can easily fill with groundwater. Other depressions are the product of depositional processes that leave dams and closed basins behind. In this respect, morainic ridges left as dams by glaciers as they recede are particularly effective. Another group of lakes are formed by blocks of ice that are separated from the main glacier that eventually melt and leave behind depressions (holes) that fill with water to become kettle lakes. As a sub-chapter, depressions formed by the impacts of extraterrestrial objects (asteroids, meteorites, comets), undoubtedly also exogenic in nature, are also discussed.

2.1 Lake Basins Formed by Endogenic Processes: Tectonics, Structural Control and Volcanism

2.1.1 Lake Forms Depending on Tectonics and Structural Control

Tectonics explains processes of movements in the Earth's crust that belong to the category of inner dynamics (endogenic) and form *structures* as a framework for *sculpturing* (exogenic processes) by weathering, gravity and the many transport processes by water, wind, or glacier ice and ground frost on the surface.

The term "structural control" sums up all features which have been formed directly or indirectly (visible after erosional processes) by sub-surface (endogenic) forces including vertical tectonic movements, plate tectonics, faults, folds or joint patterns and closed depressions that have been filled by water to form lakes, ponds, or pans. Endogenic features with a volcanic origin are presented in a separate sub-chapter.

Established as a model in the 1960s, plate tectonics or continental drift is the floating of the rigid crust of Earth (from 5 to over 50 km thick) on viscous molten rock called magma, heated to 1200 °C or more by radioactive processes within the Earth's interior. Continents floating with a rock density of approximately 2.5 g/cm^3 (similar to granite) rise high and dip—like a loaded ship—deeper into the magma zone (than the ocean bottom) that has a basaltic composition of density 3.2 g/cm^3. Magma cools in contact with the colder crust and this contact has different levels—deep below continents, high under the oceans—by magma cooling beneath, the continents descend slowly down. Magma rises under the oceans in a circular manner. The result is diverging movement beneath the oceans that can spread the thin magma crust to both sides in the direction to neighbouring continents with the effect of more and more outward spreading of the ocean bottom. Along the spreading zone magma raises, cools down and basaltic rocks fill the gap forming a rise or ridge over the spreading zone. Should only spreading occur our globe would consequently enlarge its surface (and diameter), however this is not the case: where oceanic crust collides with continental crust the low lying and denser section (the ocean floor), is subducted below the continent into the magma zone where it liquefies again.

In a zone where both ridged crusts are moved alongside each other, friction is high and movement occurs in small steps when the moving forces overwhelm friction. These stepwise movements are earthquakes. Friction may also fracture the crust so that magma rises and emerges at the surface as lava—as the result of collision volcanoes are born on the continents or on island chains. Volcanic islands also form in ocean-bottom spreading zones or over "hot spots"—beginning under water, occasionally rising to the surface as new islands (Hawaii, Azores, Canaries, Galapagos and many others). A hotspot remains constant while the plate continues to move over it leaving a trail of volcanoes behind with older volcanoes moving away and younger volcanoes forming over the hotspot. All these movements (dipping down at subduction zones, moving along each other as transform faults, compression of the crust, pushing up of mountains or section of the crust in general) are slow from a human viewpoint of millimetres to tens of centimetres/year maximum. Nevertheless, the resulting formations (as mountain ranges, folds, faults, basins etc.) may survive weathering and erosion/denudation at the surface, apparent with the main structures in the relief around the earth.

Fractures and joints are small internal lines and planes of weakness in rock created by contraction during cooling, or by stress during folds. Most fractures and joints are imperceptible in deeper sections of the rock but open to the surface—as exposed quarries can indicate. The results of weathering are consequences of climatic influences on rock-like extensions by heating during the day (expansion) and cooling at night (contraction), or through water filtering into invisible joints that break the rock through freezing (expansion). Over long time periods (thousands of years or more), joint patterns are visible but too small to give any precedence toward lakes or ponds described in this book. Their significant sizes, forms and patterns are mainly the result of erosion by rivers and in particular glacier ice. In landscapes of repeated glacial erosion with alternating ice-free periods of intensive weathering, variances in rock resistivity and joint patterns are exposed and dominate the number, clustering and individual forms to lakes. These landscapes (as Scandinavia, Canada, Alaska, northern Russia, southernmost South America) exhibit the highest number of lakes and in fact represent 80–90 % of the more than 300 million lakes in the world.

Most of the large structurally controlled basins are shaped by the collision of plates (continental drift), where crustal deformation occurs by compression, leading to folding and faulting. Worldwide, the widest region with this pre-requisite to large structurally controlled basins is in central Asia (Fig. 2.1). A more than 3000 km wide zone from Lake Baikal in the north to the Himalayas in the south demonstrates numerous tectonic basins and depressions of many sizes—the largest is the Tarim Basin with the Taklamakan Desert to the west and the Gobi Desert basin to the east. The lake districts between the Altai Mountains in the north and the vast Tibetan landscapes in the southern section of this compressional zone present many lakes exceeding 50 km in diameter/length, often situated in basins of altitudes over 4000 m (Figs. 2.2, 2.3, 2.4, 2.5, 2.6, 2.7, 2.8 and 2.9).

The second largest region of crustal compression in the world ("basin-and-range" region stretching 1000 km) is the wide undulating landscape found in North America between the Pacific Ocean, central California in the west, and the Wasatch Mountains of eastern Utah (Fig. 2.10). This zone comprises the long wide central valley of California that lies between the Coast Range and the Sierra Nevada and especially the light-coloured basins now filled by shifting sand (or by silty salt crusts from evaporative processes), interrupted by dark coloured ranges often rising 2000–3000 m above the light-coloured basins. Lakes have filled many of the basins during times of colder climates and less evaporation, particularly in the Late Pleistocene epoch from 14,500–12,000 BP when mountain glaciers melted and delivered huge amounts of meltwater (see example of Lake Bonneville/Great Salt Lake in Chap. 1: Introduction). Even today several of the basins can produce shallow lakes in summer after strong precipitation (thunderstorms), or from glacial meltwater in springtime. The light colour of the salt crusts results from evaporation of high amounts of fresh water.

Analogous to North America but with less uniformity, the central Andes in South America exhibit a compressional zone (approximately 750 km wide) with high mountain ranges including dormant and active volcanoes (reaching over 6000 m high), interrupted by wide salt basins (called Salar) and large lakes (Fig. 2.11). These basins are situated on the "Altiplano"—high-elevated flat with wide depressions at almost the same altitude between about 3600 m asl (as in

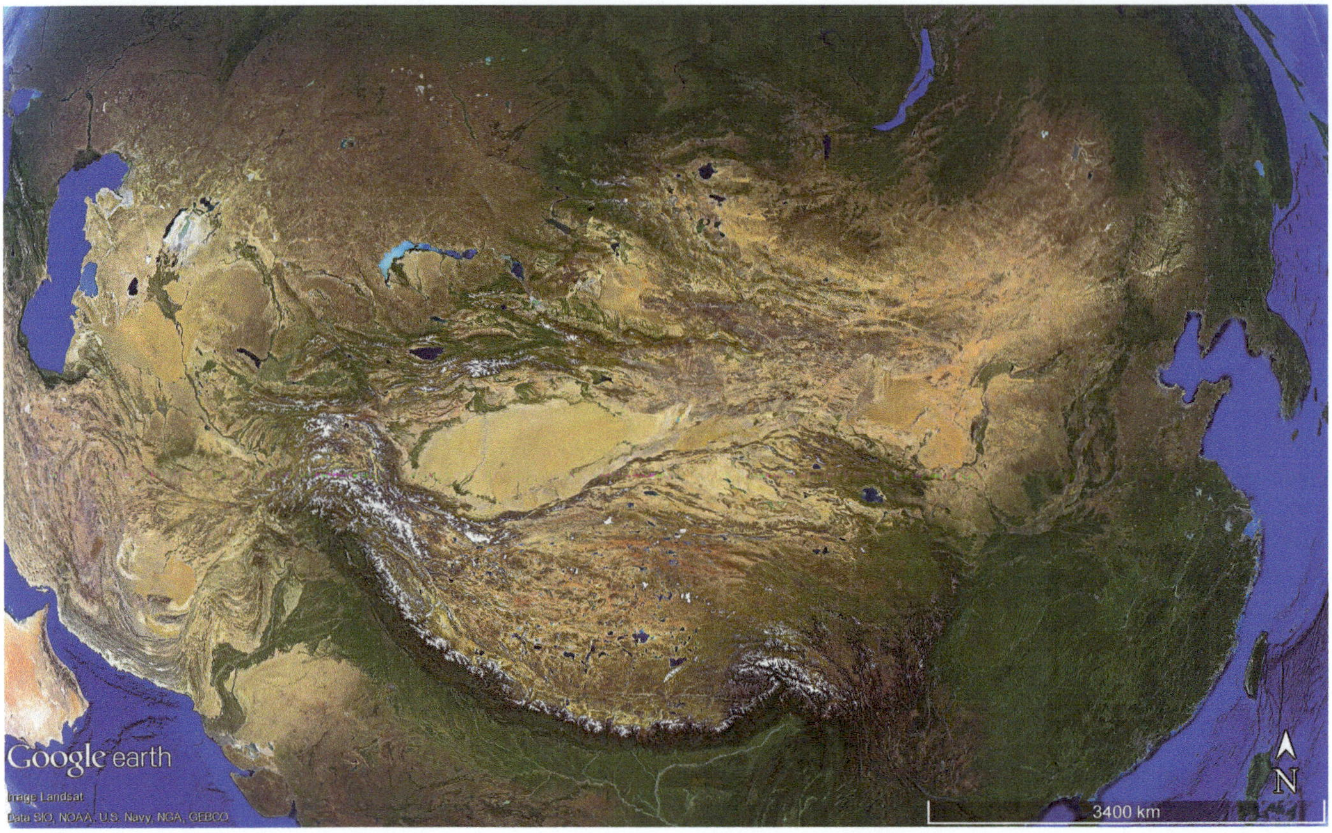

Fig. 2.1 From the Himalayas to the south end of Lake Baikal, structurally-controlled basins extend for more than 3000 km as the Tarim-Basin and the Taklamakan desert to the west, Gobi desert in the east, the Tibetan-Plateau determining the wide southern lake complex, and the western Mongolia lake district in the north, fringed by the Altai-Mountains (Image credit: ©Google earth 2013)

Fig. 2.2 A group of large lakes in structural depressions of the Cenozoic Era rest on the Tibetan Plateau in northern China (central co-ordinates *35°14'N, 91°24'E*—lake levels between 4800 and 5000 m asl). Scene is 225 km wide (Image credit: ©Google earth 2013)

Fig. 2.3 (**a**) Lake fringed by folds of sedimentary rocks of northeast Tibet (*35°14'N, 91°24'E*—lake levels approximately 4800 m asl) in a 52 km wide scene. Lake to the southeast is Lake Cuorendejia where usage has changed its surface area (and water content) significantly: in 1976 Lake Cuorendejia covered 137 km^2 and in 2009 the surface area increased by 50 % to 202 km^2. (**b**) In Northern Tibet lakes lie at 4913 m asl in folding structures fed from snow and glaciers from high mountains (up to 5474 m asl) in the east (approximately *35°32'N, 90°15'E* and 15 km south of Lexie Wudan Lake) (Images credits: ©Google earth 2015)

Fig. 2.4 In northeast Kirgizstan the Altai Mountains encompass Lake Yssykköl (lake length 177 km at 1603 m asl, approximately *42°24'N, 77°16'E*) the second largest mountain lake on earth (after Lake Titicaca). Lake Yssykköl has 6200 km^2 of surface area, 1740 km^3 water volume, and a max. depth of 668 m with an average depth 270 m (Image credit: ©Google earth 2013)

Fig. 2.5 Lake Sayram in northwest China (*44°35'N, 81°12'E*, 30 km long W-E), sits at 2074 m asl, max. depth 90 m, surface area 458 km^2 with volume of 21 km^3) (Image credit: ©Google earth 2013)

Fig. 2.6 Taiyang Lake, Tibet, *35°55'N, 90°38'E*, 4886 m asl, 15.5 km long, fed by glacier meltwaters. Mountains in the south are >5900 m high, in NE up to 6800 m asl. The magnitude 8.1 Kunlun-earthquake of Nov. 14th, 2001, activated the rupture which forms the lakes basin, and beach ridges along its southern shoreline have been displaced (Image credit: ©Google earth 2013)

Fig. 2.7 A 2.1 km long lake in Tibet (*34°36'N* and *92°01'E* at 4696 m asl) embedded in a syncline of red sandstone. Image shows ice cover on lake (Image credit: ©Google earth 2015)

Fig. 2.8 A 6 km long lake (4953 m asl) in northwest Tibet (*34°46'11.06"N* and *88°35'11.79"E*) follows a fault line. The smaller lakes are fringed by cuestas of sedimentary rocks. Smooth surface formations point to former glaciation of the area (Image credit: ©Google earth 2015)

Fig. 2.9 A 10 km wide lake scene in Tibet (approximately *34°45'N, 89°03'E*). Elevation nears 4850 m asl. Lakes follow the least resistant sections of the sedimentary rocks (Image credit: ©Google earth 2015)

Fig. 2.10 Around 1000 km wide structural elements (the Basin-and-Range-Province), from the Coast Range of California in the west to the Wasatch Mountains east of the Great Salt Lake in Utah (USA) (Image credit: ©Google earth 2013)

the largest Salar of the satellite image Salar de Uyuni) and the level of Lake Titicaca in the north at 3810 m asl. In the colder Late Pleistocene period and most probably in former times of the Ice Ages the flat relief of the Altiplano allowed the basins and very large lakes to connect.

In contrast to the compression of crustal sections produced by continental drift leaving wide and often shallow basins, the spreading of the crust creates a rift valley (graben) narrow and with steep flanks (as in East Africa, compare Fig. 2.12). Depressions in the rift contain the long lakes Albert, Edward, Kiwu, Tanganyika, Malawi/Nyassa and Turkana where the discharge area from inner-tropical high precipitation areas is wide. In small catchment areas, ponds or salt flats occur (Fig. 2.13).

Other depressions with a similar outline (i.e. long and narrow) may appear as a result of erosion/denudation along fault lines (e.g. transverse faults) caused by tectonic forces having fractured the rocks through constant pressure and movement. Frequently strong erosional forces form deep basins (with lakes). For instance, glacial erosion is evident in examples from Loch Ness in Scotland on the Caledonian Fault/Great Glen Fault (Figs. 2.14 and 2.15) and in the straight parallel lake forms in the Russian mountains (Fig. 2.16). Strong erosion by Ice Age glaciers often result in unearthing primary patterns of rock resistivity (Figs. 2.17, 2.18, 2.19, 2.20, 2.21, 2.22, 2.23, 2.24, 2.25, 2.26 2.27 and 2.28), whereas fault lines (often accompanied by a band of fractured rock from crustal movements and friction) appear from all types of weathering (Fig. 2.21). Sedimentary rocks that are heterogeneous from stratum to stratum as a result of changing depositional conditions (in deep or shallow waters/oceans, on land by rivers, by evaporation, or from clasts cemented into solid rock), tend to exhibit these differences in cuesta-landscapes (a ridge with a gentle slope on one side and a cliff on the other) with numerous variations of elongated mountain chains and long depressions (Figs. 2.29 and 2.30). Round structural patterns in crystalline rock may indicate batholiths and plutons, old and eroded bulge- or dome-like forms rising from magma bodies (Figs. 2.31, 2.32 and 2.33). The polygonal patterns and intensely crossing of lines found in the landscape (by chance filled with lakes and ponds) are based on fractures and joints in the rock (Figs. 2.34, 2.35 and 2.36) and are usually significantly smaller than those from tectonics like faults or folds.

2.1.2 Lakes Formed by Volcanic Processes

Volcanic deposits as clasts or lava flows are typically permeable due to an intensive jointing process during rapid cooling

Fig. 2.11 In the central part of South America's Andes (Peru and Bolivia), from Atacama Desert along the coast of the Pacific Ocean to Amazon's green mountains and lowlands in the east shows parallel folds from the attachment of a crustal section to the crystalline rocks which date back to the Gondwana supercontinent (approximately 180 Mio years). The main white area is Salar de Uyuni, with Salar de Coipasa directly north of it. The light blue area is Lake Poopo and the large black area to the north is (deep) Lake Titicaca. Structural compression into many parallel folds is best viewed in the eastern section of the Andes. Rivers/valleys back-cutting from the Amazon lowlands open and may empty basins in between the mountain chains—seen in the north central part of the image. The scene is 750 km wide (Image credit: ©Google earth 2013)

and from the high concentration of open pores located between clasts (Fig. 2.37). However over time, the floor of the basin dammed by lavas can become impermeable by the deposition of clay and silt particles from weathering. Irregular-formed lakes occur where lava flows block a part of the pre-existing topography (Figs. 2.38, and 2.39). Water sources for crater lakes may derive from precipitation (including snow or ice melts from high summits), or from ground water when the depressions are situated in a low flat topography. Whether fresh water or saline waters is present depends—as in other lakes—on the balance between the input of water and the rate of evaporation. Lakes situated close to shafts of volcanic activity can have mixtures of different minerals/elements, such as hot vents and other exhalations that often contain sulphur. Small ocean islands where volcanoes shape its main elevation can also have salt water or brackish waters seep through the rocky rims of the craters.

Endogenic processes caused by magma movements below the rigid crust of the lithosphere, develop wide (and often elongated) depressions by tectonic activity. The volcanic group of endogenic processes tend to shape round and small lakes at the place of craters as openings for lavas or pyroclastics (ashes, lapilli, volcanic bombs, pumice) during active phases. Their size is limited because of limited space on top of a mountain. (Figs. 2.40, 2.41, 2.42, 2.43 and 2.44). Round (or oval) volcanic lakes can be larger if volcanic edifices have collapsed into a caldera (Figs. 2.45, 2.46, 2.47, 2.48, 2.49 and 2.50). Calderas develop when a very strong eruption empties the magma chamber below the volcanic edifice and its summit area implodes into this chamber leaving a wide opening (a caldera). Lakes in these depressions may also take alternate forms if calderas or craters partially fill with younger lava, tephra, or rock falls from the crater walls. Not all volcanic eruptions lead to mountain building. Occasionally volcanic eruptions are simply gas explosions of a very short duration (called "maar" from German examples) forming more or less round openings in the ground. (Figs. 2.51, 2.52 and 2.53). If a maar is deep enough and located in a humid climate, it will fill with groundwater and

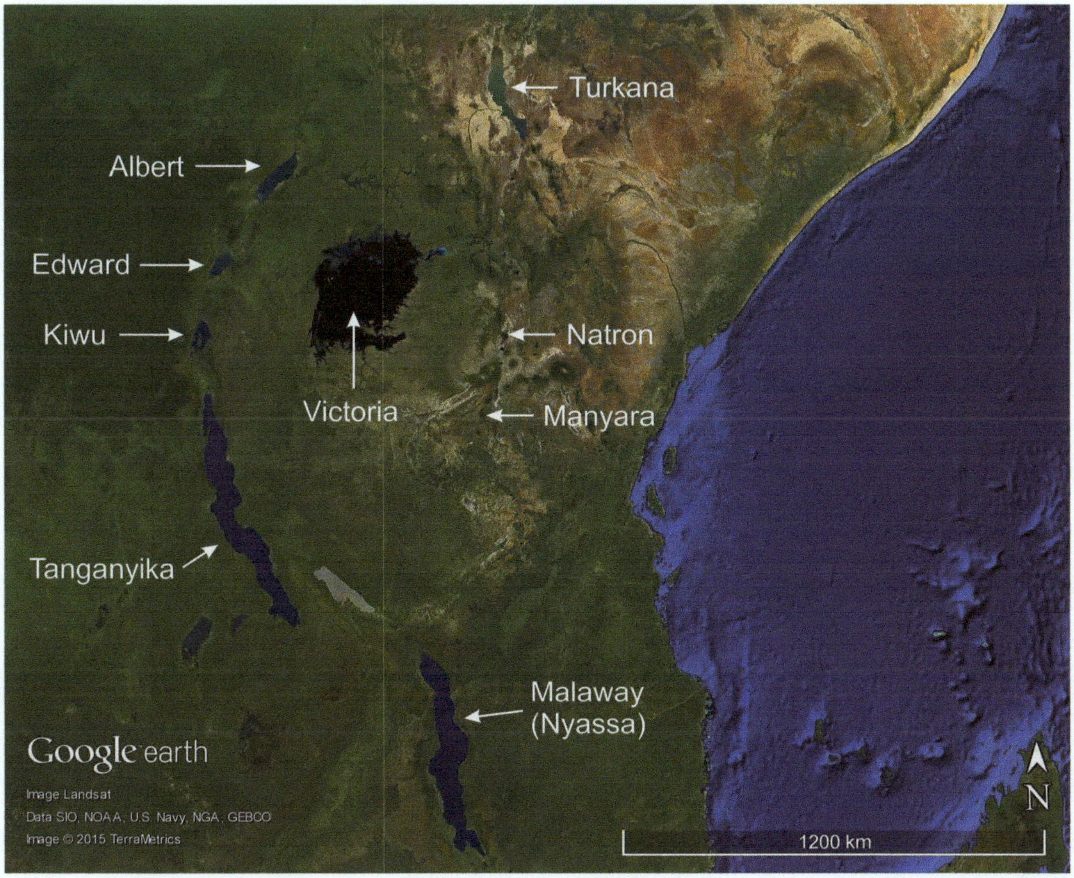

Fig. 2.12 Large East African Rift Valley lakes (compare Table 2.1, from different sources). The five lakes in the western rift valley branch (Albert, Edward, Kiwu, Tanganyika and Malawi/Nyassa) are large and mostly deep (Tanganyika 1470 m, Malawi/Nyassa 706 m, Kiwu 480 m). The lakes in the eastern branch present smaller and at least Lake Natron and Lake Manyara are very shallow. Lake Turkana (Lake Rudolph) is the World's largest alkaline lake (although the water is potable) and Lake Natron has an alkaline brine with high content of different salt minerals (pH is over 12!) and yet is rich in fauna and flora, in particular to the south where fresh water springs occur (Image credit: ©Google earth 2015)

Table 2.1 Approximate data for East African Rift Valley lakes (compiled from various sources)

	Max. length	Surface	Volume	Average/Maximum depth	Surface elevation
Name	km	km²	km²	m	m
Albert	168	5,300	132	25/51	615
Edward	85	2,320	40	17/112	912
Kiwu	98	2,700	500	240/480	1,460
Tanganyika	670	32,900	18,900	570/1,470	773
Malawi (Nyassa)	560	3,000	8,400	292/706	500
Turkana (Rudolph)	285	6,400	203	30/109	360
Natron	52	(variable)	?	<3	600
Manyara	?	230	?	shallow	955

Fig. 2.13 Tadjoura-Basin in Djibouti eastern Africa (*11°55'N* and *42°24'E*), one of the many parallel narrow graben or rift features filled with saltpans or episodic lakes. With of scene is 11 km (Image credit: ©Google earth 2015)

Fig. 2.14 Crossing of two main structural lines in northwest Scotland (*56°57'N, 5° 22'W*): W-E-structures are old and the younger SW-NE-lineament is the Caledonian Fault also known as the Great Glen Fault was active during the Silurian Period about 430–390 Mio years ago. The northern section has moved northeast and the southern section southwest. Scene is 80 km wide (Image credit: ©Google earth 2015)

Fig. 2.15 The Great Glen Fault (Caledonian) in Scotland, with Loch (lake) Ness at the northeast end (lake centre *57°01'N, 4°29'E*). Scene is 112 km wide (Image credit: ©Google earth 2013)

Fig. 2.16 Lakes carved out by glacial erosion along two faults in northeast Russia (around *67°51'N, 66°14'E*). The largest lake is 13 km long (Image credit: ©Google earth 2013)

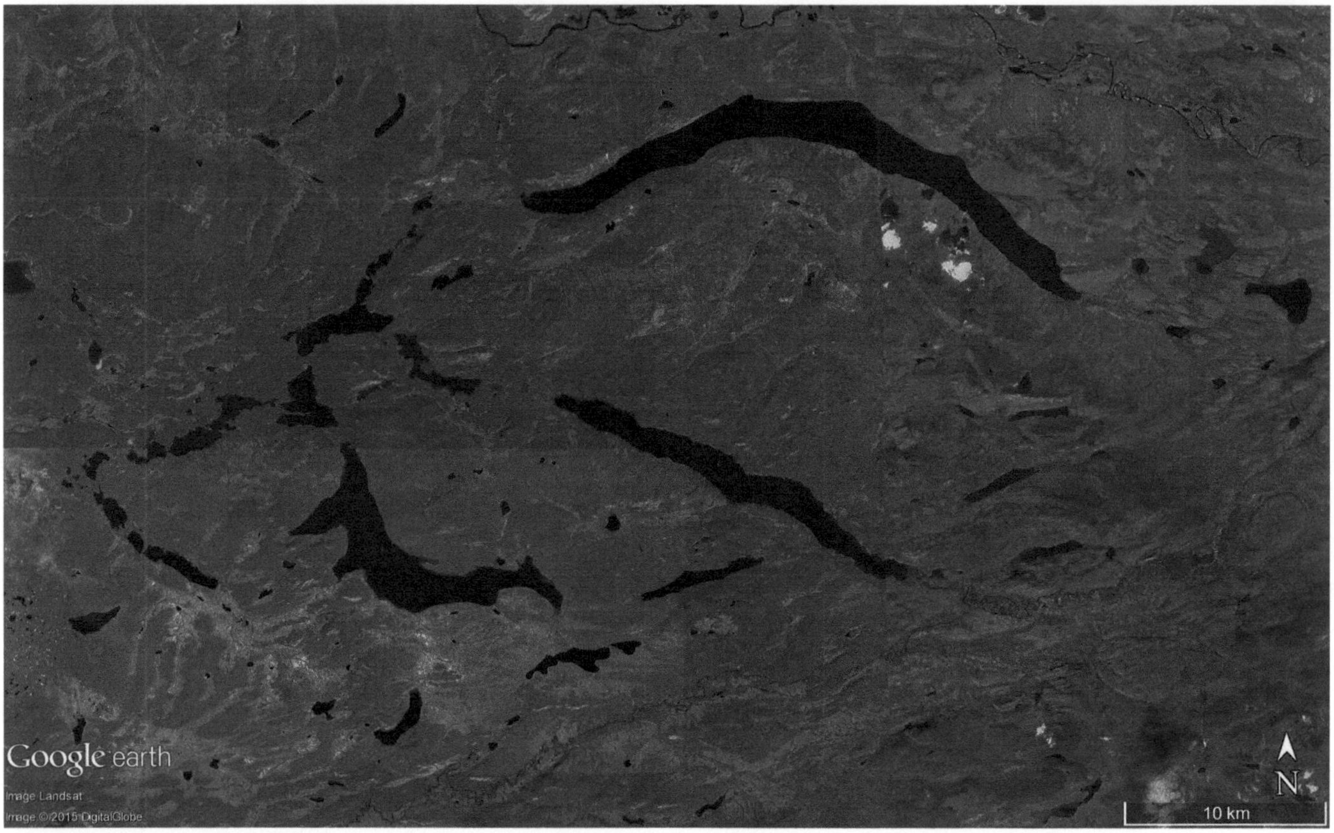

Fig. 2.17 The Tuva Province of south central Russia (around *52°38'N, 97°00'E* and 1100 m asl) exhibits structural elements resulting from the compression between India and Eurasia, exposed by the direction of lake basins (southwest—Many-Khol'Lake, centre—Lake Kadysh, northeast—30 km long Noyan-Khol'Lake) (Image credit: ©Google earth 2015)

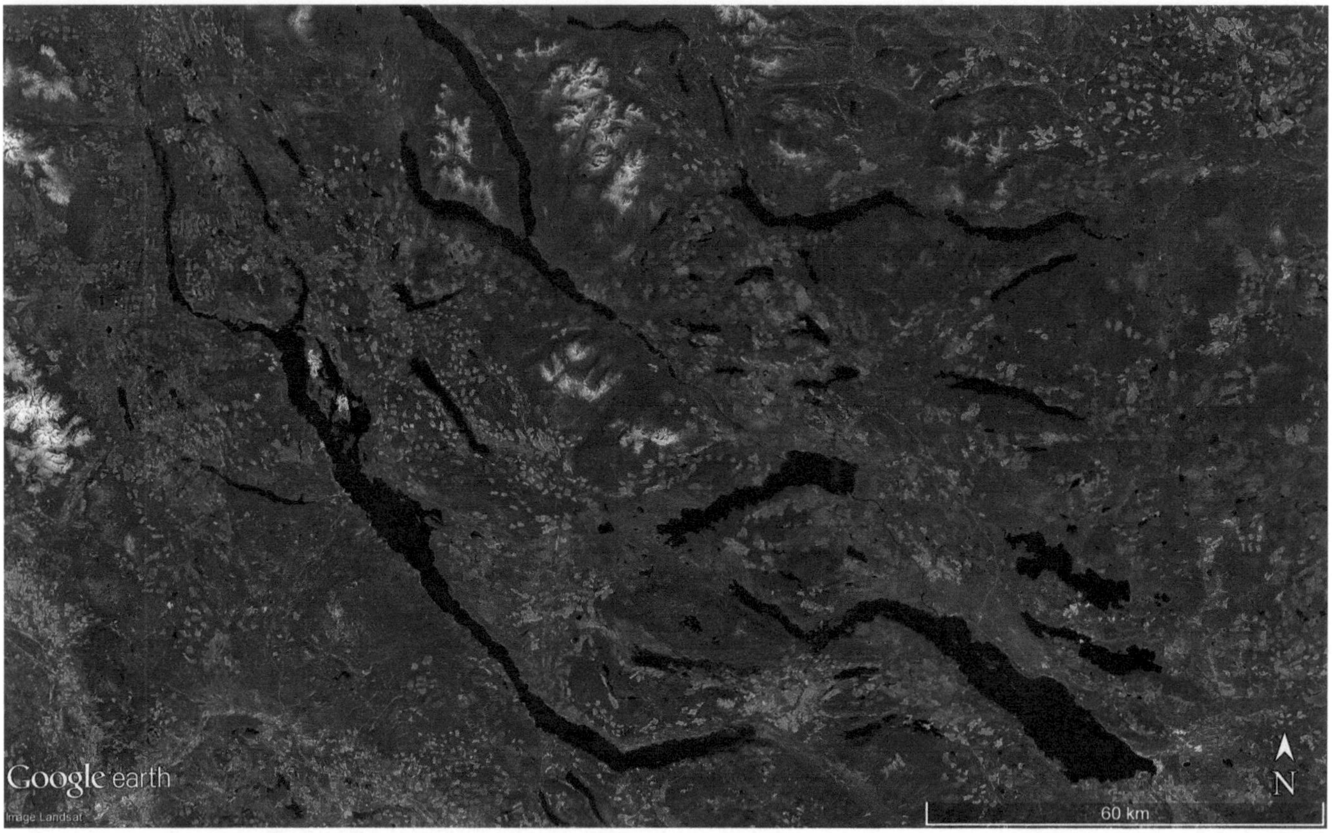

Fig. 2.18 Long lakes in Canada have filled old glacier beds eroded along zones of less resistant crystalline rocks. The 200 km wide scene lies approximately *54°56'N, 125°18'W* (Image credit: ©Google earth 2013)

Fig. 2.19 Lake Wakatipu on New Zealand's South Island (lake centre is *45°05'S, 168°35'E*) has two knick-points following a SW-NE-trending fault. Scene shown is 100 km wide (Image credit: ©Google earth 2015)

Fig. 2.20 Star-shaped Presidente Rios Lake in southern Chile *(46°28'S, 74°12'W)* scoured by glacial erosion following structural weakness during several Ice Ages. Scene is 175 km wide (Image credit: ©Google earth 2015)

Fig. 2.21 In eastern Mongolia *(48°03'09.85"N, 90°10'22.28"E)* at 2579 m asl, a 15 km scene shows a fault line that forms the northeast coast of this small mountain lake (Image credit: ©Google earth 2013)

Fig. 2.22 One of many familiar scenes in northern Scotland *(58°22'N, 4°29'W)* of rock sculpturing, resulting from several glaciations. The different resistivity of the rocks formed basins and ridges; the basins are filled by lakes from the last deglaciation (approximately 13,000 BP). The scene is 5 km wide (Image credit: ©Google earth 2015)

Fig. 2.23 Lakes in Ireland (*55°02'N* and *8°01'W*) oriented by various structural forms and directions in a landscape eroded by several glaciations. Scene is 15 km (Image credit: ©Google earth 2015)

Fig. 2.24 Crossing structural elements in Swedish rocks made visible by the lake contours. Scene is 17 km wide at around *58°48'N, 15°50'E* (Image credit: ©Google earth 2015)

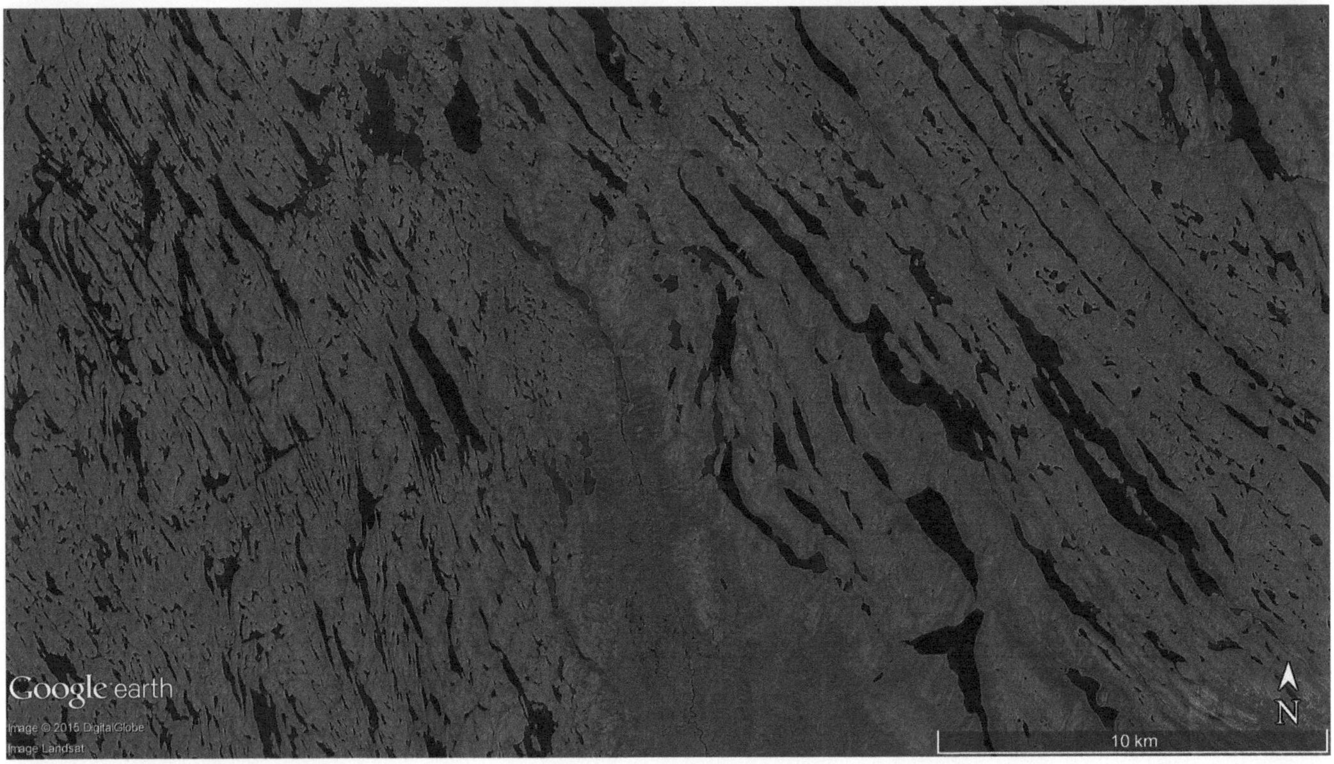

Fig. 2.25 A group of parallel lakes sit in folded and glacially polished rock in Labrador, Canada. Scene is 34 km wide at *58°01'N, 69°41'W* (Image credit: ©Google earth 2013)

Fig. 2.26 The Glaskogen Lake in west Sweden (centered at *59°29'N, 12°28'E*). The lake's shorelines are dominated by variations of resistant and less resistant crystalline rocks. The structure was uncovered by glacial erosion during the Ice Ages. An 18 km wide scene (Image credit: ©Google earth 2015)

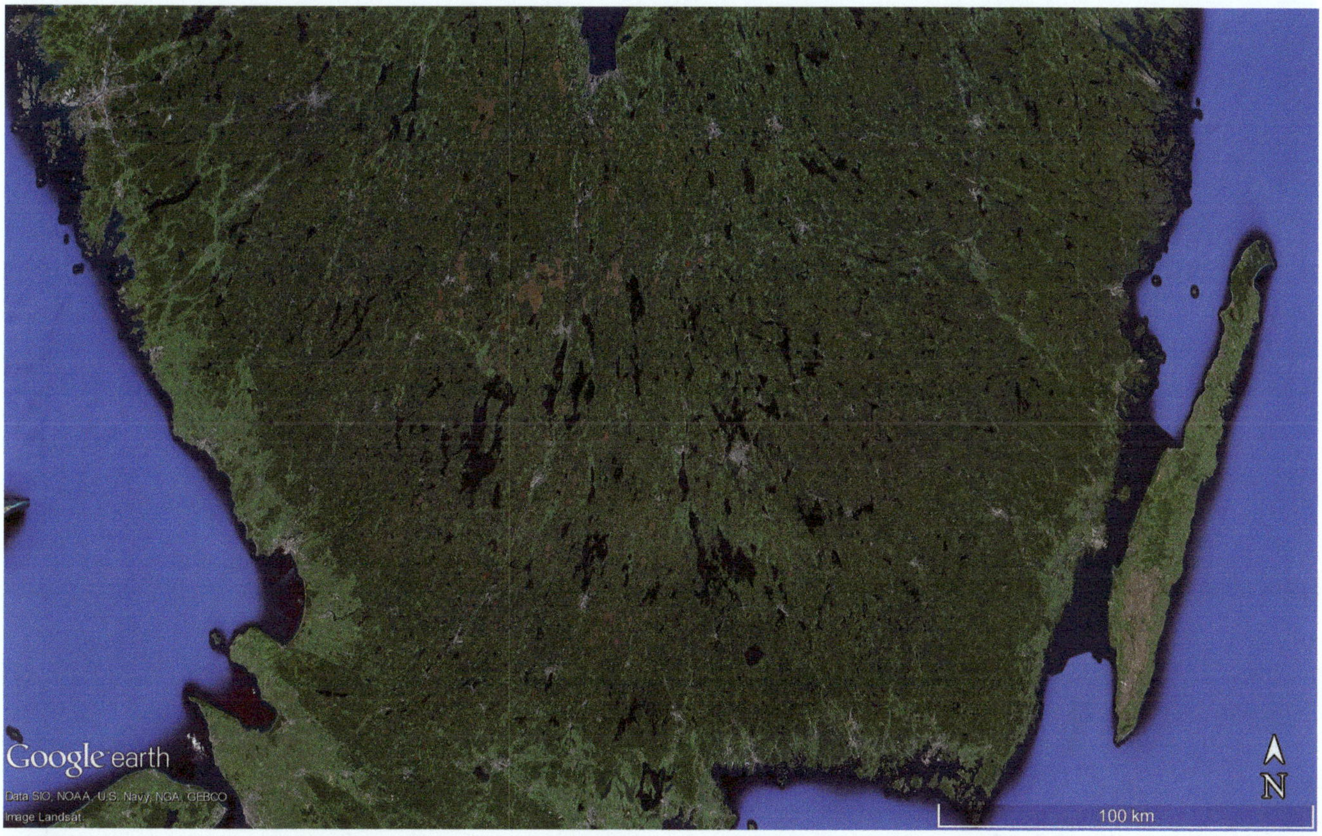

Fig. 2.27 Orientation of the long axes of lakes in southern Sweden demonstrates the direction of glacier divergence from the Ice Ages, and their erosive influence along the less resistant structures of crystalline rocks. Centre of the scene (350 km wide) is at about *57°06'N, 14°14'E* (Image credit: ©Google earth 2013)

wide clusters of maar-lakes occur—as in southern Patagonia of Argentina.

When effusive (lava-producing) or explosive (clast-producing; called pyroclasts signifying fire-particles) volcanism becomes inactive, occasionally post-volcanic activities remain as an exhalation of gas, hot springs, and even geysers. These hot springs are often ponds resembling nearly vertical shafts many tens of metres deep and some exhibit boiling water. The hot water percolates through rock, transports and deposits minerals (as geyserite a siliceous material) where the spring's water evaporates in the thin sheets of overflow around the rim of the pond (Figs. 2.54, 2.55 and 2.56b). Vivid colours may occur in and on these mineral deposits from bacteria (yellow to red colours) or from algae (greenish to black colours) and depending on water temperature; bacteria dominate from 94 to 76 °C and in lower temperatures algae start to grow (Figs. 2.56a, b).

2.2 Lake Basins and Lakes Formed by Exogenic Processes

2.2.1 Impact Crater Lakes

Depressions or basins shaped by the impact of extra-terrestrial objects (meteorites, asteroids, comets) are rare on earth. Only about 200 are known so far, but certainly more wait for detection, in particular using satellite images scanning remote regions of the globe. The earth, moon and other planets are widely covered by impact craters of many sizes suffered from the "Great Bombardment" in the early development periods of the Sun System up to approximately 3.8 billion years ago. In contrast to these cosmic bodies, volcanism and plate tectonics constantly renew Earth's surface and older features disappear through these processes (e.g. by subduction into magma), or by weathering and erosion under

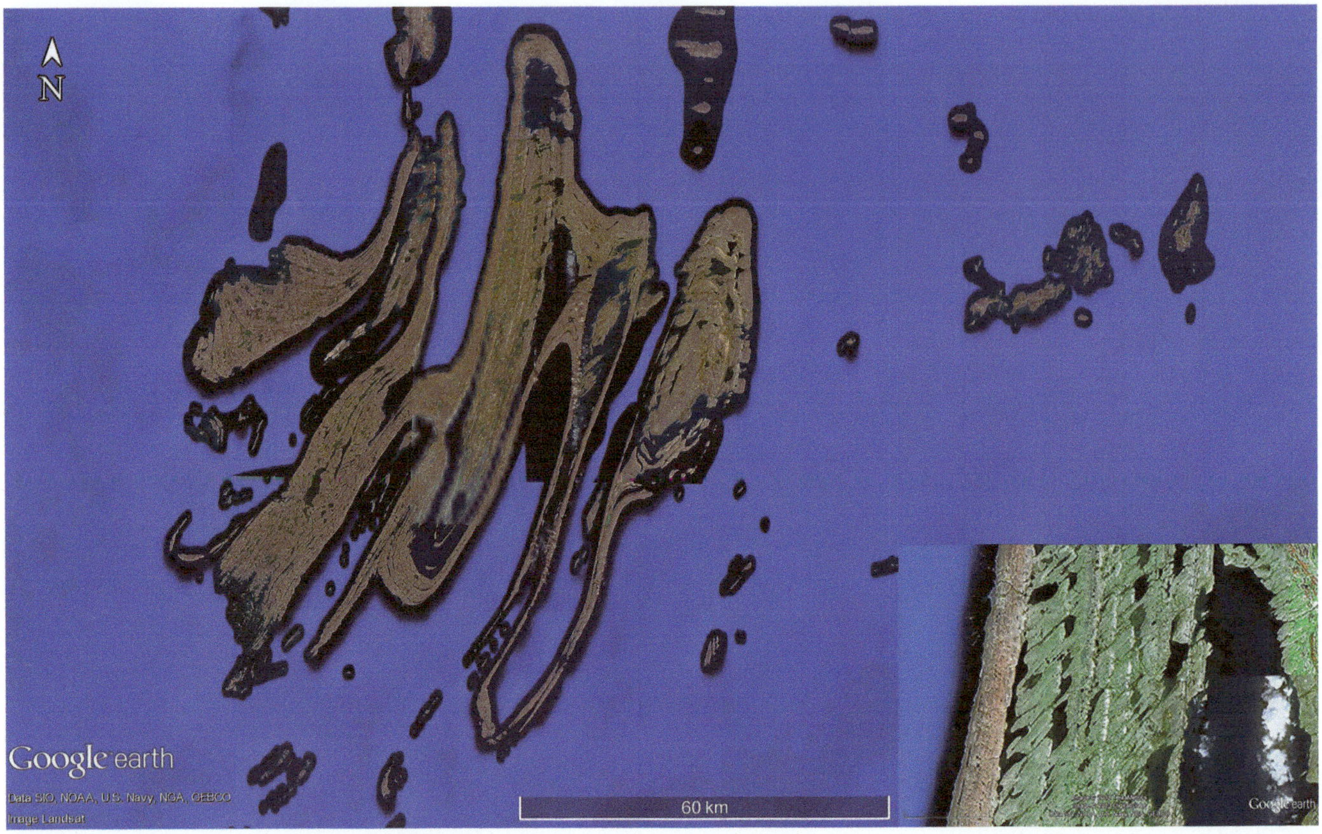

Fig. 2.28 The Belcher Islands (centred at around *56°11'N, 79°22'W*) in southern Hudson Bay of Canada present an excellent example of wide folded structures just above sea level. Lakes on the islands replace the strata of less resistance resulting from glacial erosion of the Ice Ages. Main scene is 210 km wide. Insert (lower right corner) shows a 19 km wide scene from the southwest archipelago with distinct glacial sculpturing from a northeast direction (Image credit: ©Google earth 2013)

Fig. 2.29 Stretching along the northern part of Lake Titicaca (3810 m asl) at the Peru/Bolivia border (*15°19'S* and *69°42'W*), coastlines are controlled by folding structures and cuestas from sedimentary rocks. Width of scene is 33 km (Image credit: ©Google earth 2013)

Fig. 2.30 (**a**) An ephemeral lake approximately 700 m long in western Algeria (*31°55'29.31"N* and *2°01'52.91"W*). The depression is obstructed with debris delivered from both cuestas (Image credit: ©Google earth 2013, image taken April 6th, 2009). (**b**) Ephemeral shallow lake (about 1.5 km long) situated in a syncline of folded sedimentary rocks in Morocco (approximately *29°25'N, 8°02'W*) (Image credit: ©Google earth 2015)

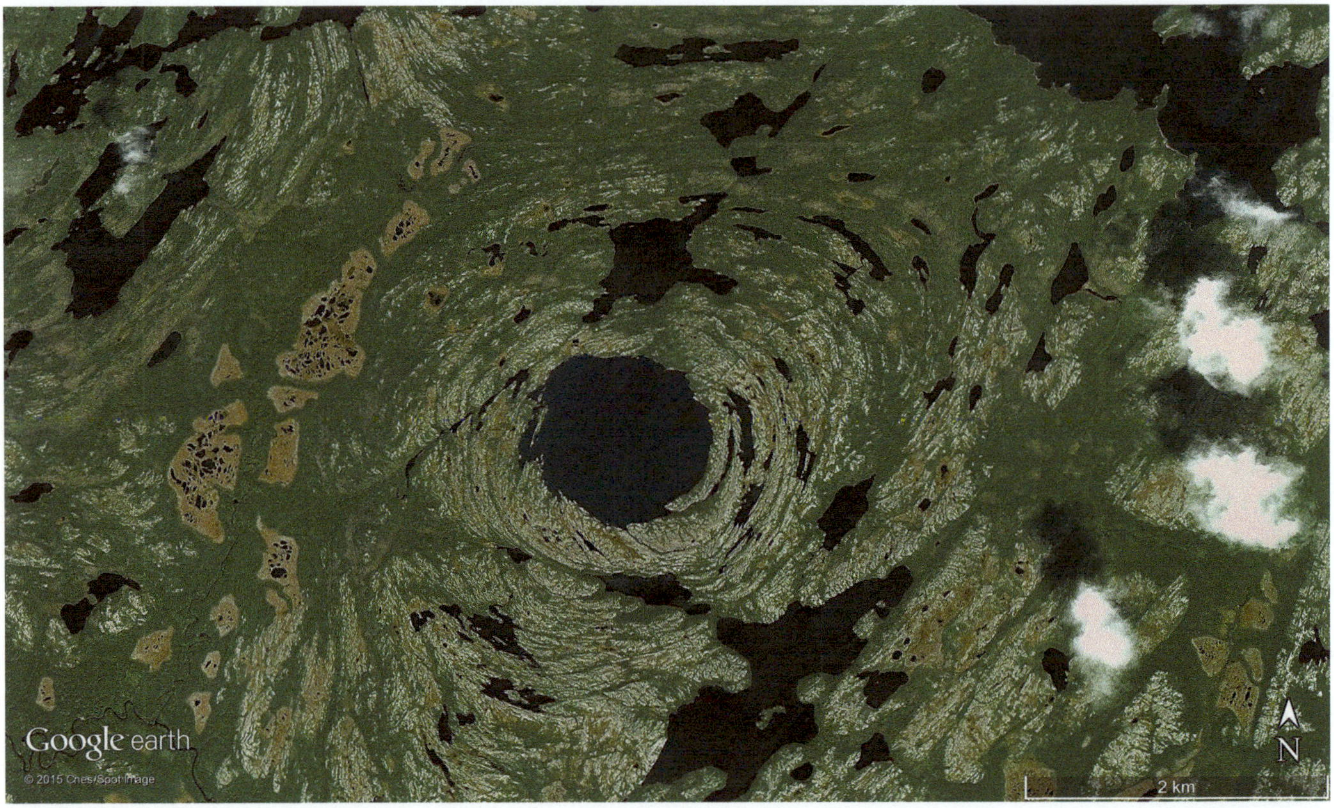

Fig. 2.31 An intrusive body of rock less resistant in the centre has been eroded through several glaciations and now exhibits a lake surrounded by circular structures. The scene stretches 10 km across in eastern Québec (Canada) at *50°50'N, 59°5'W* (Image credit: ©Google earth 2013)

Fig. 2.32 By glacial erosion, a small pluton of crystalline rocks has been unearthed in the Labrador landscape of Québec, Canada. Scene is 5.8 km wide at *50°24'N* and *61°27'W* (Image credit: ©Google earth 2015)

Fig. 2.33 A bulge of hard rock from a crystalline batholith at *50°11'N, 67°51'W* in Québec (Canada) fracturing in all directions. The less resistant parts of the bedrock are eroded by ice streams of the Ice Ages. Scene is 58 km wide (Image credit: ©Google earth 2013)

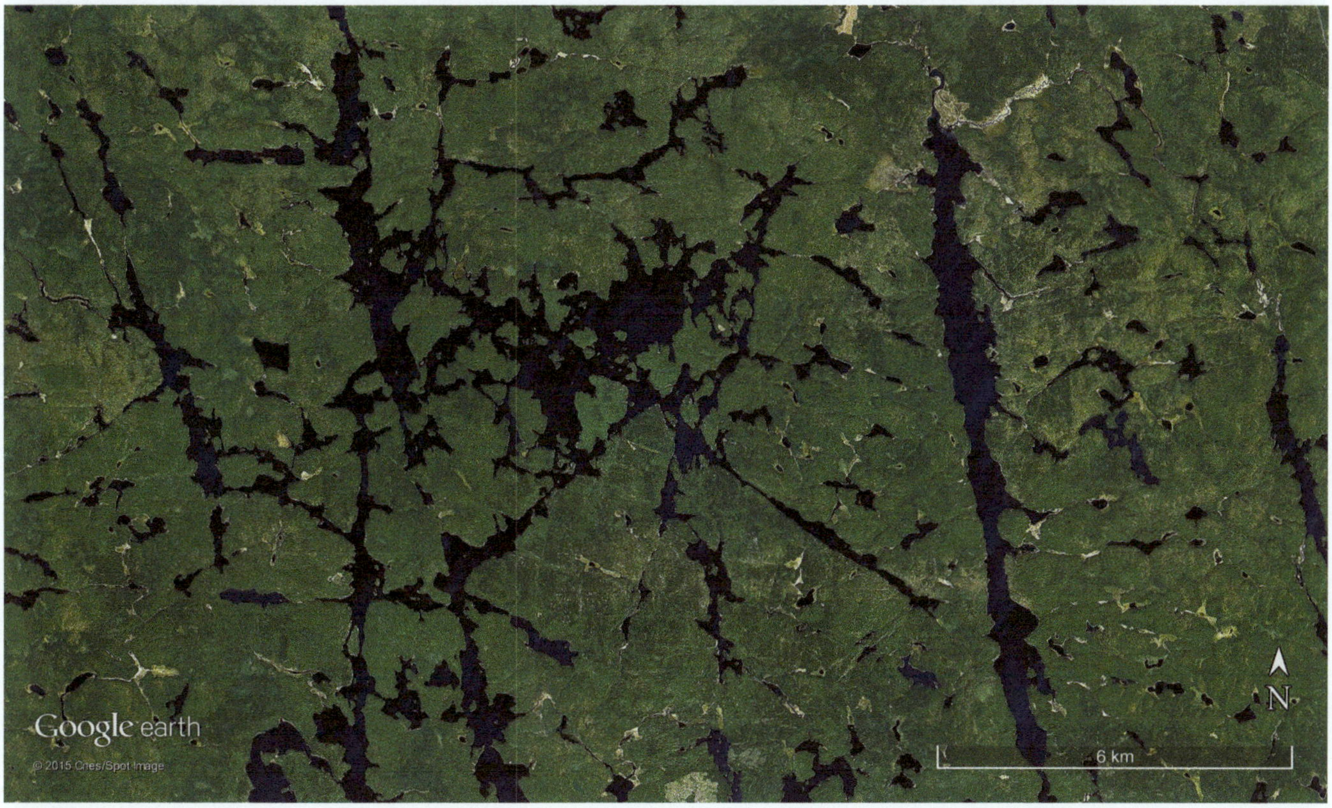

Fig. 2.34 Crossing and eroded joints filled by lakes in southern Canada (approximately *47°05'N, 81°28'W*). Width of scene is 18 km (Image credit: ©Google earth 2013)

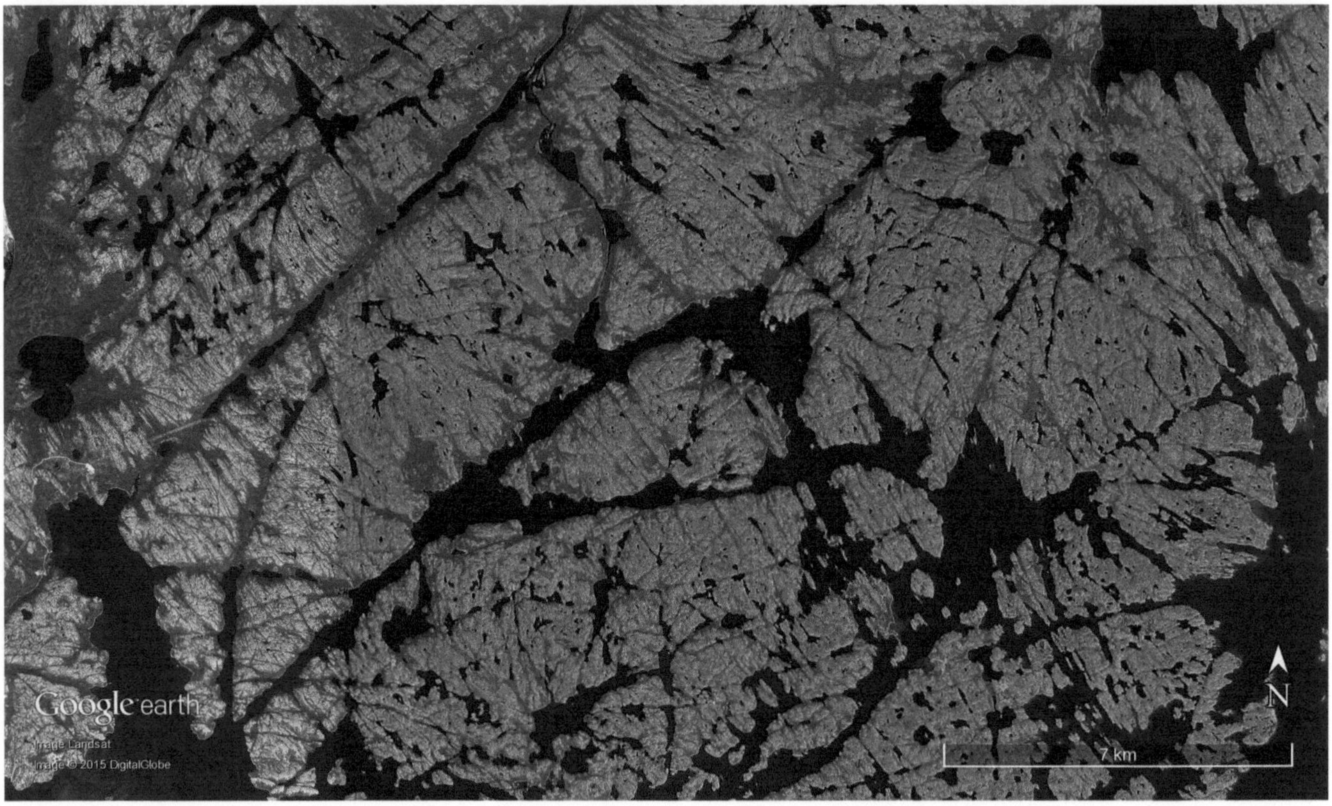

Fig. 2.35 The satellite image shows strong control of lake clusters by pre-existing joints in crystalline rocks, excavated in Québec (Canada) by glacial erosion in a 23 km wide scene. The region is around *51°17'N, 58°27'W* (Image credit: ©Google earth 2013)

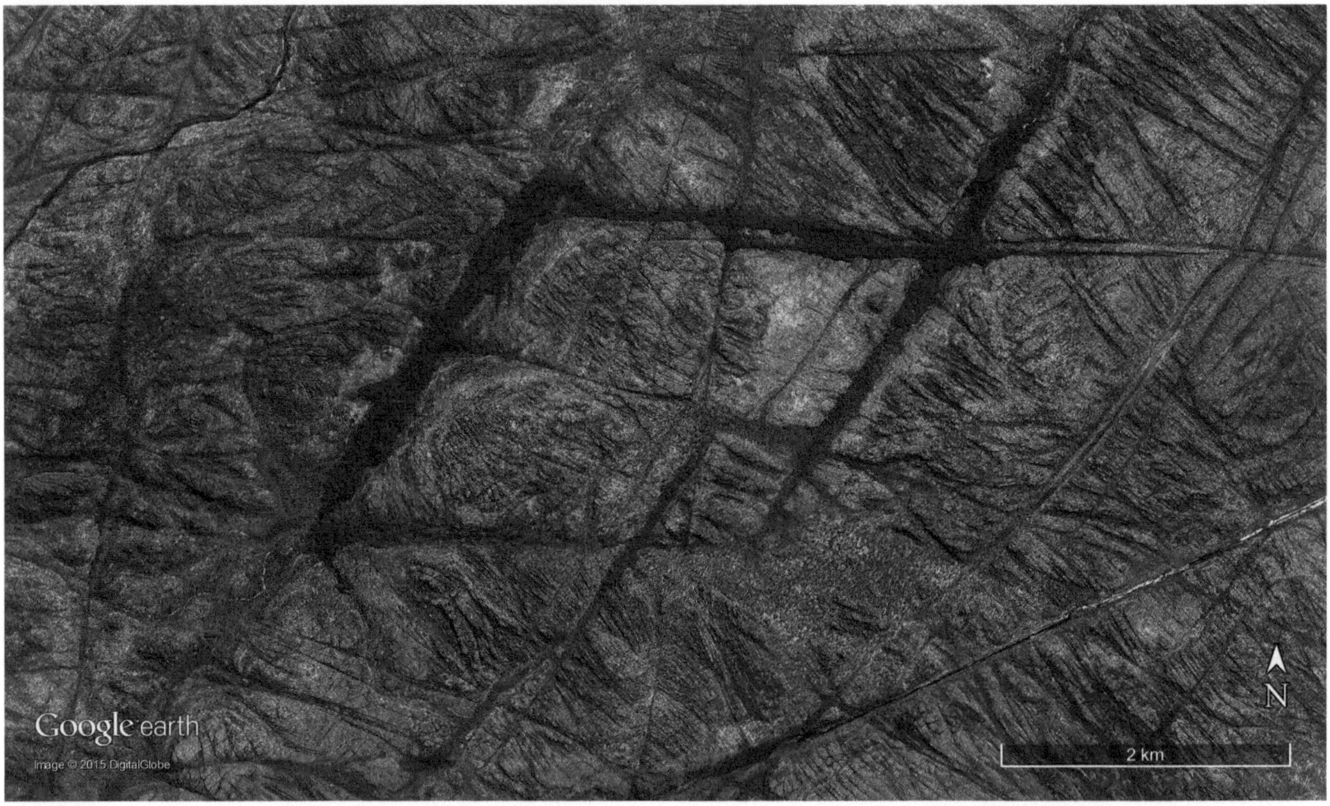

Fig. 2.36 In the Northern Territory of Australia (approximately *12°21'S, 133°54'E*) a rhomboid lake 5 km long has been formed by the filling of deep joints in sandstone/quartzite rock most likely deposited at least one billion years ago (Image credit: ©Google earth 2012). Argentina (*26°06'39.09"S, 67°25'08.08"W* at 3323 m asl). Scene is 2 km wide (Image credit: ©Google earth 2015)

Fig. 2.37 Lake (Laguna Alumbrera) surrounded with swamps created by young lava from Antofagasta Volcano in western

Fig. 2.38 A 9 km long pyroclastic flow from the eruption of Descabezado Volcano in 1932. The flow dammed a valley in the Chilean Andes (*35°26'54.62"S* and *70°48'07.77"W* at 1500 m asl), creating quite a small lake (Laguna Mondaca) on account of the permeability of the pyroclastic material up to 50 m thick (Image credit: ©Google earth 2015)

Fig. 2.39 Meltwater lakes blocked by lava (4500 m asl) near Cerro Parinacota (6312 m asl) at the border of Chile and Bolivia *(18°10'S, 69° 09'W)*. The largest lake is about 7 km across (Image credit: ©Google earth 2014)

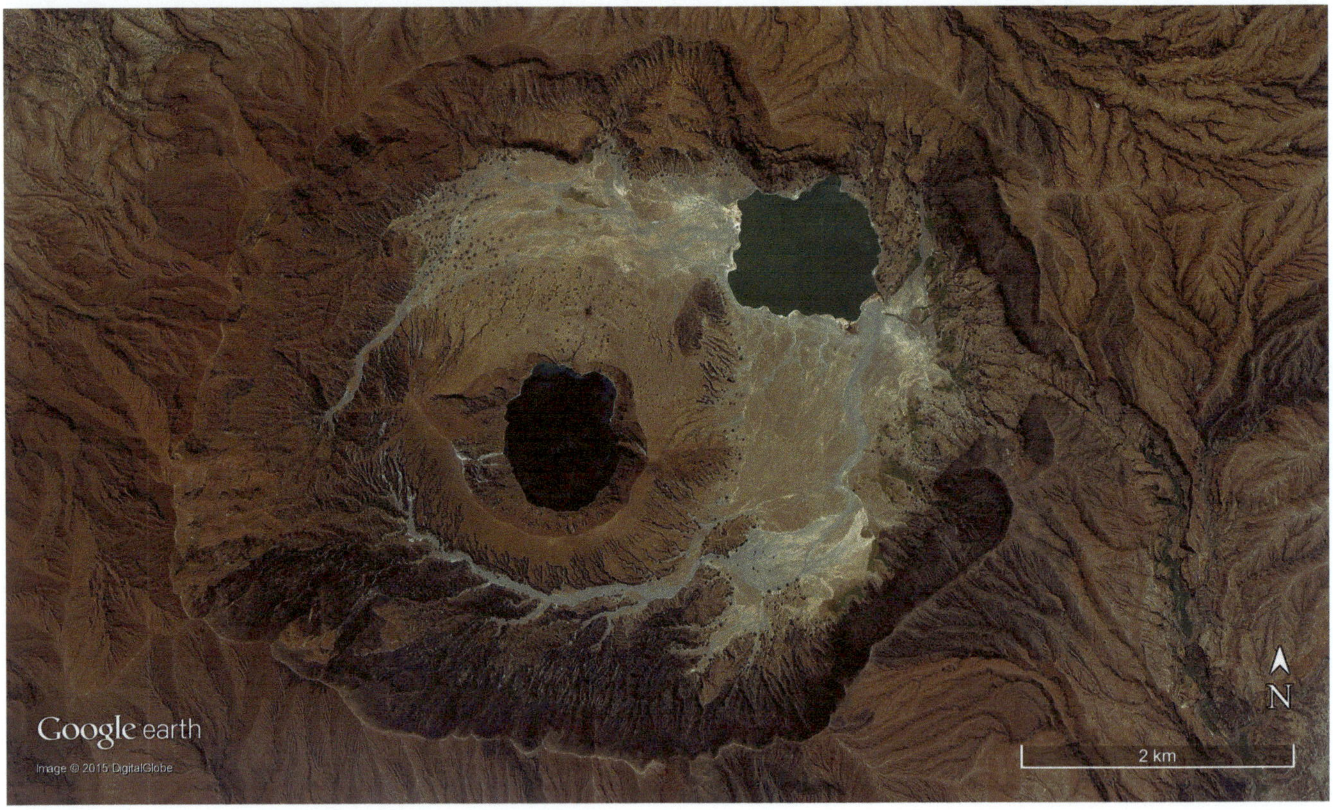

Fig. 2.40 Deriba Crater (5–8 km across) forms the highest part of Jebel Marra (approximately *12°57'N, 24°16'E* at 3042 m asl) in Darfur the western part of Sudan, and displays two crater lakes in shafts of an explosive volcano from around 3500 years ago (Image credit: ©Google earth 2015)

Fig. 2.41 A 6.6 km wide crater lake on Mt. Nemrut in eastern Anatolia (Turkey, *38°37'N, 42°15'E* at 2252 m asl) is half-filled from younger lava flows (Image credit: ©Google earth 2015)

Fig. 2.42 In Puebla Province of southeast Mexico (approximately *19°08'N, 97°32'W*) an 80 m deep crater with a central ash cone exhibits two saline lakes. Scene is 1.4 km across (Image credit: ©Google earth 2013)

Fig. 2.43 Mt. Kelimutu (*8°46'S, 121°48'E* at 600 m asl) on Flores Island (Indonesia) is recognized for three crater lakes of differing colours. The lakes have approximate diameters of 400 m, and that in the southwest can sometimes express a vivid red colour (Image credit: ©Google earth 2013)

Fig. 2.44 Lago Bracciano (approximately *42°07'N, 12°15'E*) is a Pleistocene explosion crater (9 km across with an area 57.5 km^2 and max. depth 160 m) filled by groundwater in the Sabatini Hills south of Rome (Italy) (Image credit: ©Google earth 2013)

Fig. 2.45 Toya caldera lake at Usu volcano on the southern Hokkaido Island of Japan (approximately *42°36'N, 140°51'E* and 11 km across with area 70 km^2 and max. depth 180 m) is the result of a summit collapse from 8000 to 7000 years ago. After several thousand years of inactivity the volcano awoke again with nine eruptions between 1663 and 2000 AD. The island in the centre of the lake is Nakana-shima (Image credit: ©Google earth 2015)

Fig. 2.46 Close to Manila capital of the Philippines, the large caldera lake of Taal volcano (approximately *13°59'N, 121°E*) has an area of 267 km^2 and is up to 160 m deep. The volcano is still very active and dangerous with 34 eruptions between 1572 AD and 1977 AD. Here is a lake on an island in a lake on an island in the ocean. The scene is 55 km wide (Image credit: ©Google earth 2015)

Fig. 2.47 Young volcanic island in a 7 km wide caldera with Ozero (lake) Kol'tsevoye on Sachalin Island, Russia (approximately *49°20'N, 154°43'E*) (Image credit: ©Google earth 2013)

Fig. 2.48 Crater Lake of Oregon USA (approximately *42°56'N, 122°06'W*). The caldera of the ancient Mt. Mazama exploded around 7000 years ago. The lake at 1883 m asl is 9.5 km across, has a size of 53 km^2 with a max. depth of 594 m. The lake is thought to be one of the clearest lakes in the world due to a very small discharge area (only from the caldera walls) and minor amount of sediment input. The small island to the west is Lizard Island exhibiting a tephra cone and lava tongues (Image credit: ©Google earth 2013)

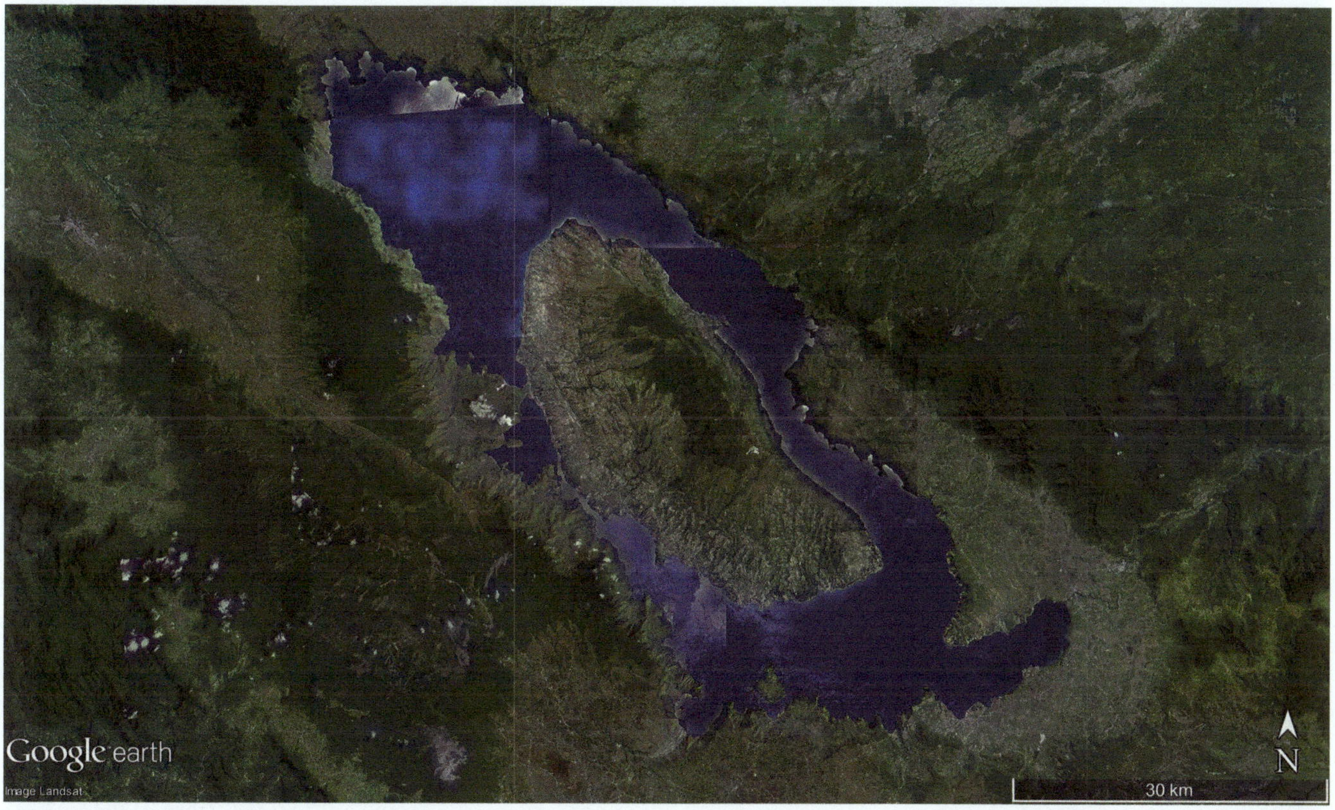

Fig. 2.49 Lake Toba, the largest lake in South East Asia is situated in Sumatra, Indonesia (100 km long and 30 km wide, depth of 505 m at *2°35'N, 98°48'E*). The lake is a caldera from the ancient super-volcano Toba which exploded around 74,000 years BP. Believed to be the largest volcanic eruption within the last two Mio years, Toba produced 2800 km^3 of airborne matter producing a sudden temperature fall of many degrees and possibly a bottle neck for the development of mankind. The size of the lake is presently 1103 km^2 with a volume 240 km^3 (Image credit: ©Google earth 2013)

Earth's climatic conditions. The impact of a fast moving body (on average 17 km/s of meteorites and up to 51 km/s of comets) originating from space consistently ends in a strong explosion (where most of the body mass is transferred into energy). More or less round depressions appear regardless at what angle the object impacts earth.

Small object (less than about 1 km across) mostly leave a simple crater, often with a higher rim of ejected rock or base rock pushed up during the central explosion. The impact of significantly large bodies (with a much higher energy) year-shock the rigid crust of Earth down to its base, and counter-reaction from the magma forms a central bulge in the crater through a rebound process (compare Figs. 2.57, 2.58, 2.59, 2.60, 2.61, 2.62, 2.63, 2.64, 2.65, 2.66, 2.67 and 2.68). The current status of known and confirmed impact craters on Earth's landmasses can be found at the Earth Impact Data Base, Planetary and Space Science Centre (PASSC), University of New Brunswick, Canada. Impact craters can be from just a few decades to more than 1 billion years old. It is astonishing that impact craters older than some tens of millions of years are still visible even though the craters have moved by continental drift and certainly undergone weathering and erosion under different climates. The lakes in these craters often are much younger. Tropical swamps and rain forests, wide uninhabited regions, large glaciers (i.e. Greenland or Antarctica), or former glaciers of the Ice Ages may explain the distribution pattern of impact sites as in Fig. 2.57. Water in impact craters can occur in humid climates from precipitation, in low-lying regions from groundwater, and in arid climate zones that may have transformed impact crater lakes into saltpans through evaporation.

2.2.2 Lakes Formed by Karst Processes

A group of sedimentary rocks that are comparatively easy to erode by solution are rocks from chemical deposition (evaporates like salt/halite and gypsum or in the shallow waters of oceans as carbonates/limestone). Although the rate of solution on limestone is slow from a human perspective or lifetime (i.e. 0.1 mm to 0.01 mm per year), the resulting karst landscapes exhibit numerous closed depression with diame-

Fig. 2.50 The largest lake in Central America, Lake Nicaragua (161 km long with an area 8264 km^2 at 31 m asl, centre approximately *11°32'N, 85°20'W*), is very shallow with a maximum depth of only 26 m. The twin islands in Lake Nicaragua consist of two high and steep volcanoes: Concepcion (1610 m) and Maderas (1340 m) (Image credit: ©Google earth 2013)

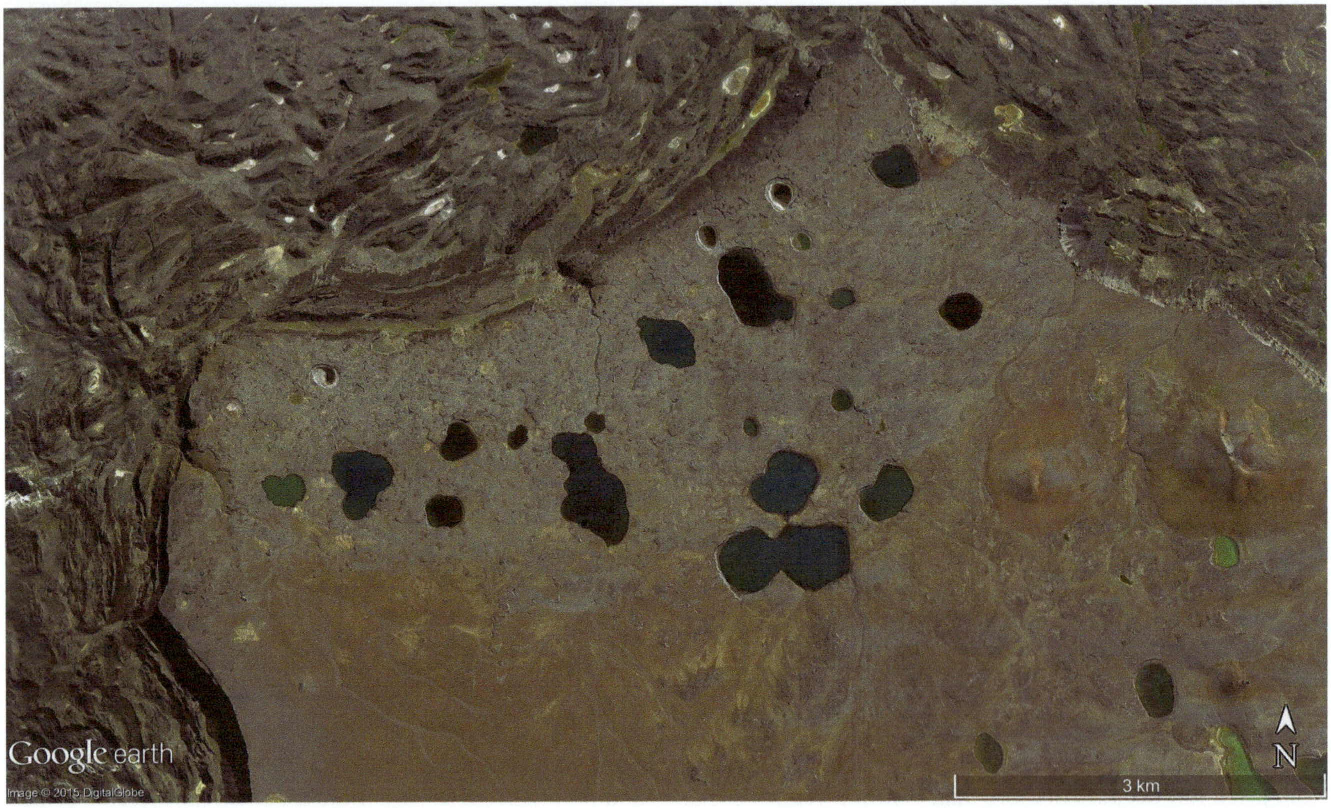

Fig. 2.51 Explosive maar-craters (the largest are 1 km across) on and within a lava plateau of Patagonia (Argentina) (approximately *46°43'S, 71°29'W*) (Image credit: ©Google earth 2015)

Fig. 2.52 A field of small explosive maar-craters in a 3.3 km wide scene in Patagonia, southern Argentina *(48°23'S, 71°01'W)*. Scene of the insert is 850 m wide and positioned *41°11'S, 67°33'W* (Image credit: ©Google earth 2013)

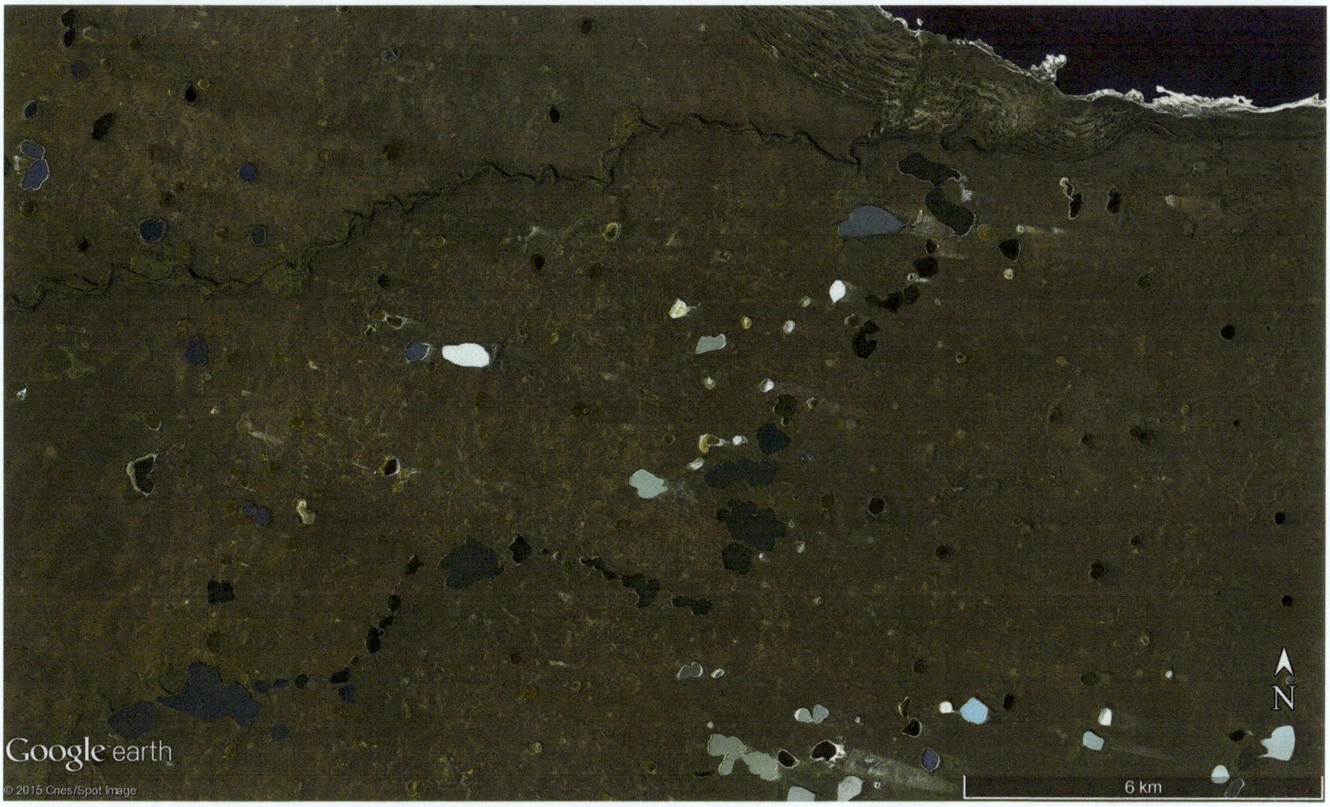

Fig. 2.53 A 20 km line of small lakes explains a fissure with former explosive volcanic outbursts under an older basaltic plateau in SE-Patagonia (Argentina) (around *48°32'S, 71°18'W*) (Image credit: ©Google earth 2013)

Fig. 2.54 Many hot springs/ponds in West Thumb Geyser Basin of Yellowstone National Park (Wyoming, USA), in a 280 m wide scene and 2370 m asl *(44°24'59.67"N, 110°34'18.92"W)* (Image credit: ©Google earth 2013)

Fig. 2.55 A hot spring beauty pool in the Upper Geyser Basin, Yellowstone National Park (Wyoming, USA). Scene is 200 m wide and 2240 m asl *(44°28'06.30"N, 110°50'20.29"W)* (Image credit: ©Google earth 2013)

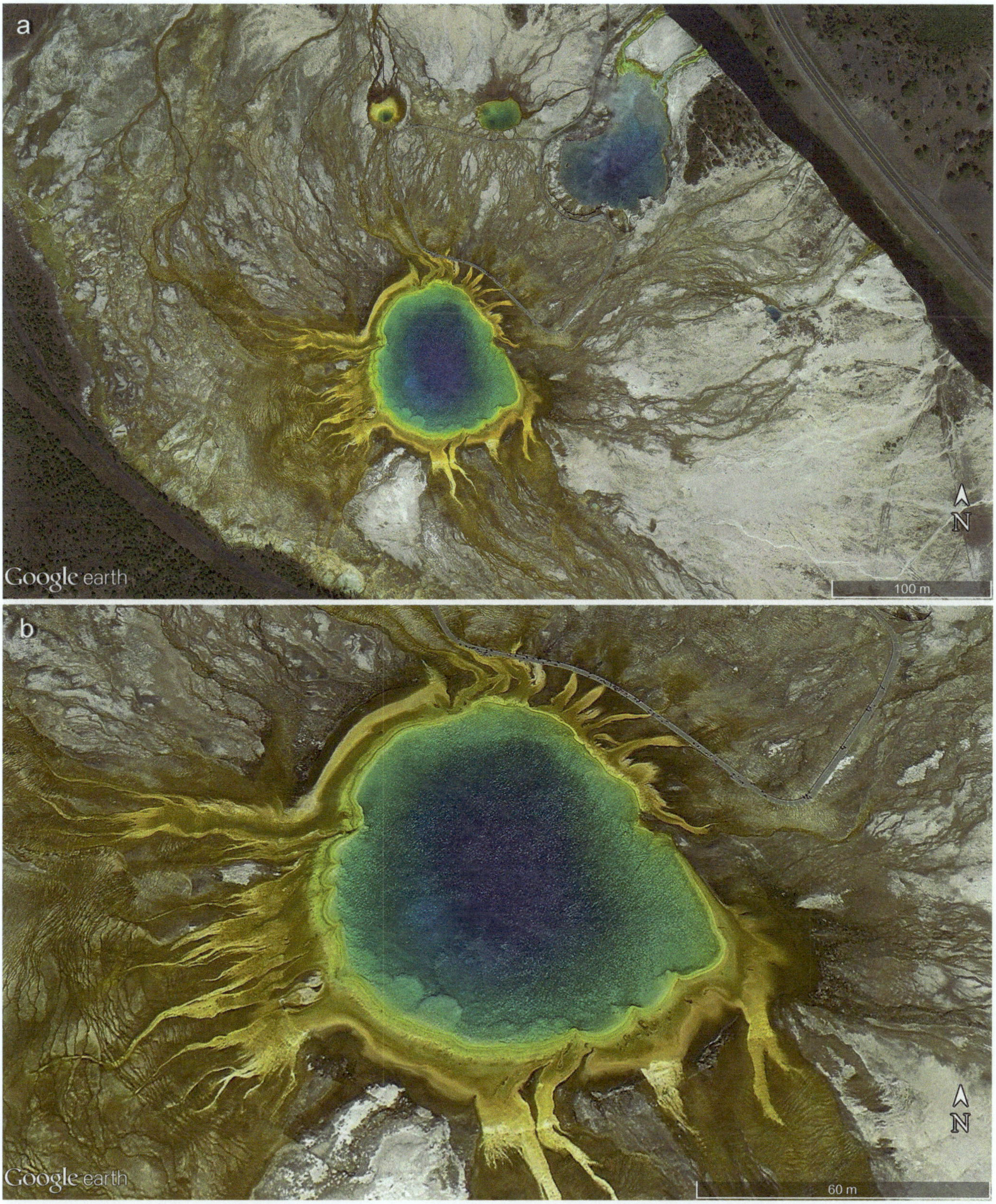

Fig. 2.56 (**a**) Grand Prismatic Spring (*left*) and Excelsior Geyser Crater (*right*) in Yellowstone National Park (Wyoming, USA), at 2220 m asl and *44°31'31.01"N, 110°50'16.24"W*. Both springs have diameters of about 100 m. (**b**) Grand Prismatic Spring is the largest hot water pool in Yellowstone National Park (Wyoming, USA) with a diameter of 101 m. The colourful fringe of the basin is the highest part of a post-glacial (after the last Ice Age from 11,000 BP onward) shield-form (16 m high) developed from the deposition of geyserite (a siliceous mineral called opaline; having neither definite form nor apparent structure). Geyserite is extracted by the evaporation of soluble minerals precipitated from the volcanic base rock. *Yellow* to *reddish* colours indicate the presence of bacteria and the dark (*black*) colours further away from the boiling basin with cooler temperatures of overflow waters are algae (Images credits: ©Google earth 2013)

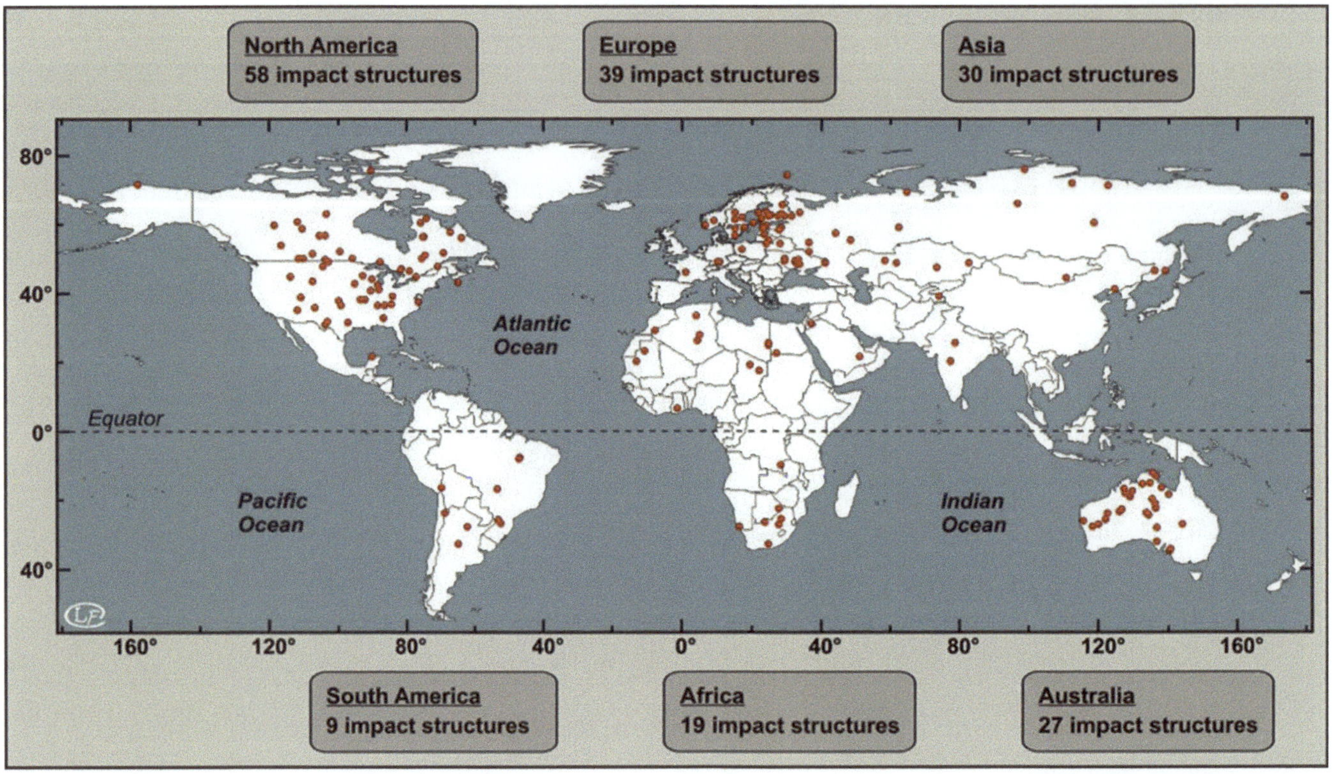

Fig. 2.57 World map of 182 sites with cosmic impact craters on the continents. Several more impacts are under investigation—the large empty areas in the northwest of North America (under glaciers for long time spans), the northern section of South America (wide swamps and dense river networks under the cover of tropical rainforest), central Africa (with similar conditions to northern South America), including central and east Asia—seem to reflect more a lack of information than a lack of terrestrial impacts (Credit: Earth Impact Data Base 2010, Planetary and Space Science Centre [PASSC], University of New Brunswick, Canada. Graphics: SM May)

Fig. 2.58 Perfect circular impact crater in Québec, Canada *(61°16'40.39"N, 73°39'43.08"W)*. Called Crystal Eye, New Quebec or Pingualuit crater, the lake is 3 km across (or 3.44 km including the higher rim), which to the north is around 70 m above the lake level. Age is estimated to be about 1.4 Mio yrs (Image credit: ©Google earth 2013)

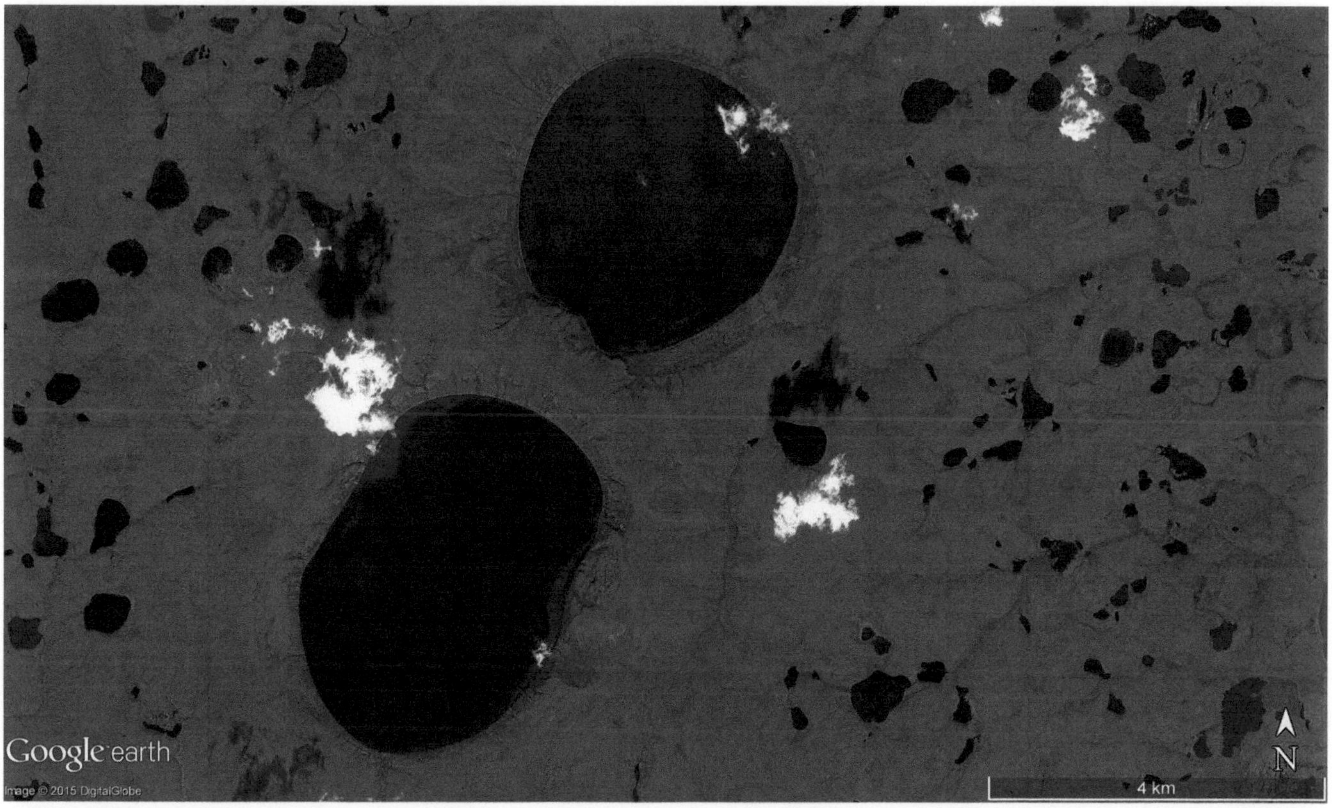

Fig. 2.59 Two impact craters indicate a young age as the forms have never been smoothed by inland glaciers and are located west of Kotzebue Sound in NW Alaska *(66°21'N, 164°04'W)*. Long diameters are 4.4 and 3.5 km (Image credit: ©Google earth 2015)

Fig. 2.60 Lonar impact crater (2 km wide and 150 m deep) in Maharashta, India, from around 52,000 years ago with a saltwater lake 1.2 km across (centre of the lake is at *19°59'N, 76°30'E*). Scientists discovered the impact was caused by an asteroid because of the presence of maskelynite a glass only formed by extremely high-velocity impacts (Image credit: ©Google earth 2013)

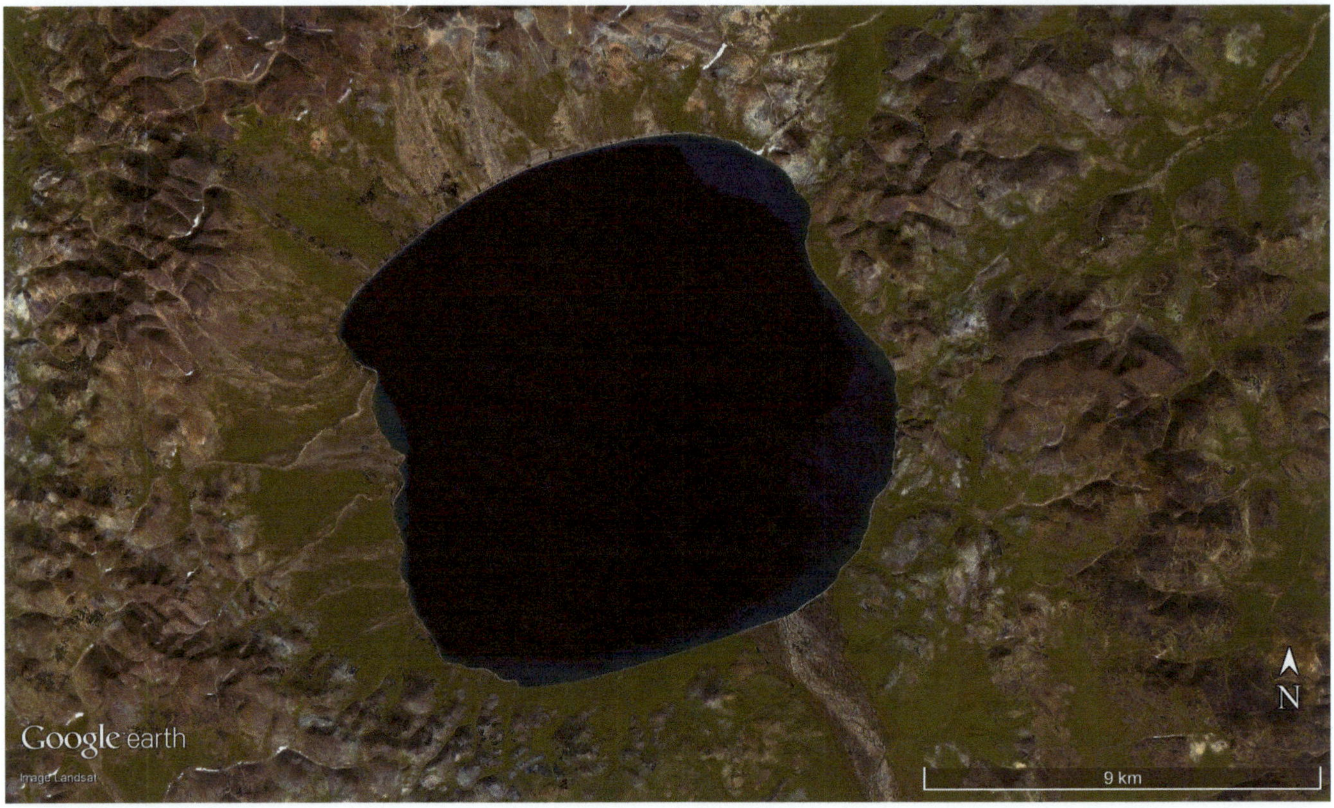

Fig. 2.61 Elgygytgyn impact crater lake, Russia (approximately *67°29'N, 172°05'E*). N-S-diameter of lake is 13 km and max. depth 170 m. Age of the impact has been dated to 3.6 Mio years and because Ice Age glaciers have never covered the lake, there are undisturbed lake sediments from the time of impact, a valuable archive of environmental conditions in these high continental parts of northern Asia (Image credit: ©Google earth 2013)

ters ranging metres to hundreds of metres (sinkholes; Figs. 2.69, 2.70, 2.71, 2.72, 2.73, 2.74, 2.75, 2.76 and 2.77 or uvalas; Figs. 2.78 and 2.79 if developed in a line) to over hundreds of kilometres in the tectonically adapted poljes formed by the coalescence of several sinkholes. Many of the smaller sinkholes Figs. 2.69, 2.70, 2.71, 2.72, 2.73, 2.74, 2.75, 2.76 and 2.77 are simply collapsed openings over subsurface solution caves (disappearing streams).

Limestone is one of the most widely distributed sedimentary rock on Earth, stemming from former shallow oceans often folded and lifted above sea level. Even high mountains (parts of the eastern European Alps) are built from limestone, including parts of Southeast Asia from Thailand to China and many islands of the Caribbean. Consequently, landscapes with hundreds of thousands of karst forms (sinkholes, uvalas, poljes) exist on all continents. However, lakes in these limestone and karst landscapes are rare because the rocks are permeable along their joint and stratum plains/lines and the water disappears underground. Exceptions are where fine sediments like silt and clay seal these waterways and can only occur when the topography is low and flat with groundwater near the surface. This is particularly the case with thousands of lakes formed on the limestone plateau of the Florida peninsula, USA.

The largest carbonate forms of the Mediterranean region are karst features known as poljes (Serbo-Croatian meaning: field). Poljes are wide limestone depressions (up to and exceeding 100 km) positioned within mountainous reliefs and exhibit flat floors overlain with fine sediments, as a base for agriculture. These sediments seal the ground from water contact, and solution in contact to the limestone is only effective along the side slopes. This may enlarge the basins outwards over long time scales and often define relatively straight contours; at least on one side. This process leads to the original formation of depressions as tectonic basins or as part of a rift/graben structure with subsidence in the central section. Moreover, poljes are a combination of structural forms as the rough outer frame, and sculpturing by limestone solution as a special process of weathering, i.e. an exogenic geomorphological factor. Figures 2.80, 2.81 and 2.82 present examples from the Balkan region.

Fig. 2.62 The 72 km wide impact crater Manicouagan in Québec (Canada) has slowly transformed into its current state. What was originally the rim of the crater has since eroded away due to Ice Age processes. The impact occurred around 212 million years ago and the diameter of the asteroid is estimated to have been approximately 5 km. The central island's bulge is 500 m higher than the ring-like lake as a counter reaction of the Earth's crust to the strong impact. The bulge appeared immediately after the impact but since has transformed through multiple glaciations and long periods of weathering. Centre of the image is at *51°24'N, 68°42'W* (Image credit: ©Google earth 2013)

2.2.3 Lakes Occurring Along the Coastlines of the World

Located along the coastlines of the World's open oceans are numerous lakes in flat, depositional and low-lying landscapes (at sea level) where water is always available to fill the depressions. In this chapter, current coastal lakes, their form and infill as a result of littoral/marine processes are discussed. By chance, lakes from other origins (e.g. volcanic, glacial) can also occur near the coast, but are not characteristic of the special processes in coastal environments. The sea level history from the low stand of the last Ice Age (22,000 to 16,000 BP) was around 120 m below the current sea level, and its rise to the modern levels (around 7000–6500 years BP) determines the maximum potential age of coastal lakes to be no more than 7000 years.

A lagoon is the principal form and genesis to coastal lakes. A lagoon is a body of water usually positioned parallel to the coastline with its longest axis separated from open waters by a barrier of a beach, or between a sequence of beach ridges or additionally, by a belt of coastal dunes. These barriers isolated former open embayments from the sea by the accumulation of sand (including shell hash and sometimes with pebbles). To seal off a bay, the barrier needs to accumulate on the near-shore seafloor that is only possible in relatively shallow water; otherwise the time span available (not more than 7000 years) is too short, or waves and currents are too strong for a final deposition. Here we present, amongst other coastal lakes, a short series of the main forms. Many more can be found in Scheffers et al. (2012): The Coastlines of the World with Google Earth—Understanding our Environment.—Coastal Research Library 2, Springer, Dordrecht, 293 pp.

The making of a barrier necessitates the accumulation of sediments from outside, from the sea and by movement of currents and waves perpendicular to the coast; however, in most cases sediments are transported by *longshore drift,* a process in which prevailing winds trigger both currents and waves approaching from an angle to the coastline, forming oblique waves. The waves break at an angle to the shoreline and push sand forward along the coastline in the surf zone (swash) depending on the dominant wave direction. Should

Fig. 2.63 In Labrador (Québec, Canada) the twin craters of Clearwater West and Clearwater East (36 and 26 km across) have a total lake area of nearly 1400 km² with maximum depth of 178 m. The twin craters were determined by an impact around 290 Ma. The site is approximately *56°10'N, 74°20'W* (Image credit: ©Google earth 2015)

Fig. 2.64 The ring structure in the Democratic Republic of Congo (36 km across, centre at *3°38'S, 24°31'E*) appeared after forest clearing in recent years. The centre of the ring is 50–60 m higher than the outer ring, which resembles something between a river and the bend of a narrow lake. Confirmation (e.g. analysis of impact minerals) is needed for this ring feature as an impact signature and the age of (potential) impact is still to be determined (Image credit: ©Google earth 2015)

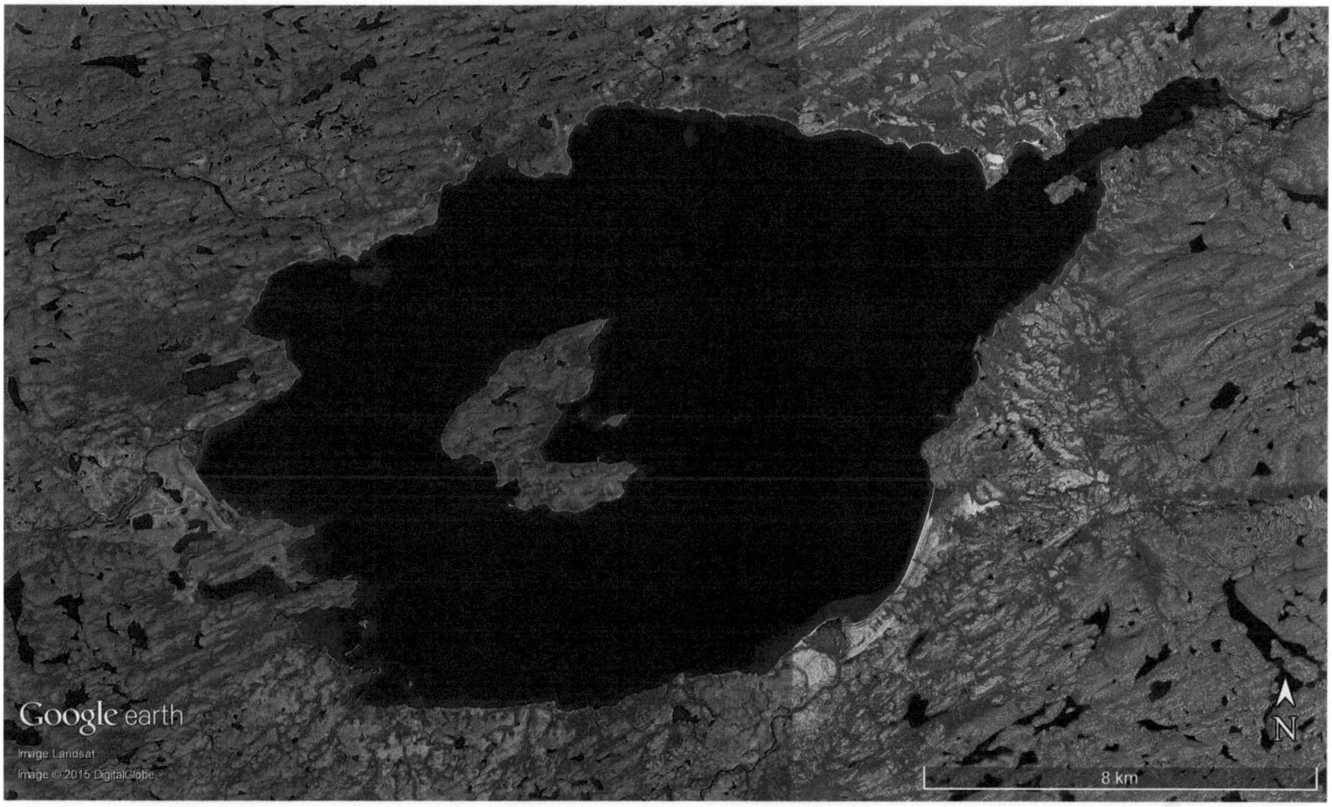

Fig. 2.65 A meteorite crater dating 32–40 Ma contains Mistastin Lake in Labrador, Canada (max. 22 km long, centre at *55°53'N, 63°18'W*). The island in the centre (Horseshoe Island) is a central bulge similar to Clearwater West impact crater and Manicouagan Crater, both in Labrador. The elongation of the lake and central island is from glacier movement broadly from the west. Previously believed to be a volcanic structure, the occurrence of rocks melted to glass and iridium verified the impact character of the site (Image credit: ©Google earth 2015)

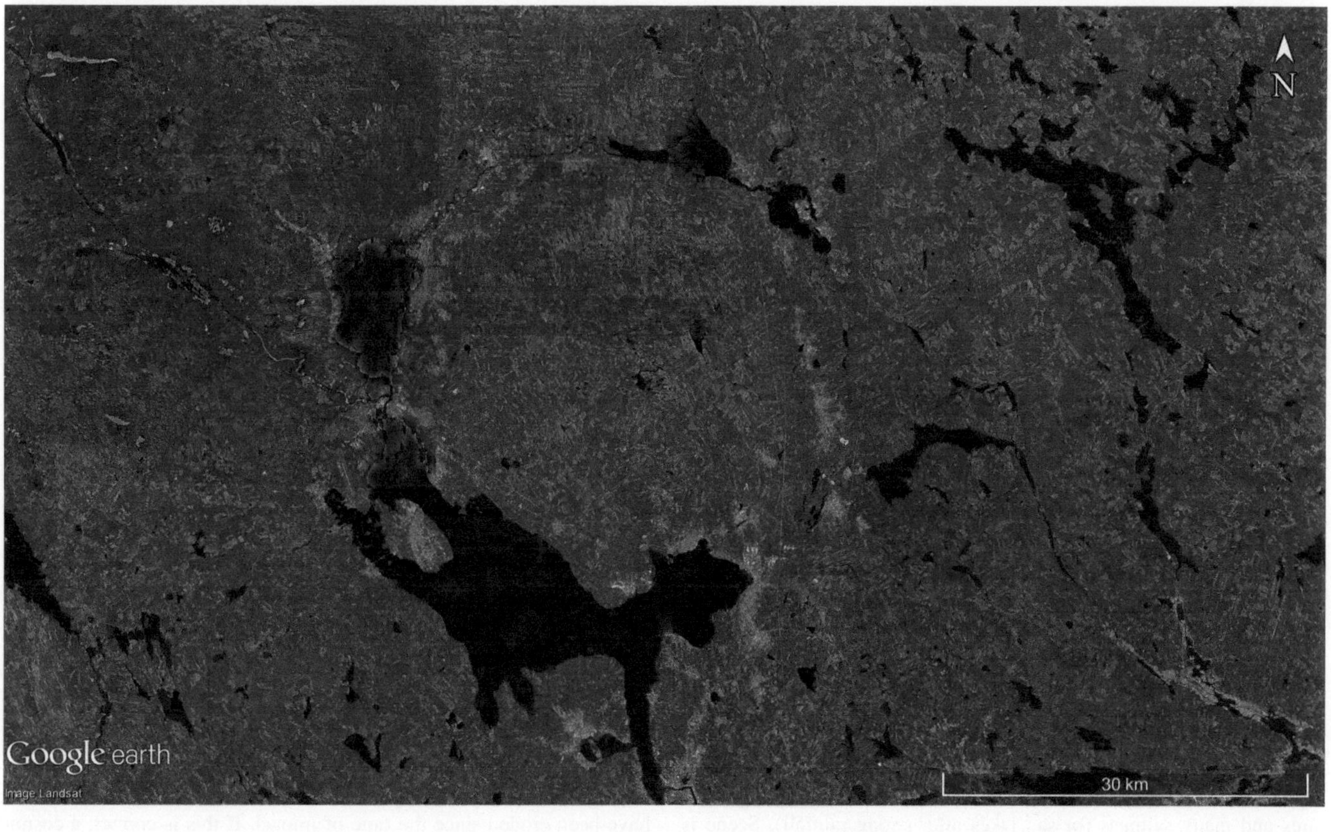

Fig. 2.66 A probable meteorite impact (377 Ma) Siljan structure 50 km across in Sweden *(61°01'N and 14°55'E)*. Several lakes appearing around the outer margin may be the result of repeated glacial erosion during the Ice Ages (Image credit: ©Google earth 2013)

Fig. 2.67 Created from an impact less than 400 Mio years old, Nicholson crater lake located east of Great Slave Lake in central northern Canada is up to 12.6 km wide (Centre is *62°40'N,102°42'W*) (Image credit: ©Google earth 2013)

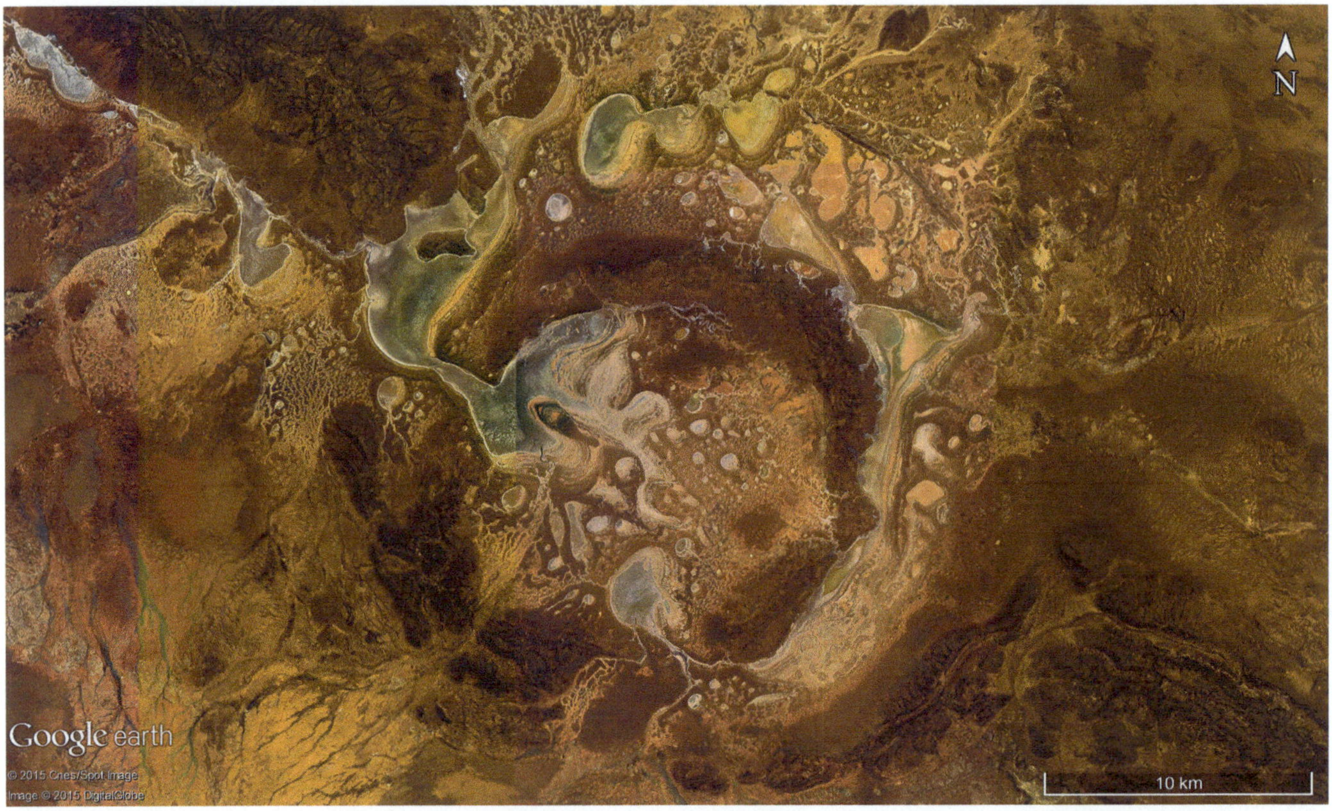

Fig. 2.68 Shoemaker impact crater in Western Australia with ring structures and many saltpans (or salt lakes after strong rainfall). Scene is 50 km wide at *25°50'S* and *120°55'E*. Age of the impact is under debate and estimated to be 2.6 Ga, 1.63 Ga or 568 Ma. Here, there are deep rock imprints of the very old impact and it is estimated that 2–3 km of rock have been eroded since the time of impact. If this is correct, a cosmic impact of an asteroid or meteorite of 2–4 km diameter is able to alter the Earth's crust several km deep (Image credit: ©Google earth 2015)

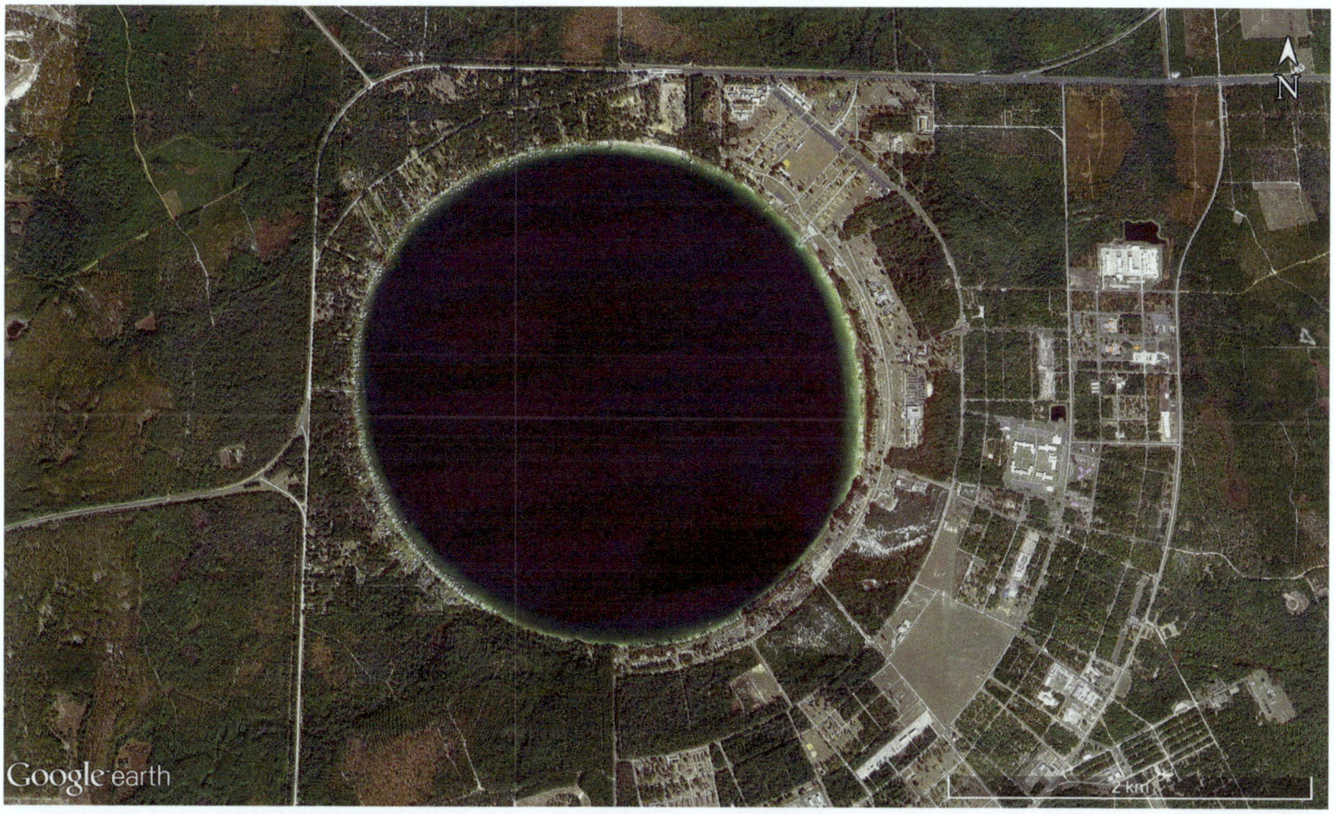

Fig. 2.69 Many Florida sinkholes, in particular those showing lakes, are almost perfectly round deep shafts connected to underground water levels with cave systems. This sinkhole at *29°58'N* and *82°00'W* has a diameter of 2.9 km (Image credit: ©Google earth 2013)

Fig. 2.70 A Croatian sinkhole in a valley bottom at *43°49'N* and *16°01'E*, diameter only 180 m (Image credit: ©Google earth 2015)

Fig. 2.71 Two sinkholes connected to a twin lake with a length of 960 m. Florida at *27°16'N* and *81°22'W* (Image credit: ©Google earth 2013)

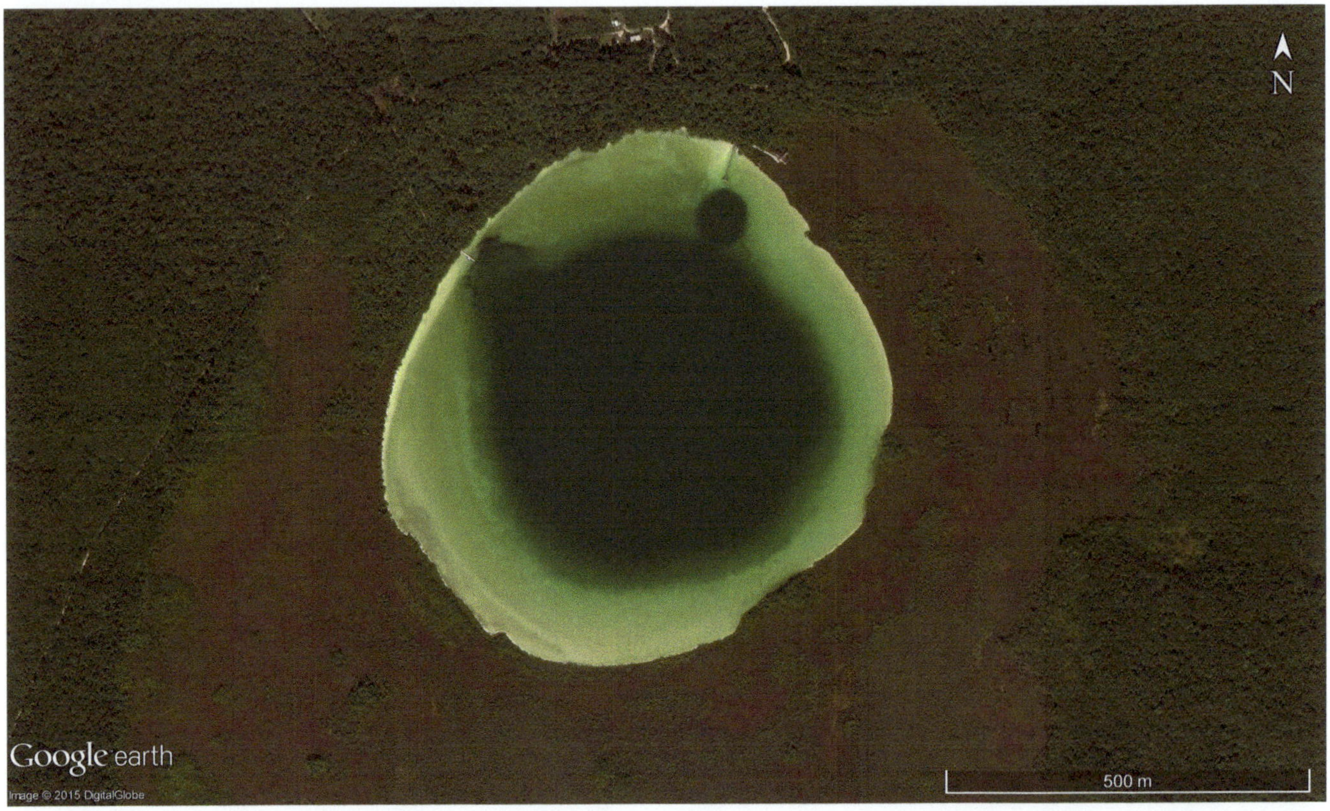

Fig. 2.72 Sinkhole (locally known as cenote) in Yucatán southeast Mexico merged into at least three original parts (one large and two quite small holes). Cenotes had been used by the Maya people for more than 2000 years either as holy places for sacrifices, or as natural wells for water supply The cenote is 750 m across at *20°09'43.94" N, 87°33'14.08"W* (Image credit: ©Google earth 2015)

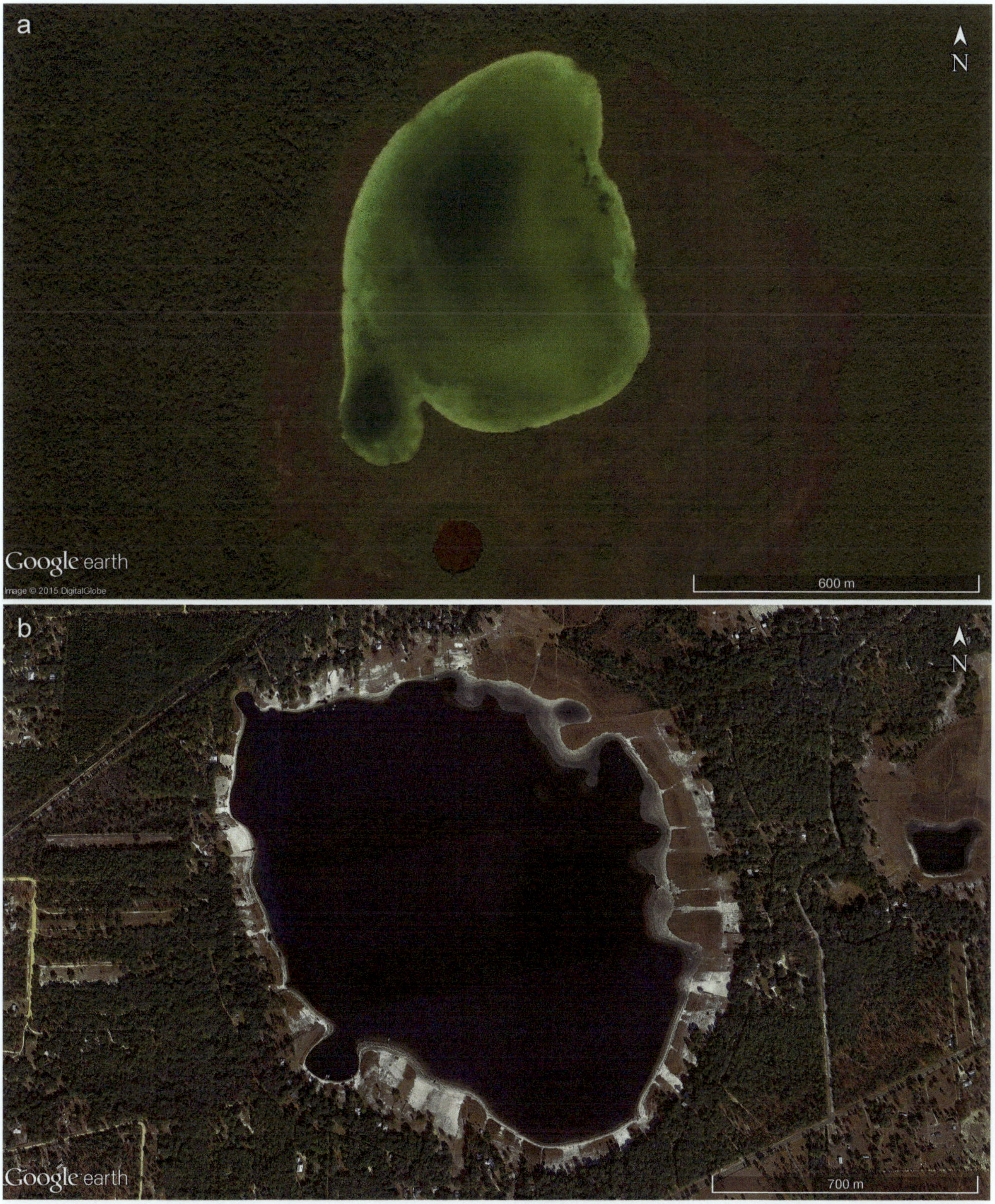

Fig. 2.73 (**a**) Three younger and more accentuated cenotes have been formed in a larger shallow sinkhole at Yucatán southeast Mexico at *20°13'N* and *87°32'W*. Scene is 2.6 km across. (**b**) A Florida sinkhole of 1.6 km diameter is decorated by at least six smaller sinkholes along its contours. The site is at *29°44'N* and *82°01'W* (Images credits: ©Google earth 2015)

Fig. 2.74 Isolated and combined sinkholes in limestone from the Florida Peninsula at *29°33'02.86"N, 81°57'44.57"W*. Scene is 2.5 km wide (Image credit: ©Google earth 2013)

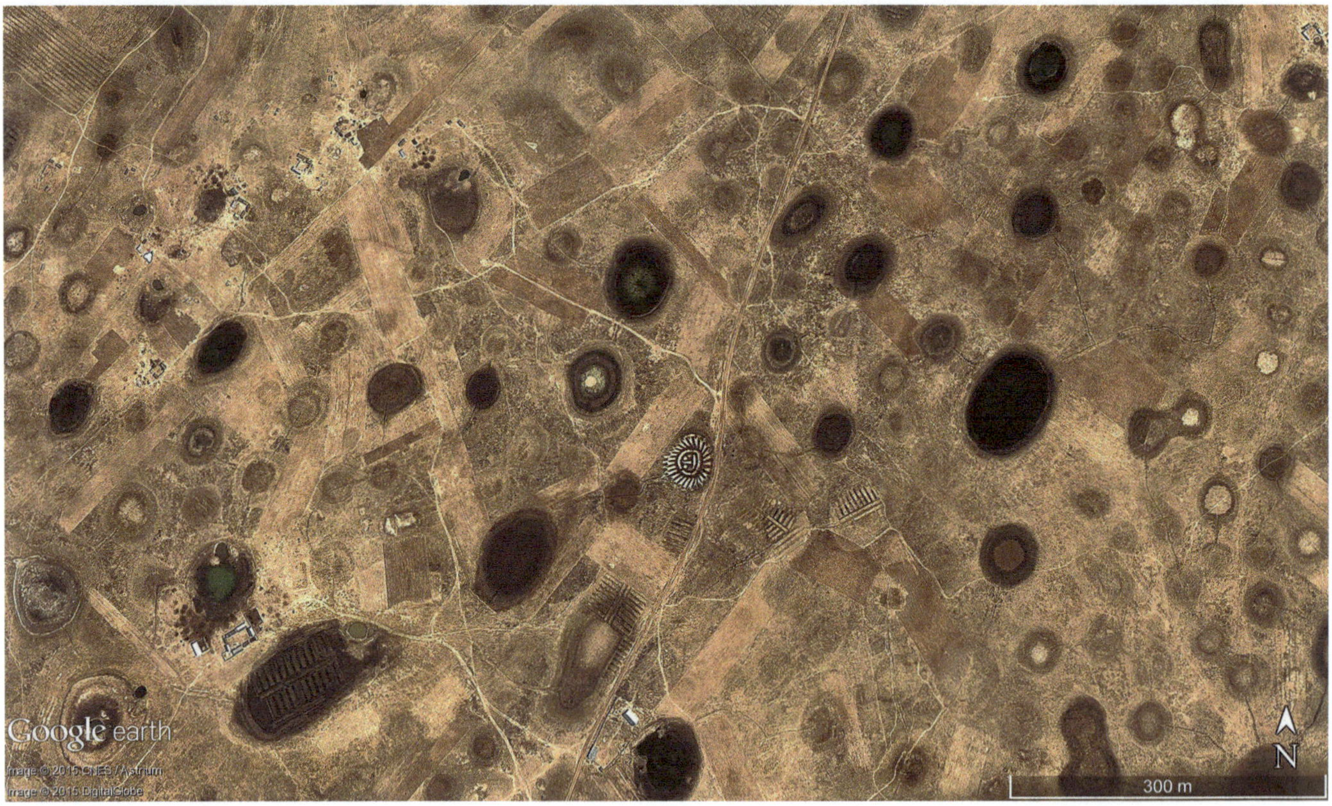

Fig. 2.75 Northeast of Lake Titicaca in Peru (3810 m asl) clusters of shallow sinkholes have developed on a plain, some with water content and others as little swamps. The scene at *15°11'S* and *70°17'W* is only 1.3 km wide (Image credit: ©Google earth 2013)

Fig. 2.76 In southern coastal Florida USA *(30°07'37.60"N und 84°10'33.56"W)* groundwater rises as springs from sinkholes and deliver water into the Gulf of Mexico. This scene is just 1.2 km across (Image credit: ©Google earth 2013)

Fig. 2.77 Karst lakes set in limestone filled by groundwater from a riverbed (to the right of the image) in northernmost Greece. Scene 2.3 km wide is at approximately *40°58'N* and *19°58'E* (Image credit: ©Google earth 2015)

Fig. 2.78 Sinkholes merged in lines with a valley-like appearance. These features are called uvala taken from examples in Serbo-Croatia. The 9 km long uvala consists of at least three sinkholes from Yucatán, Mexico (*20°03'N* and *87°35'W*) (Image credit: ©Google earth 2015)

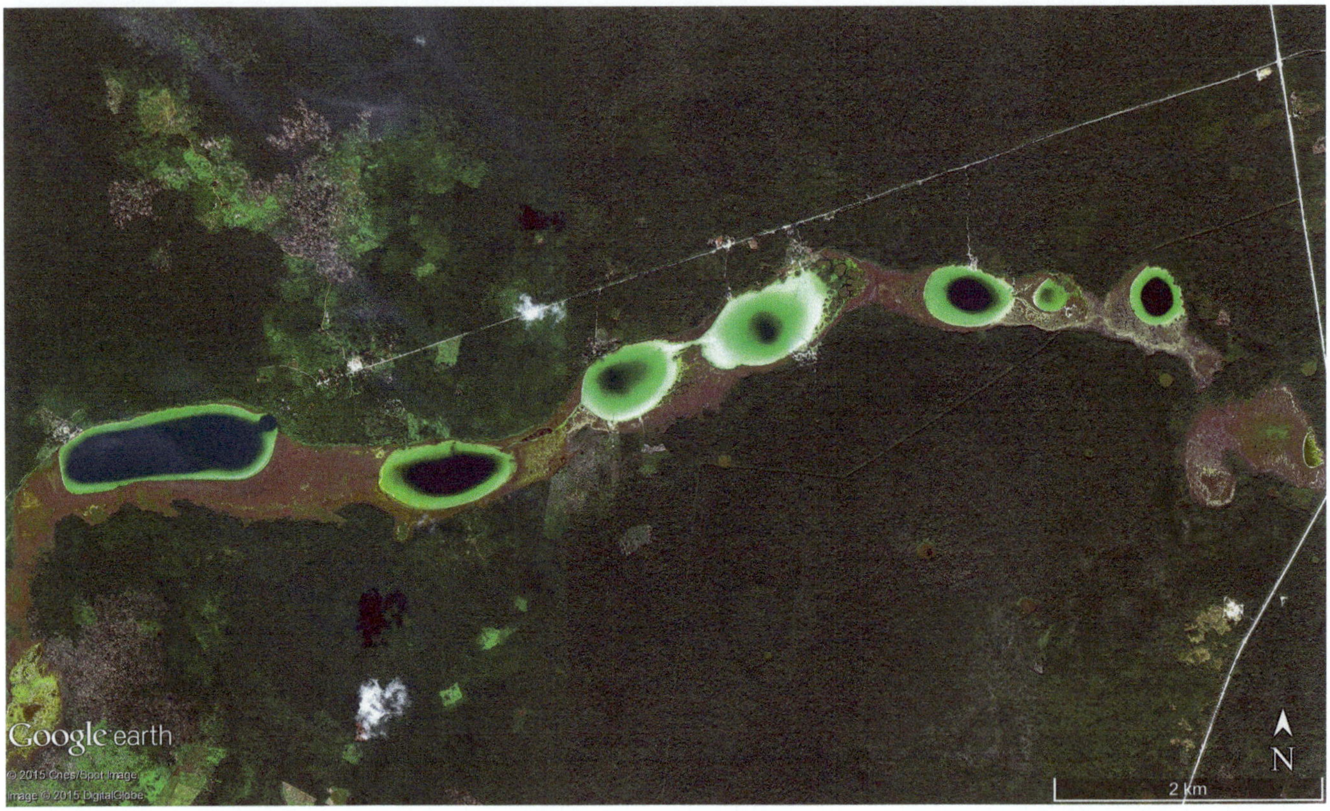

Fig. 2.79 Seven sinkholes are connected to a fine example of an uvala over 9 km long on the Yucatán Peninsula of Mexico (approximately *19°28'N, 88°05'W*). The uvala clearly demonstrates multiple collapses of an underground cave (Image credit: ©Google earth 2015)

Fig. 2.80 The polje of Maslenica Lake in Croatia is connected to the Adriatic Sea by a flooded canyon-like valley to the north (*44°12'N* and *15°32'E*). The 11 km long basin shows parallel coastlines along faults in the southwest and northeast (Image credit: ©Google earth 2015)

the coastline change directions, sediment transport may lose contact with the shoreline and continue further in the original direction where a sub-littoral bar develops and later matures into a spit. Over time the spit extends and grows until finally coming into contact with the next coastal promontory, which can seal off the bay whereby a lake appears at the place of the former embayment.

In most cases the direction of longshore drift can be observed through satellite images, as in the Italian example (Figs. 2.83, 2.84, 2.85, 2.86, 2.87, 2.88, 2.89, 2.90, 2.91, and 2.92). In this case sediments are delivered by the River Fortore west of the lagoons and moved forward to the east. The starting point to the barrier is at the lee side of the first promontory (Fig. 2.83b) and through the process of wave refraction, the barrier adapts asymptotically to the next promontory between the two lagoons. Similarly, the starting point to the barrier closing Lago di Varano is set leeward and smoothly approaches the east rocky shore by wave refraction. The Lago di Varano barrier is narrow because of diminishing sediment supply from the distant Fortore River and westerly winds are clearly the main force for the longshore drift. In numerous publications the process of *longshore drift* is reduced to *longshore currents*. Since the topography of the barrier consists of beach ridges and some low dunes (Fig. 2.83b) situated above sea level and outside the influence of currents, the effective methods of barrier formation are almost certainly by wave action, pushing sediments above the water level in wave run-up. Winds can transport sand from a dry beach and form dune belts lagoon-wards of the beach, as evident in the example of southern Brazil (Fig. 2.84). A special group of lagoons that develop in the same way by longshore drift are valleys opening into the sea, which have lower courses submerged by post-glacial sea level rise. The outward extension of these lagoons is more or less perpendicular to the coastlines. Considering valley floor profiles are deeper toward their end, constructing a straight barrier across is difficult because of deeper water and consequently stronger wave energy in the centre of the opening, and for this reason these lagoons are rare (Figs. 2.93, 2.94, 2.95, 2.96, 2.97, 2.98, 2.99, 2.100, 2.101, and 2.102). The scientific term for this lagoon type is "liman" (Greek meaning "harbour") and stems from examples in the Black Sea (north of the Danube Delta) and the northern part of the Dardanelles (Aegean Sea, just west of Istanbul). Distinctive types of lakes or enclosed water basins are lagoons on atolls, i.e. surrounded by coral reefs. The lagoons contain seawater, and if totally enclosed become hyper-saline; if corals can survive in these lagoons they may form secondary patterns (Fig. 2.103a, b).

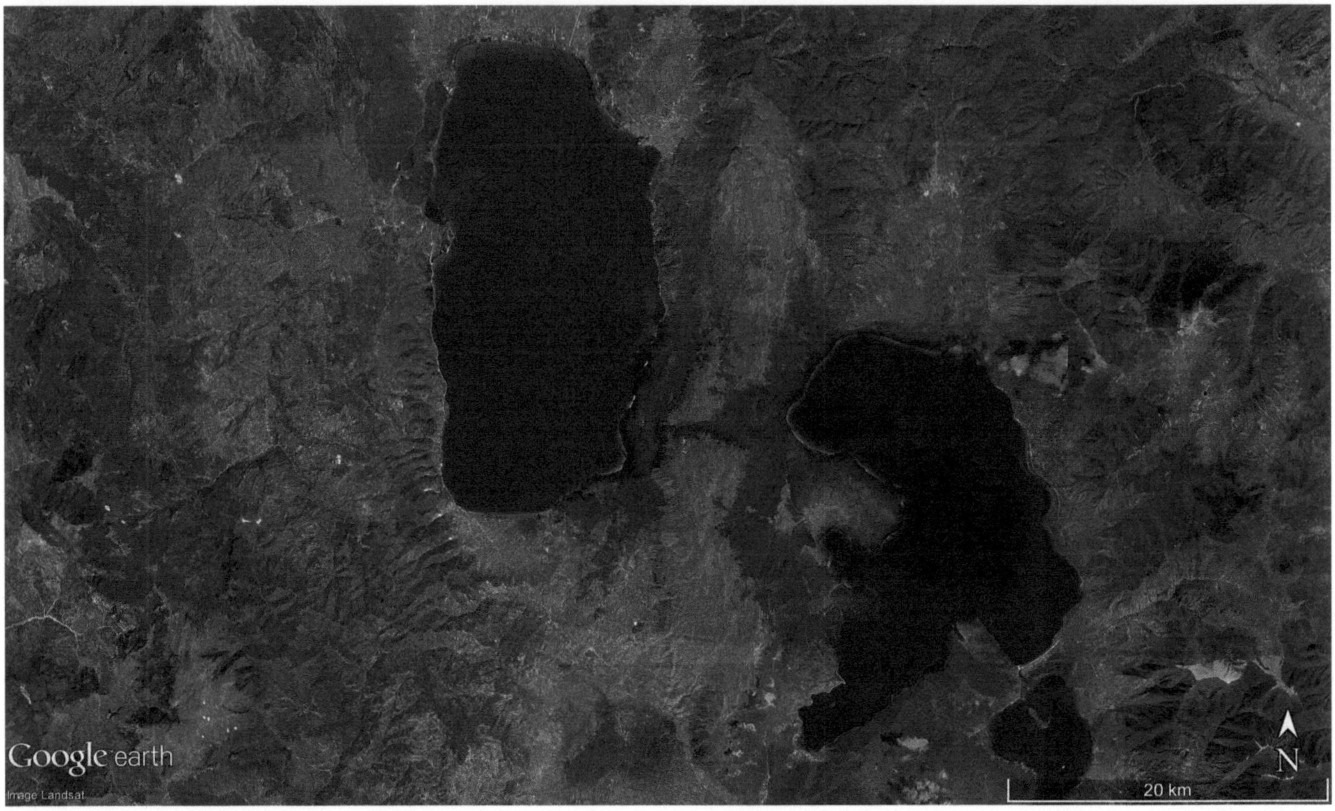

Fig. 2.81 Lake Ohrid at the border of Albania (in the west) and Macedonia (in the east) is one of the deepest lakes in Europe. The lake with a maximum depth of 288 m and average depth of 155 m is 30 km long and up to 15 km wide. The western shoreline tracks parallel to a fault active since Pliocene, giving the lake basin an age of around 5 Mio years. For a lake 5 Mio years is extremely old; the basins tend to fill with sediments ending the life of a lake. In the case of Lake Ohrid, ongoing subsidence of the graben feature has evidently overcome sedimentation also the discharge area of the lake is small, and is mostly fed by sediment-free karst springs along eastern fault lines. The polje is larger than the lake itself and extends to the north and south (marked by agricultural features) with a maximum length of 44 km. Because of the many endemic species found in the lake, Lake Ohrid is a World Heritage Site of UNESCO (lake centre at *41°02'N* and *20°43'E* and lies 690 m asl.) (Image credit: ©Google earth 2015)

Nearly all of the morphological features distinctive to the World's oceanic coastlines can also be found along (inland-) lakes, from cliffs to rock platforms, deltas to beaches, dune belts to beach ridges, barriers, spits, tombolos and many other forms (compare examples Fig. 2.104a, b).

2.2.4 Lakes Formed by Rivers

Within the science of limnology (study of inland waters) rivers and lakes belong to different categories: rivers have flowing (moving) waters and lakes have standing (still) water. Rivers flow along valleys and the resultant erosive processes cause an inclination (though not necessarily regular) to the valley beds from the source of the river to its mouth. Closed river-made depressions are non-existing with the exception of plunge pools/pot holes at the base of waterfalls and rapids (Figs. 2.105a, b). Consequently, lake-forming basins are primarily the result of dam-building processes by the gravitational deposition of material brought down from valley slopes such as rock fall or landslide. However, rivers can also actively form lakes in their beds through deposition along their banks during anabranching processes (a stream breaking off from a river and re-joining it further downstream), meandering or delta forming. These lakes are typically narrow and elongated (as riverbeds are), and usually shallow (Figs. 2.106, 2.107, 2.108, 2.109, 2.110, 2.111 and 2.112).

From the discussions above it becomes clear that river-lake formation can only be expected in the lower part of a river course and in lowland relief with fluvial deposition. Lakes will not occur in river or valley sections with erosive aspects and narrow cross-section profiles such as gorges. However, there is an exception when inorganic and organic components meet, in the case of rivers crossing limestone rocks that then carry carbonates in solution. In this instance, when river water levels are low, an underwater rocky barrier may develop upward and increases the probability for chlorophyll production if good light conditions are present—

Fig. 2.82 Vegoritida Lake in northern Greece positioned 515 m asl, and 12 km long (centre at *40°45′N* and *21°47′E*). A straight fault line marks the western shoreline and a section of the eastern shoreline (in the southern section of the lake). Similar to Lake Ohrid, the polje of lake Vegoritida is larger than the lake itself with characteristic agricultural areas to the north and south being the longer parts of the tectonic graben (Image credit: ©Google earth 2015)

Cyanophyceae (photosynthetic bacteria) and *Chlorophyta* (algae and mosses) settle on the rock and precipitate calcium carbonate through metabolism—the process of travertine building. Travertine dams can breach the surface of the river and dam sections of a gorge, as in the examples of Band-e-Pamir in Bamyan Province of Pakistan and the Plitvice National Park of Croatia (Figs. 2.113, 2.114 and 2.115b). The rivers have transformed into a series of small lakes and rocky barriers with waterfalls.

The following images will give some examples of the principle morphological groups of river made lakes.

2.2.5 Lakes Formed by Glaciers and Ice Ages

During the Pleistocene Ice Age (2.4 Mio to 12,000 years ago), large areas of all continents had been covered by glacier ice and in particular the higher latitudes and expanding thousands of kilometres from the high Arctic toward the equator, e.g. in North America. Many of the world's high mountain chains had also developed glaciers (and many still have). Moving glacier ice is a particularly suitable instrument to shape the surface of bedrock and carve depressions of any size, or deposit drift material as ground moraines and terminal moraines. The deposition of drift material is normally a mixture of rocks, sand, silt and clay that can prevent water from filtering into the ground where lake formation in humid climates will occur (Figs. 2.116, 2.117, 2.118, 2.119, 2.120, 2.121 and 2.122). In fact, formerly glaciated regions belong to those with the highest concentrations of lakes of any size (see Table 2.2 for the largest lakes of formerly glaciated areas in Canada) in the World. Additionally, some of these areas have been further sculptured by permafrost patterns presenting a multitude of closed depressions holding water in the form of lakes, ponds, swamps or bogs. The following satellite images present a catalogue of lake forms resulting from glacial processes, either destructive glacial scouring (Figs. 2.123, 2.124, 2.125, 2.126 and 2.127), or barrier construction of morainic ridges within valleys (Figs. 2.128, 2.129, 2.130, 2.131, 2.132 and 2.133). Lakes often occur where glaciers retreat and leave exposed ground behind a terminal moraine (Figs. 2.134, 2.135, 2.136, 2.137 and 2.138). Small meltwater lakes may even occur on the glacial surface, although only short living with some only for just days or weeks.

Fig. 2.83 (**a**) Two lagoons (Lago di Lesina to the west and Lago di Varano in the east) along the Adriatic coast of Italy (centre of the 100 km wide scene is approximately *41°49'N* and *15°34'E*). Sediments are delivered by the Fortore River where its truncate delta is visible to the west. (**b**) Beginning of the Lago di Lesina barrier on the Adriatic coast of Italy *(41°53'59.22"N, 15°21'53.58"E)*. The sand bars initiate formation from the western end southward on the lee side of the promontory and then follow eastward. The barrier is made up of a series of beach ridges in the northern part, and toward the lagoon side are low dunes covered by partly cultivated vegetation. Scene is 5.5 km wide (Images credits: ©Google earth 2015)

Fig. 2.84 A group of lagoons in southern Brazil (approximately *30°23'S* and *50°20'W*). The north is to the upper right, where strong northerly winds shift parabolic dunes inland, and the southern shores of the lagoons have transformed into smooth curved beaches with barriers and spits created by wave attack from the open (unprotected) lagoon. Scene is 25 km wide (Image credit: ©Google earth 2015)

Fig. 2.85 The 14 km long Lagoon de Bilene, Mosambik (*25°17'S* and *33°17'E*) with tongue-like dunes as a result of southerly winds. Liman forms can be seen north of the main lagoon (Image credit: ©Google earth 2015)

Fig. 2.86 The Lagoa de Araruama near Cabo Frio in eastern Brazil (approximately *22°53'S, 42°12'W*) is sectioned by internal spits created by longshore drift from easterly winds (trade winds). The lagoon is over 30 km long (Image credit: ©Google earth 2015)

Fig. 2.87 Lagoons along the south coast of West Australia (approximately *33°38'S, 123°48'E*). Although the lagoons lie alongside the cool waters of the Circum-Antarctic Current (also known as West Wind Drift), the climate in this region is semi-arid and so the lagoon lakes are transformed into saltpans by evaporation. Scene is 32 km wide and the western lagoon is named Lake Daringdella (Image credit: ©Google earth 2015)

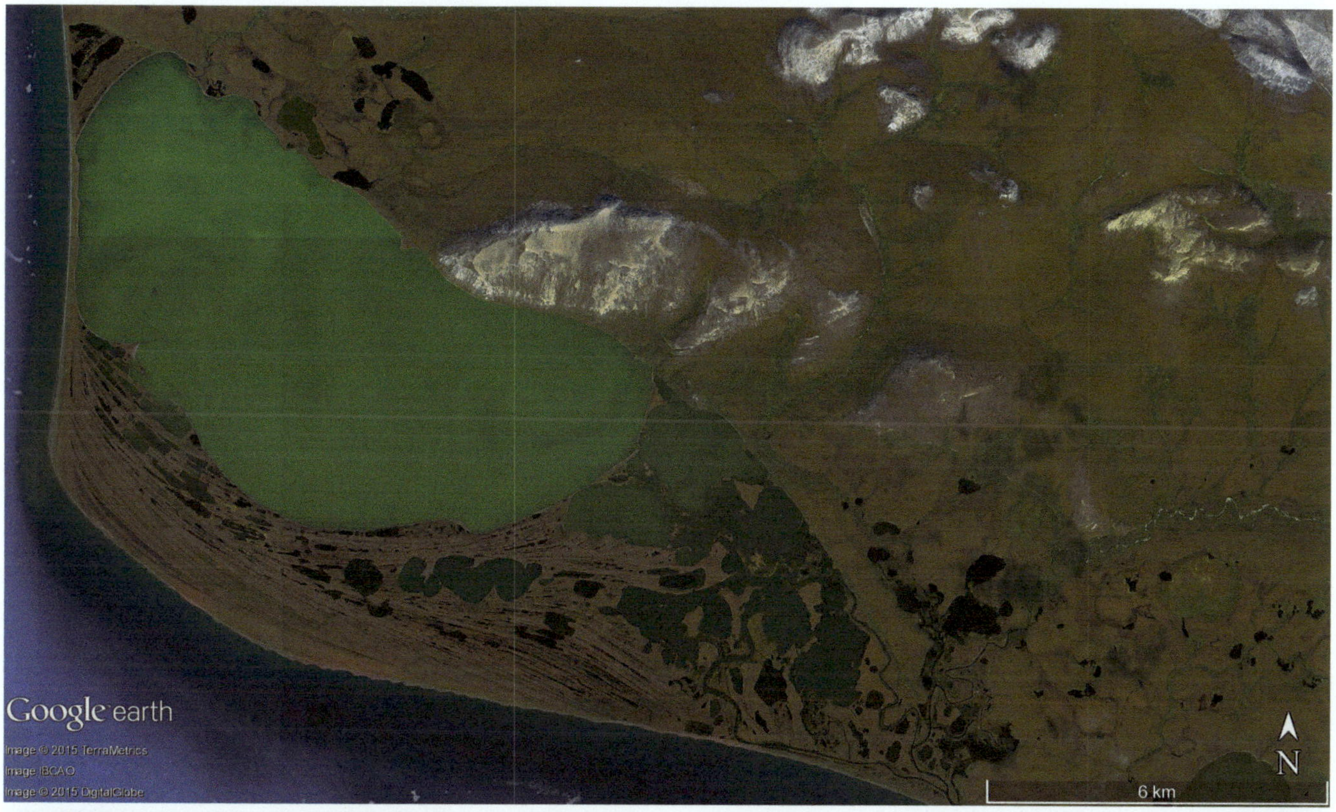

Fig. 2.88 At Cape Krusenstern in southwest Alaska (approximately *67°09'N, 163°35'W*) the large main lagoon has been dammed from the open Pacific Ocean by a barrier. In the southern part of the beach ridge system, lakes of different forms (e.g. swales within ridges, or enlarged circular-like forms by permafrost decay) occur. Scene is 23 km wide (Image credit: ©Google earth 2015)

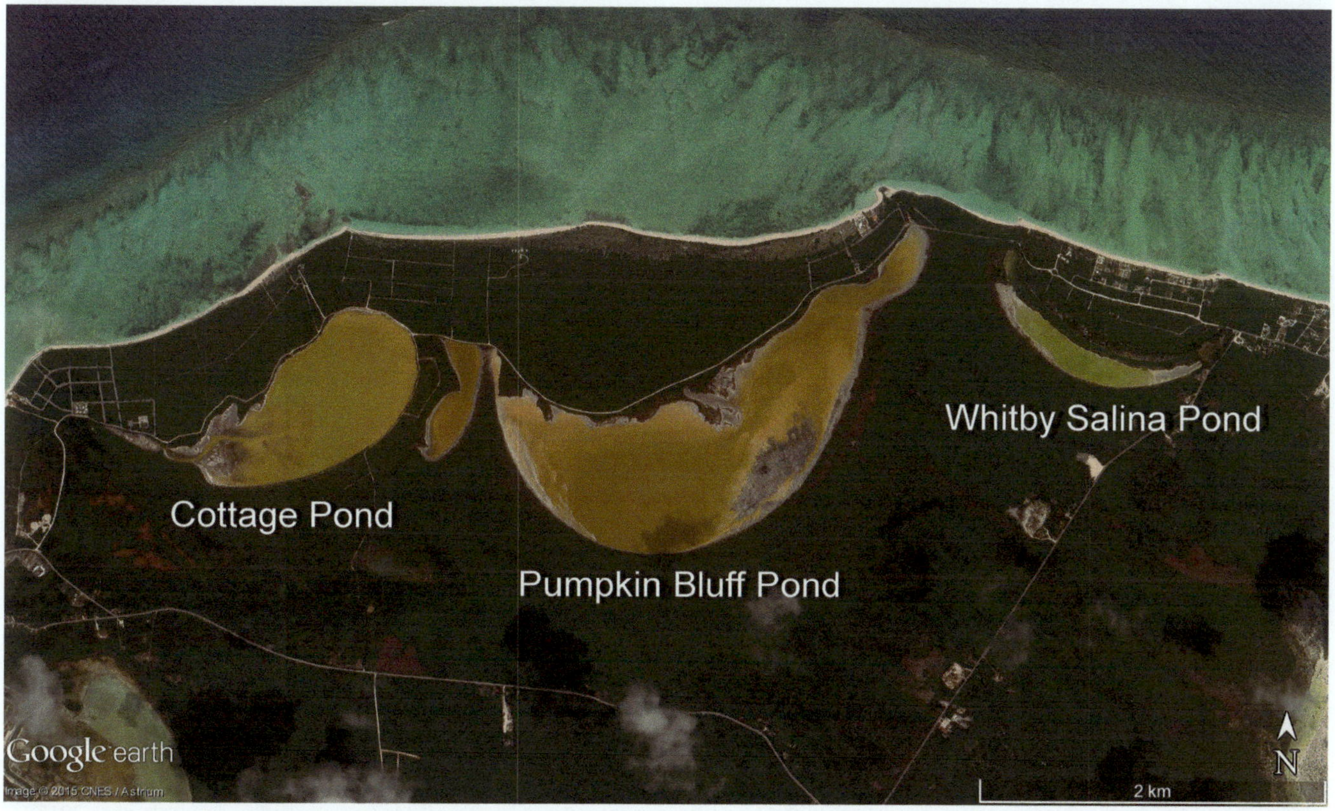

Fig. 2.89 Half-moon-shaped lagoons lie along the north coast of North Caicos Island southeast of the Bahamas (approximately *21°57'N* and *72°00'W*). The formation of these lagoons is a result of a strong hurricane attack and wave surge, eroding parts of the older and slightly cemented Pleistocene dunes further inland. Scene is 7 km wide (Image credit: ©Google earth 2015)

Fig. 2.90 (**a**) South of Perth in West Australia, transgressive sea level movements have formed three barriers and subsequently three lines of lagoons. The seaward and landward barriers now present aspects of saltpans whereby in the deeper centralized barrier a chain of lakes persists. Scene is 10 km wide with the centre at around *32°48'S, 115°44'E*. (**b**) Along the west coast of Victoria, southeast Australia (approximately *37°15'S, 139°52'E*), wide old lagoon basins are partly separated by interglacial dune belts now covered by vegetation. The lagoons transform into salt lakes and saltpans, particularly during the dry winter months. Scene is 18 km wide (Images credits: ©Google earth 2015)

Fig. 2.91 In northern Venezuela, northeast trade winds transport sand from a wide beach in the direction of the lagoon. The stepwise moving dunes are barchanoids and disappear beneath the water of the lagoon. At least 35 rhythmic steps of movement have developed, most probably representing 35 seasons of trade winds. Recent sand transport from the beach is presented in chevron forms. Scene is 2.8 km wide at around *10°58'N* and *71°36'W* (Image credit: ©Google earth 2015)

Valleys positioned below advancing and retreating mountain glaciers are exposed to danger for human populations and infrastructure: during warmer periods glaciers recede and produce large volumes of meltwater that may accumulate behind terminal moraines and easily burst through these delicate dams. Advancing glaciers may also store meltwater, either as the main valley glacier advances and blocks meltwater in side-valleys, or when side-glaciers advance and block meltwater from the main glacier (Figs. 2.139, 2.140, 2.141, 2.142 and 2.143). In all areas where glaciers occur conditions can change within a few years. The risk is, dams that are formed by (moving or stagnant) glacier ice tend to break open wide fissures by melting, allowing for fresh waterways and an outbreak. Alarmingly, the ice may also be lifted by buoyancy and an unforeseeable catastrophe may hit settlements and infrastructure.

Fig. 2.92 (**a**) Rare lagoon forms in eastern Madagascar (approximately *18°35'S* and *49°01'E*) adapted from swales positioned in between coastal dune ridges whereby the dunes are formed in a single line of sand hills. Scene is 7.7 km wide. (**b**) Lakes filling peculiar shaped basins between coastal dune ridges of eastern Madagascar (approximately *20°02'S* and *48°45'E*). Scene is 3.5 km wide (Images credits: ©Google earth 2015)

Fig. 2.93 Coastal lagoons perpendicular to the coastline as former parts of drowned valleys (limans) from the south coast of Martha's Vineyard, an island in Massachusetts, eastern USA (approximately *41°21'N, 70°28'W*). Scene is 4 km wide (Image credit: ©Google earth 2013)

Fig. 2.94 Liman-like lagoon in eastern Siberia, Russia (*60°53'N* and *171°44'E*). The closing barrier consists of a sequence of beach ridges cut by a natural outflow channel. Scene is 24 km wide (Image credit: ©Google earth 2013)

Fig. 2.95 This lagoon has been separated by a 6 km wide barrier with beach ridge sequences at the coastline of Kamchatka, eastern Russia *(60°03'N, 170°13'E)*. Several natural channels are open to the sea during springtime from meltwater discharge (Image credit: ©Google earth 2013)

Fig. 2.96 A wide lagoonal landscape along the north-easternmost coast of Russia *(69°49'N, 175°53'E)*. The outer and internal barriers contain smaller lakes of similar oval shapes with their long axes positioned in N-S direction. This pattern may originate from a dominant wind direction that also moves spring ice cover. The largest lagoon has a long axis of 11 km (Image credit: ©Google earth 2012)

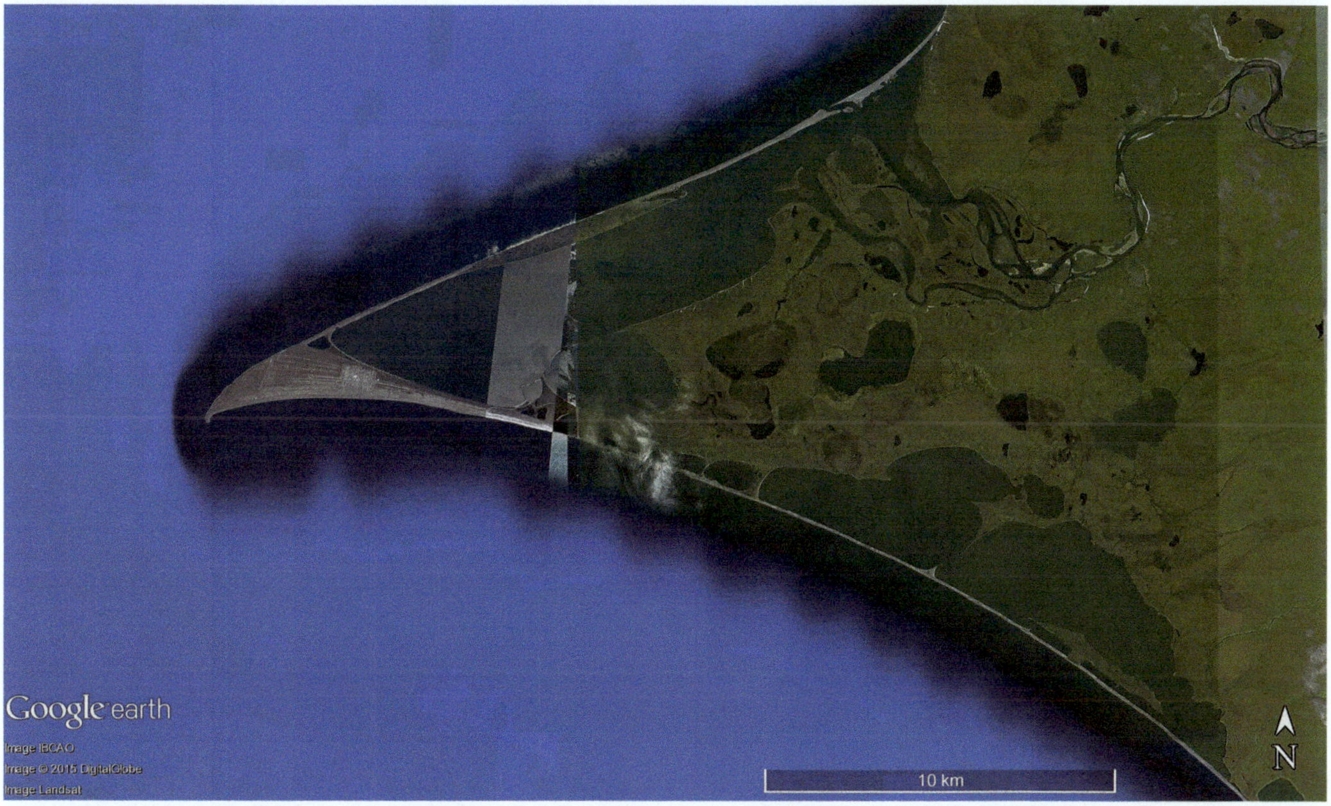

Fig. 2.97 Lagoons at Point Hope in northwest Alaska (approximately *68°21'N, 166°48'W*) are framed by two long barriers culminating into a spit. The special form of the spit and the lagoons positioned behind are created more or less simultaneously by longshore drift working from the northeast and southeast, and most probably using drifting sea ice as tools for transport of sediments and abrasion. The scene shown is 34 km wide (Image credit: ©Google earth 2015)

Fig. 2.98 A Triangular-shaped lagoon with an exposed spit in eastern Siberia, Russia (*64°25'N* and *172°20'W*) in a 20 km wide scene (Image credit: ©Google earth 2015)

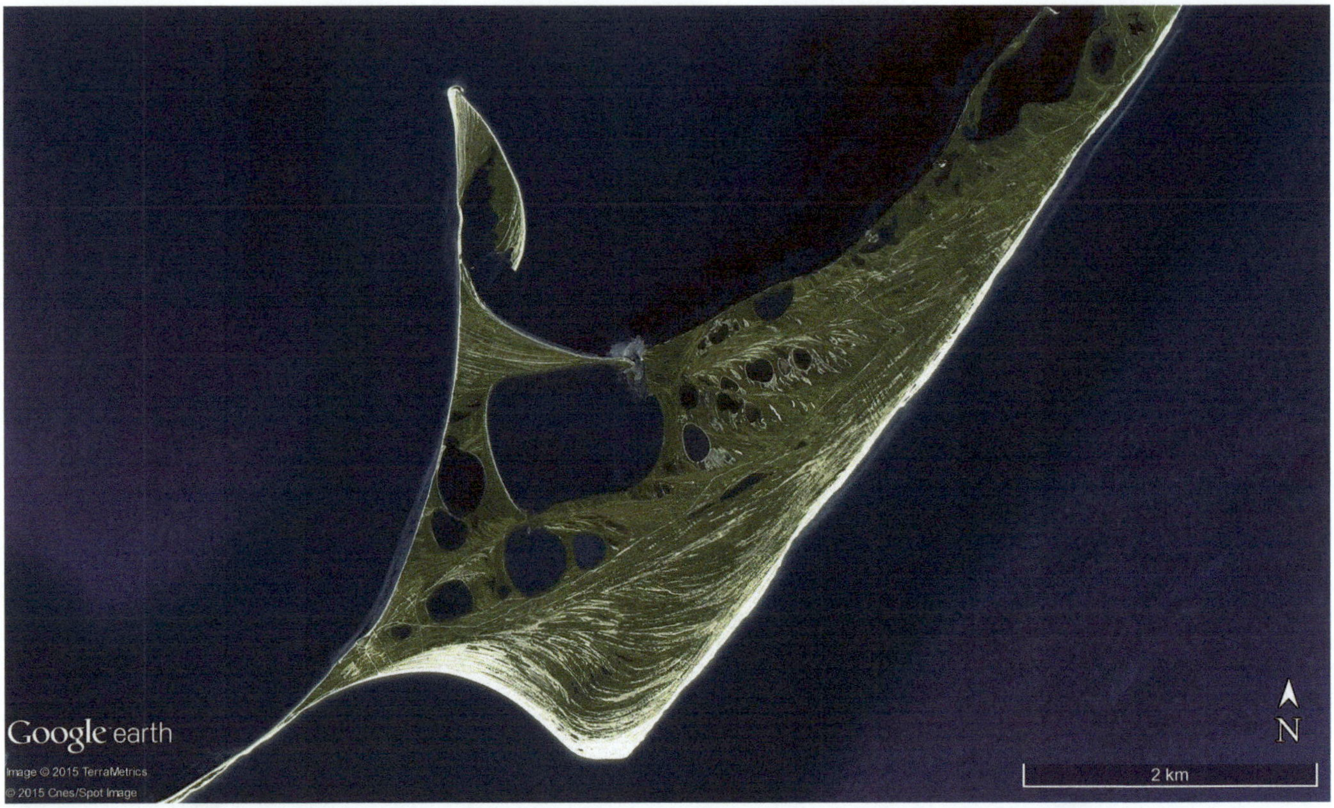

Fig. 2.99 The outer region of a long spit accommodating many small lakes situated between beach ridges (lying in different directions) can be found at the northern coast of the Azov Sea (part of Black Sea) Ukraine *(46°30'N, 36°10'E)*. The multitudes of beach ridges are an excellent archive for research in sea level variation and paleotempestology (the science and history of storms). Scene is 8 km wide (Image credit: ©Google earth 2015)

Fig. 2.100 A 9 km long section of the Dzharylhats'ka Spit in Karkinyts'ka Gulf, west of Crimean Peninsula (Ukraine south coast) along the northern part of the Black Sea. The lakes rest between beach ridges and their similar forms are the result of dominant south-southeasterly winds in summer (in winter the shallow lakes are usually frozen). Centre of the image is approximately *46°02'N, 32°51'E* (Image credit: ©Google earth 2015)

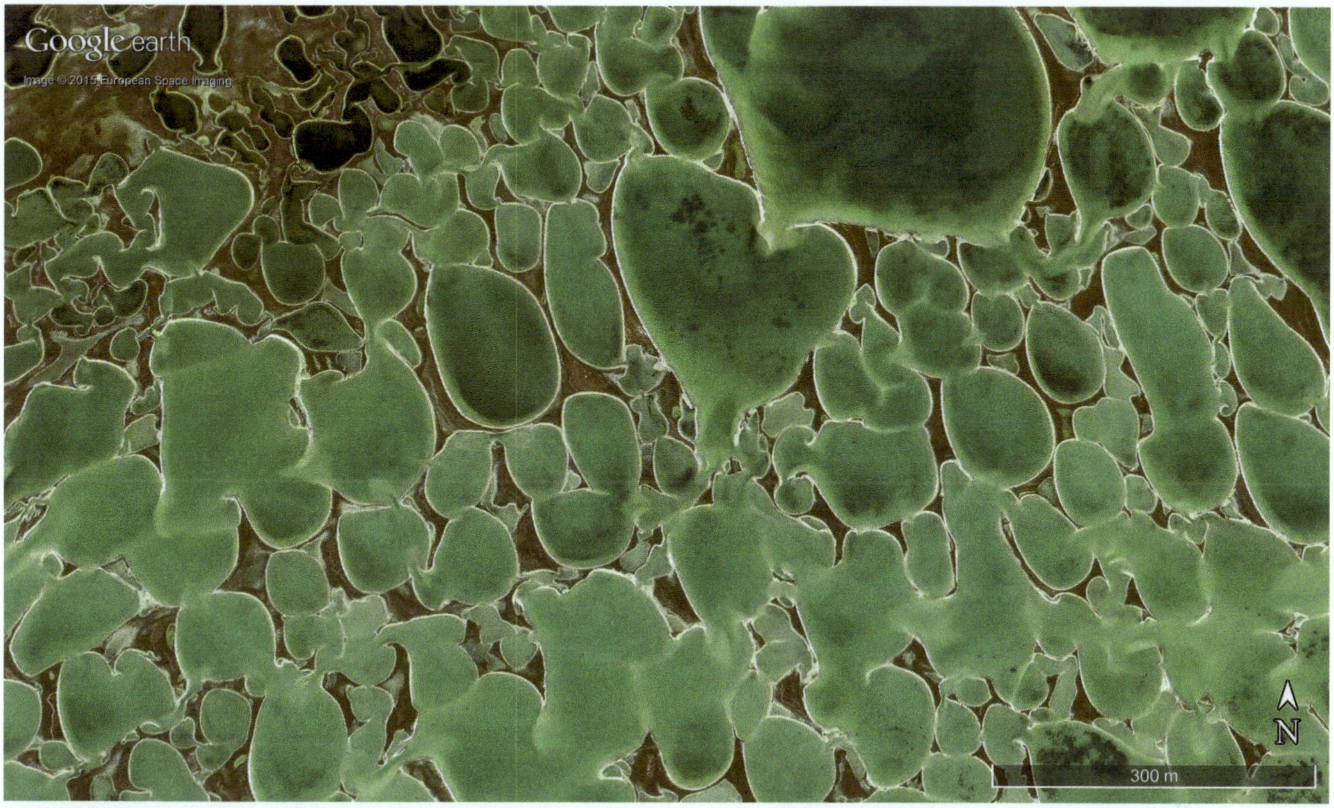

Fig. 2.101 Detail from Dzharylhats'ka Spit in Karkinyts'ka Gulf, west of Crimea peninsula along the northern part of the Black Sea in a scene 1.3 km wide. Centre of the image is *46°02'19.49"N, 32°51'17.27"E* (Image credit: ©Google earth 2015)

Fig. 2.102 Positioned on a 3.3 km wide barrier located between beach ridges along the east coast of northern Sachalin Island, Russia *(52°30'N, 143°17'E)* are large lakes the result of permafrost decay in sediments, and the small elongated lakes are just groundwater ponds between beach ridges (Image credit: ©Google earth 2013)

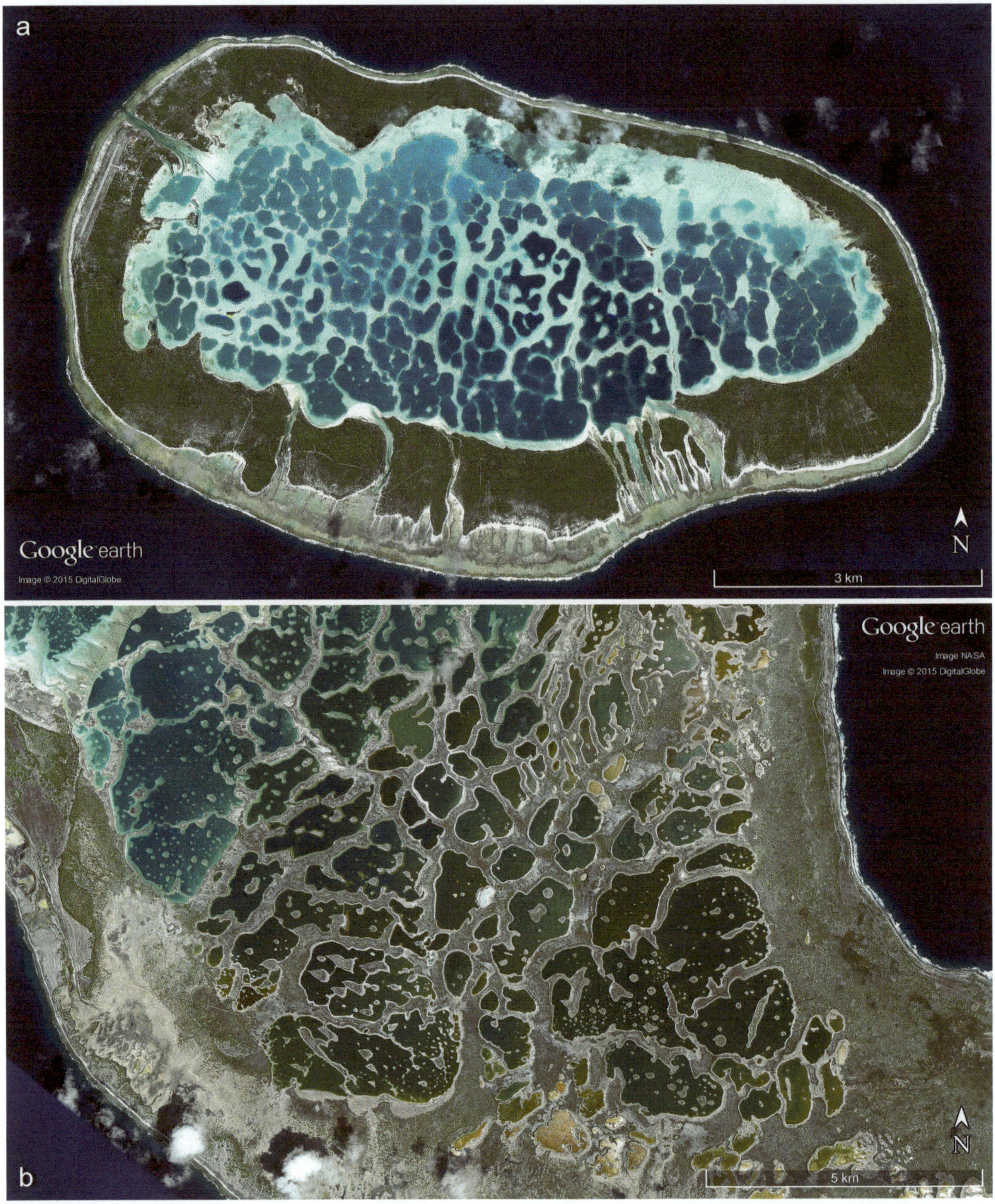

Fig. 2.103 (**a**) Atoll-shaped islands describe a more or less closed central depression containing salt water—sometimes hyper-saline caused by evaporation and limited water exchange with the open ocean. Growth of coral colonies may divide the depression into many individual basins. Scene depicts the 10 km long Mataiva Atoll in French Polynesia *(14°53'S, 148°40'W)* taken August 16th, 2006 (Image credit: ©Google earth 2013). (**b**) Basins formed by coral growth in Kiribati Atoll *(1°50'N, 157°23'W)*. Scene is 17 km wide and taken January 21st, 2006 (Image credit: ©Google earth 2015)

Fig. 2.104 (**a**) A salt lake and saltpan (Laguna Negro Francisco) located in the central Chilean Andes at 4137 m asl. *(27°28'S and 69°13'W)*. The barrier dividing the saltpan and the southern lake reaches 20 m in height. Along the southern shore of the barrier and open water waves coming from the northwest have formed a series of short spits. This scene is 10 km wide. (**b**) Detail of a barrier with spits in the southern part of Laguna Negro Francisco in eastern Chile. This scene is 2.8 km wide (Images credits: ©Google earth 2015)

Fig. 2.105 (**a**) Small lakes at the site of plunge pools/potholes resulting from sporadic waterfalls (exceeding 120 m in height being double that of Niagara Falls) resulting from more than 20 outbreaks of Lake Missoula in the late-glacial times (around 15,000–13,000 years ago). The many floods of Lake Missoula have exceeded 10 km^3/h in volume (more than all rivers in the World combined!), with speeds of more than 100 km/h and each lasting for several days. Horseshoe-shaped steps demonstrate the intensive backward erosion during these fast flows. In the collective memory of the First Nation people from Lake Nissoula region, myths are still told of a giant lake suddenly emptied and disappeared. The site is at the "Columbia Dry Falls" southwest of Grand Coulee Reservoir *(47°35'57.83"N* and *119°21'28.51"W)*. Scene is 4 km wide. (**b**) Formerly potholes and now known as "Dry Falls" from the late-glacial outburst of Lake Missoula more than 13,000 years ago in northern USA *(47°36'N, 119°21'W)*, the dry falls have transformed into a small swamp and a saline pond (saltpan when dry). The shadow reflects the 120 m high step and former waterfall site (Images credits: ©Google earth 2013)

Fig. 2.106 **(a)** A lowland lake in northern Columbia is continuously filled by river sediment. The shallow depth of the lake can be seen in the natural levees along the river courses, deposited by the resistance of vegetation that filters sediment from the flow. Scene is 11 km wide at approximately *9°35'N* and *74°42'W* (Image credit: ©Google earth 2015). **(b)** When loaded with sediments (in suspension), rivers breach their banks whereby water flow is reduced by the vegetative friction and sediments are deposited. Over the course of time wide dams form called "natural levees" on both sides of the river occur. If the river terminates in a lake, these levees develop isolated fingers of deltas and enclosed lakes of different forms; as in this image from the lowlands of northern Colombia *(8°55'N* and *74°44'W)*. Scene is 27 km wide (Image credit: ©Google earth 2013)

Fig. 2.107 During springtime in northern Russia, large rivers (setting approximately *67°25'05.20"N, 52°01'17.56"E)* carry vast sediment loads and dispense the load during riverbank overflow as natural levees (including along meander bends). Eventually narrow curved lakes develop or become irregular shallow basins at sites without significant sediment deposition. Scene is 20 km wide (Image credit: ©Google earth 2015)

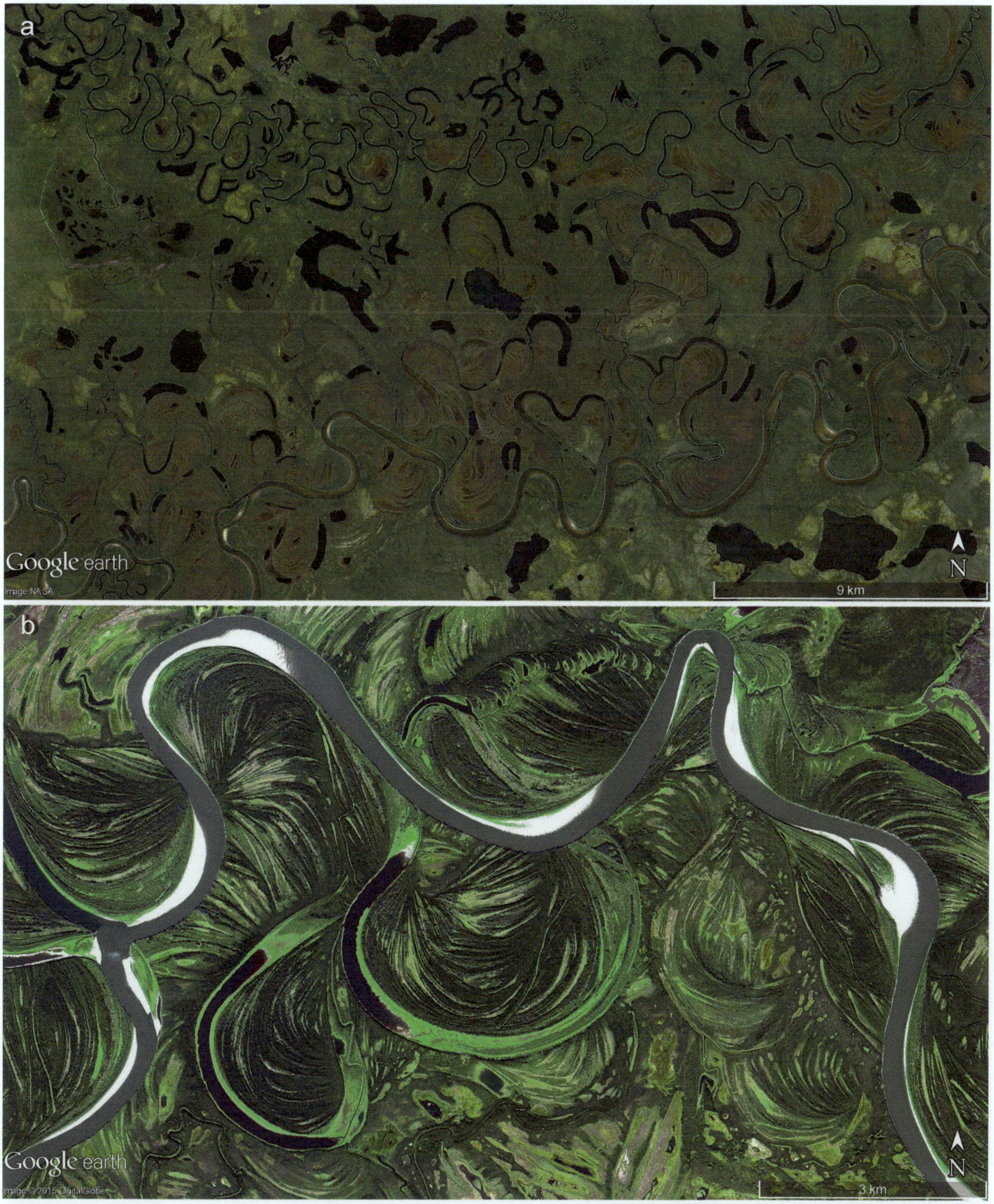

Fig. 2.108 (**a**) The principle of alongside deposition—particularly high waters flowing over river banks where vegetation filters the sediment out of the overflow—also works with meandering rivers. Natural levee formations curve and meander with the river resulting in numerous narrowly curved lakes, or ponds of shallow water in between the deposits at point bars (crescent-shaped bars located on the inside of the river bend). Oxbow lakes are stagnant and inactive meanders separated from the main river by levees. Scene is from northern Siberia, Russia *(68°30'N, 143°38'E)*. Setting is 35 km wide (Image credit: ©Google earth 2013). (**b**) Lakes replace former meander bends of an old river from Alaska *(65°31'40.94"N, 156°53'19.64"W)*. Scene is 13 km wide (Image credit: ©Google earth 2015). (**c**) Lakes lie between narrow natural levees and point bars in the bends of the Kolyma river in northern Siberia Russia *(68°44'N, 158°33'E)*. The meander of the river shifted

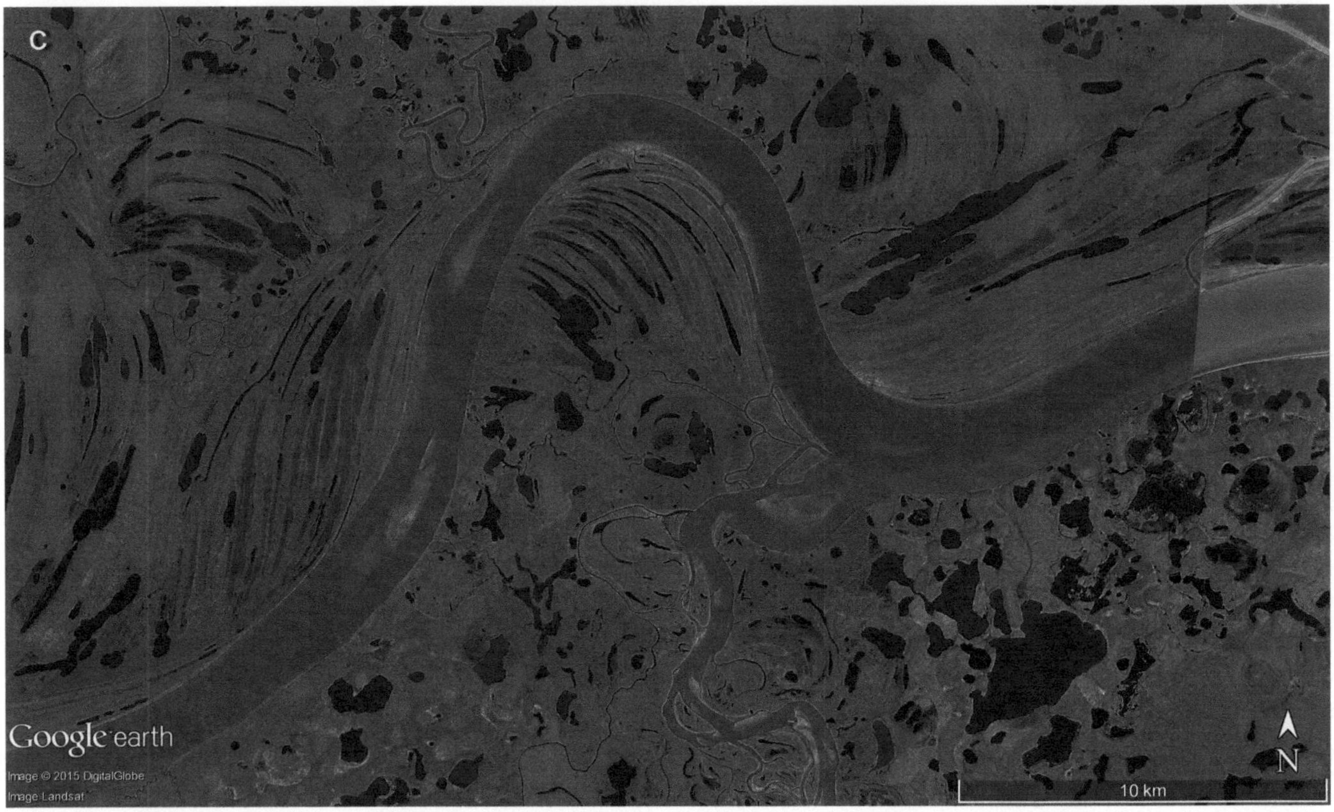

Fig. 2.108 (continued) from the main channel and left a stagnant section of the riverbed with clear, black (viewed from above) water, whereas the active river shows sediment suspension of a light colour. The small lakes with irregular contours set away from the river course are depressions created by permafrost thawing processes. Scene is 40 km wide (Image credit: ©Google earth 2015)

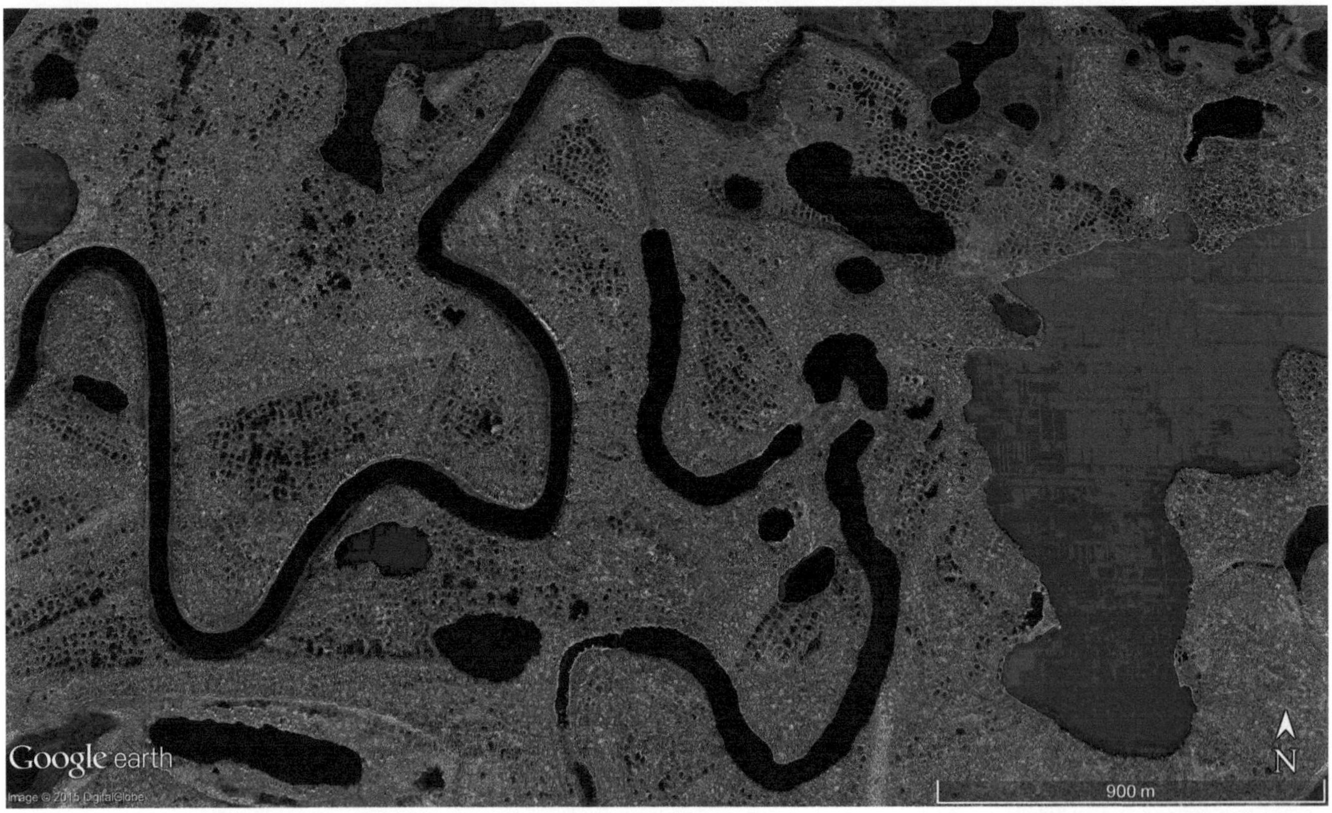

Fig. 2.109 Ground frost activities have blocked a north Siberian lowland river *(71°02'N* and *136°02'E)* and sections of the riverbed are now left as narrow-bending lakes. Scene is 3.6 km wide (Image credit: ©Google earth 2015)

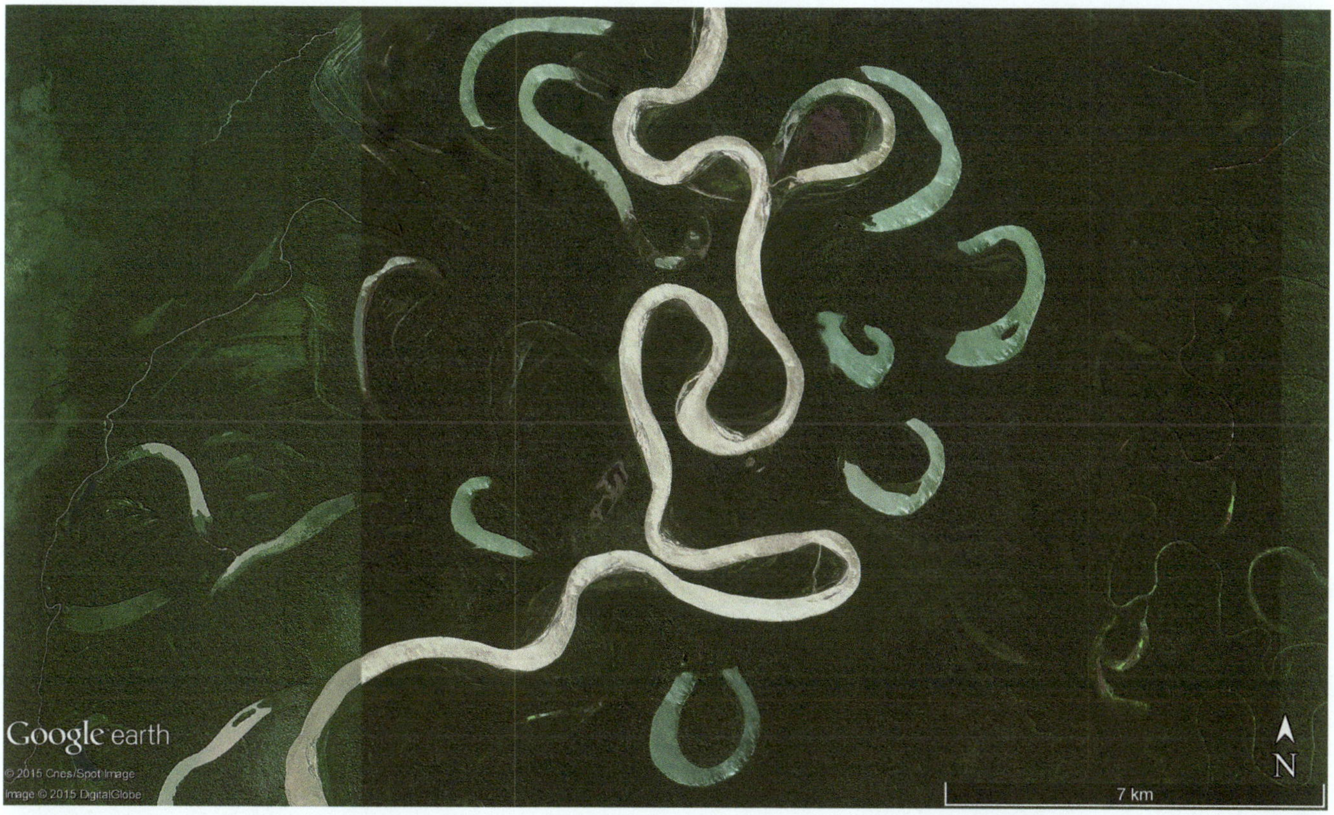

Fig. 2.110 The Rio Beni in lowlands of eastern Bolivia (approximately *13°37'S, 67°23'W)* exhibits several generations of oxbow lakes in extinct meanders dammed by younger natural levees in a 30 km wide landscape (Image credit: ©Google earth 2015)

Fig. 2.111 The Rio Purus, a northern tributary of the Amazon in northern Brazil (approximately *5°54'S* and *64°30'W)* demonstrates the continual shifting of meander bends and separation into oxbow lakes. In the oxbow lakes, sediments have been settled down and the water is clear (black shown from above) whereas the active river sediment suspension gives a lighter colour. Scene is 14 km wide (Image credit: ©Google earth 2013)

Fig. 2.112 Close to the northern shores of Lake Titicaca, Peru (approximately *15°34'S, 69°57'W)* natural levees encloses a meander lake almost 400 m in length (Image credit: ©Google earth 2012)

Fig. 2.113 Along a narrow gorge in Band-e-Pamir National Park, Afghanistan *(34°49'50.94"N, 67°12'25.86"E*, at 3000 m asl) six lakes are dammed by travertine barriers (deposition of calcium-carbonate through evaporation, triggered also by the presence of *Cyanobacteria* and *Chlorophyta* (mosses and algae). Scene is about 10 km wide (Image credit: ©Google earth 2015)

Fig. 2.114 The Plitvice National Park, Croatia *(44°52'26.65"N* and *15°36'17.50"E)*. Travertine barriers blocking a river gorge along several sites and dam small lakes that are connected via waterfalls. The gorge runs from west to east traversing through a limestone and dolomite area presenting a series of 16 lakes cascading from 636 m asl to 503 m asl. The depth of lakes varies between 1 and 47 m, and the highest waterfall measures 78 m. The travertine barriers grow upward around 1 cm/year and their age is determined to be around 6000 to 7000 years. Plitvice National Park is one of the first UNESCO World Heritages Sites dating from 1979. The scene is 2.6 km wide (Image credit: ©Google earth 2015)

Fig. 2.115 (**a**) Detail of the travertine dams from Plitvice National Park in Croatia (see Fig. 2.114). Lake levels in this image go from 623 m in the west to 609 m, to 604 m and to 565 m in the east. Scene is 1.3 km wide. (**b**) Angled detail of lakes dammed by travertine barriers in Plitvice National Park, Croatia, at about *44°52'N* and *15°36'E* (Images credits: ©Google earth 2015)

Fig. 2.116 The Dzhulukul'Lake in Altai Province, southern Russia *(50°28'54.91"N, 89°42'36.51"E* at 2201 m asl). The main lake (10.6 km in length) replaces a former glacier tongue and the more than 300 shallow kettle lakes around the region are the result of block-ice melting in late-glacial basal morainic debris (Image credit: ©Google earth 2015)

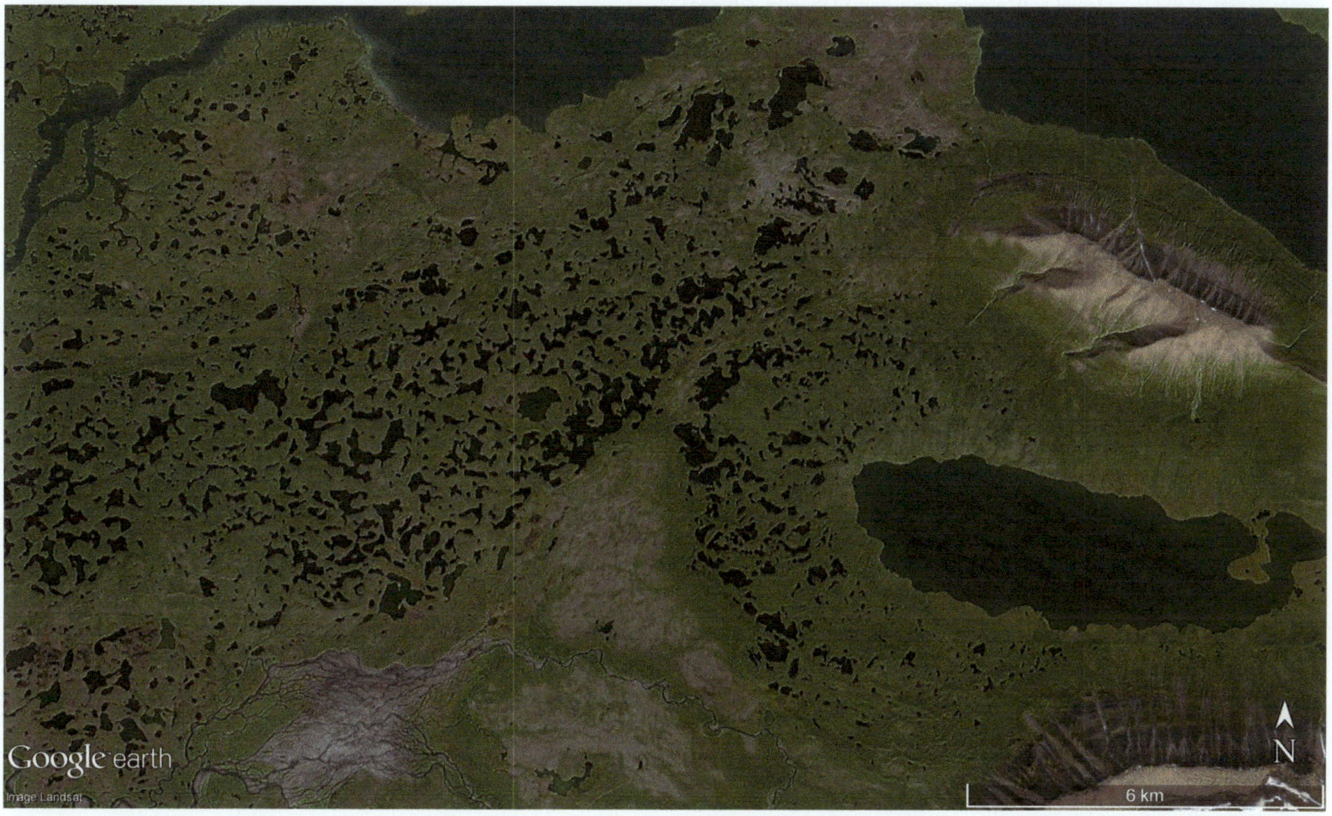

Fig. 2.117 Glubokoe Lake (to the lower right of image) replaces an Ice Age glacier tongue in central Russia (approximately *69°15'N* and *89°43'E*) and the scene exhibits a mosaic of many small lakes (kettle holes) in an undulating and chaotic ground moraine environment resulting from the Ice Age glacier advancement. Scene is 23 km wide (Image credit: ©Google earth 2015)

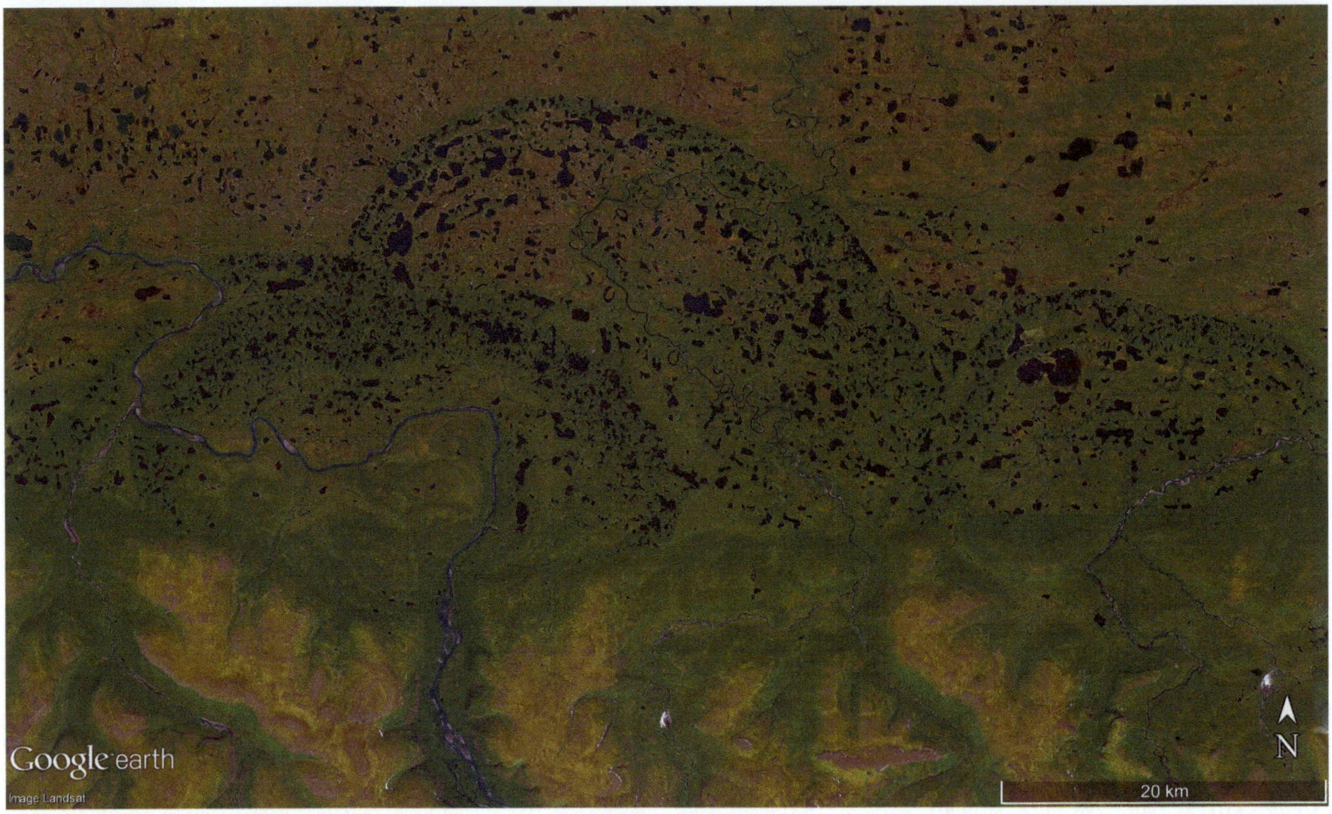

Fig. 2.118 Three semi-circular groups of lakes north of the basaltic Central Siberian Plateau, Russia *(70°34'N* and *96°08'E)* mark the maximum extent of the last glacial with glacier tongues and their re-advance morainic ridges from the south. Scene is 80 km wide (Image credit: ©Google earth 2013)

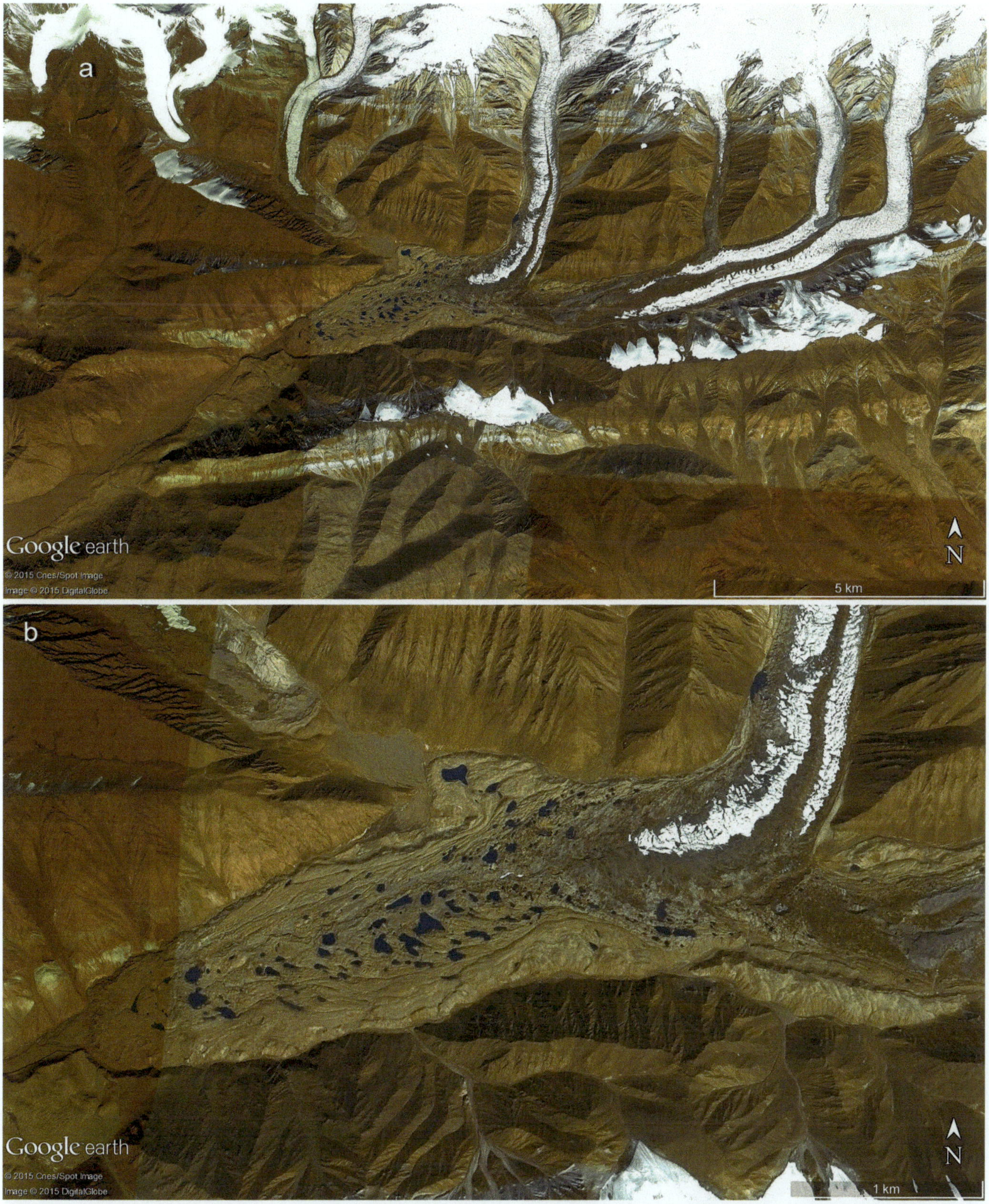

Fig. 2.119 (**a**) Kettle lakes and kettle holes on a receding glacier in the Pamir Mountains bordering the territories of Tadschikistan and Kirgisistan (approximately *39°24'N, 73°29'E*, and 4400 m asl.). Scene is 17 km wide. (**b**) Detail of Fig. 2.119 (**a**) in a 5 km wide scene (Images credits: ©Google earth 2015)

Fig. 2.120 The fast meltwater flow at the end of the last Ice Age demonstrates rhythmic changes to lakes and deposits of a similar form in Sweden (around *62°2 N* and *12°20'E)*. Scene is 13 km wide (Image credit: ©Google earth 2015)

Fig. 2.121 A 60 km long lake system on rocky ground in Québec (Canada) is surrounded by many smaller lakes. Former glaciers have undulated the rock depending on the joint patterns and resistance by which the scoured depressions are now filled with water. Scene is at *60°58'N* and *73°58'W* (Image credit: ©Google earth 2013)

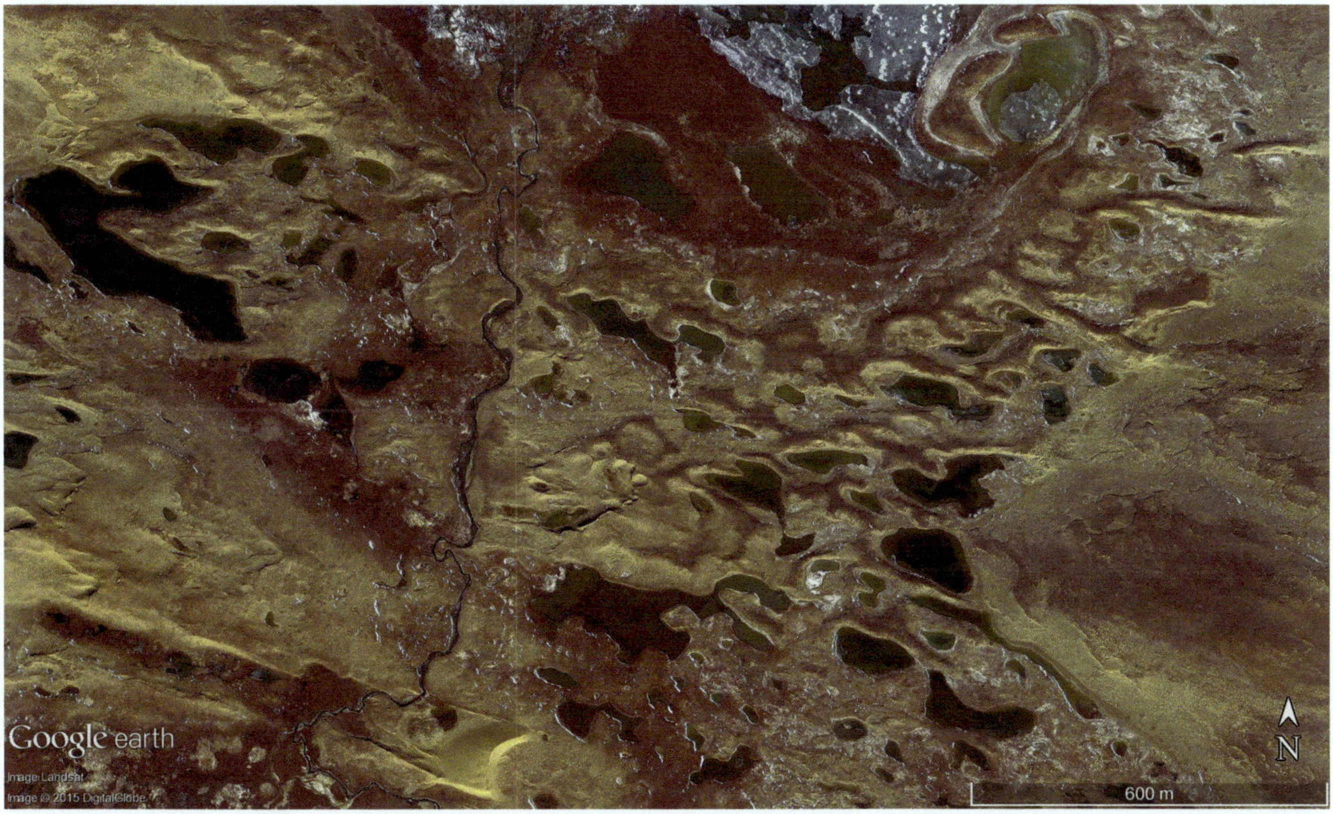

Fig. 2.122 An irregular topography with kettle lakes at the site of a block-ice melt, positioned between ground moraine hills from the last glaciation in northern China *(34°44'N, 98°17'E* at 4200 m asl). Scene is 2.6 km wide (Image credit: ©Google earth 2015)

Table 2.2 All of the largest lakes in Canada lie in a formerly glaciated landscape overriding old base rock, and not only are these lakes outstanding by their extraordinary size, but also their depth and particularly long shorelines

	Longest extension	Surface area	Average depth	Maximum depth	Shore length
Name	km	km²	m	m	km
Great Bear Lake	309	31,150	72	446	2720
Great Slave Lake	477	27,200	41	614	3050
Lake Winnipeg	412	23,550	12	62	1860
Lake Athabasca	283	7850	?	124	1900
Largest lake of Alaska (USA): Lake Iliamna	124	2622	44	301	>500

These lakes present a vivid topography and are irregularly shaped (jagged) with occasional compound features undulated by glacial processes on rocks of different resistances (data compiled from various sources)

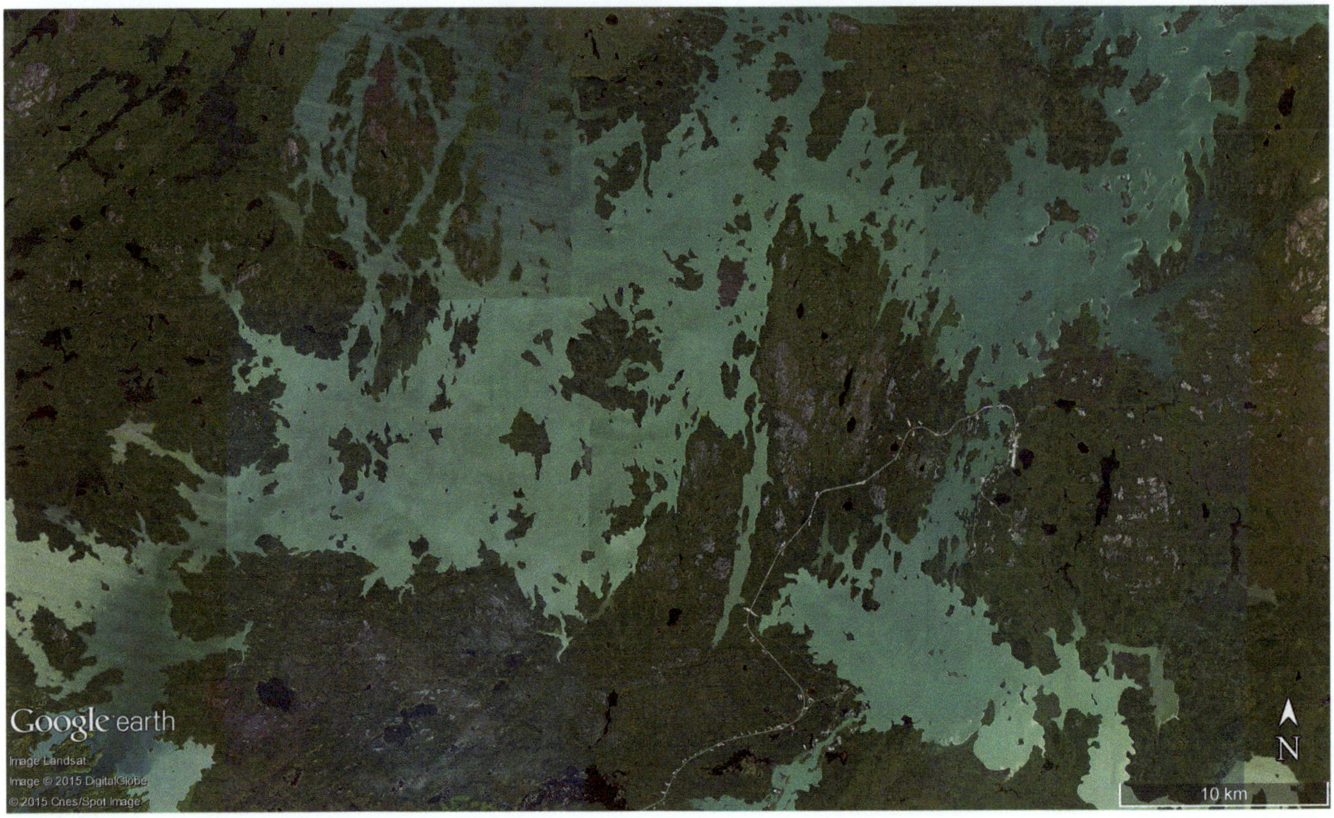

Fig. 2.123 The western section of the Southern Indian Lake in northern Manitoba, Canada (approximately *56°49'N, 99°12'W*, with a maximum length of 146 km and a surface area exceeding 2000 km^2) shows numerous skerries (small rocky islands) carved out of rock by former glaciations. The shallow lake represents a wide flooded part of the Churchill River. Scene is 65 km wide (Image credit: ©Google earth 2015)

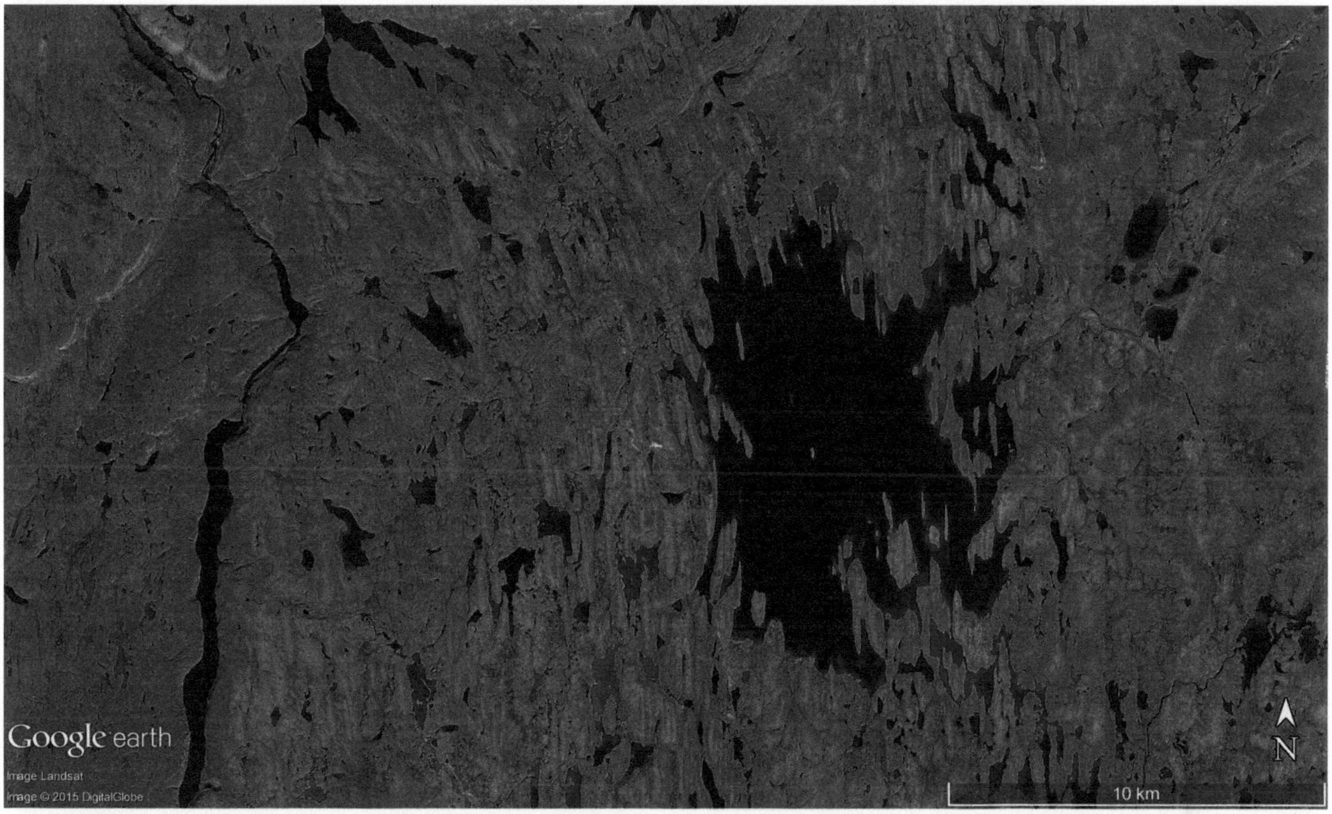

Fig. 2.124 In northernmost Norway the landscape is decorated with numerous whaleback-shaped formations resulting from the last glacial movement from north to south. When carved in rock, these formations are identified as skerries or roches moutonnées and in soft material are described as drumlins. The 35 km wide scene presents Saari Jierjärvellä (Saari means lake in Finnish) at around *69°43'N, 24°15'E* (Image credit: ©Google earth 2015)

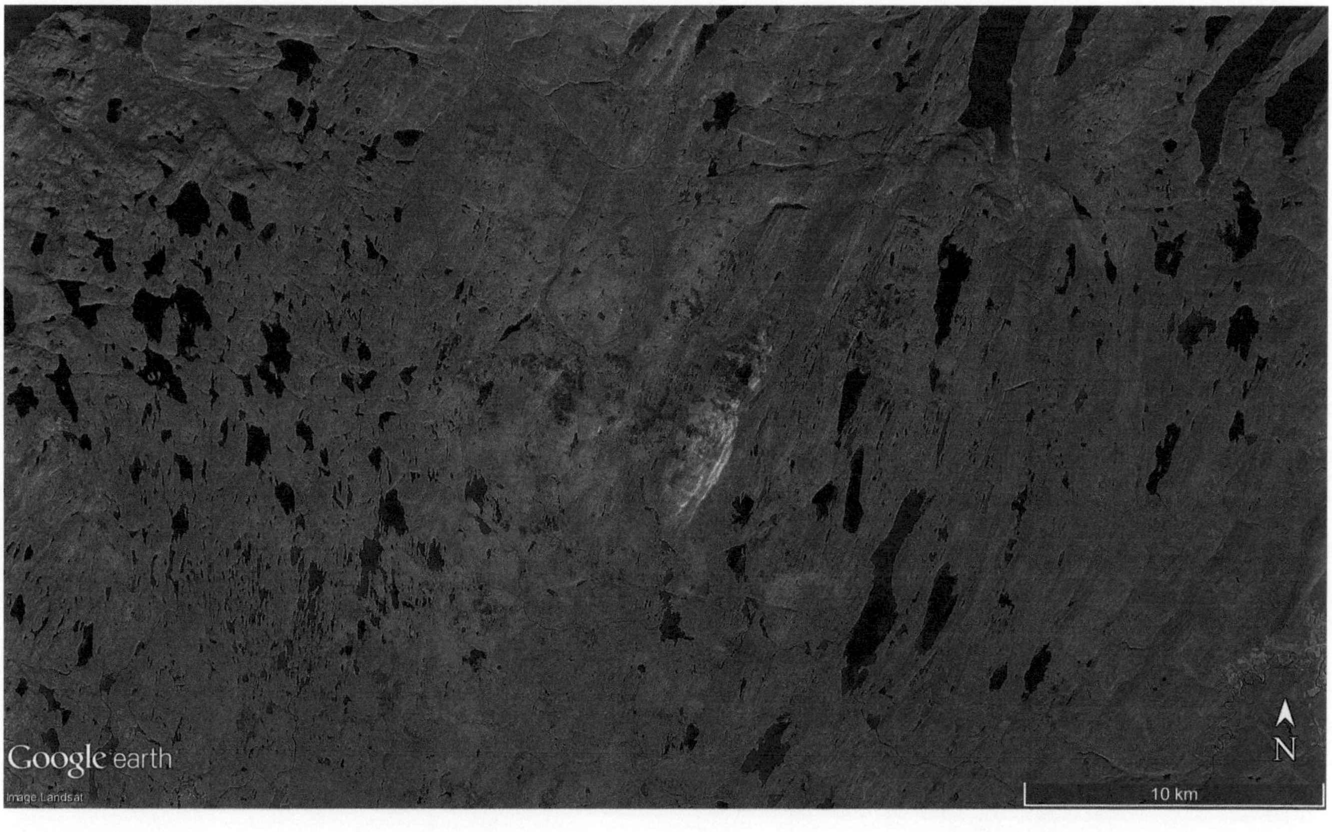

Fig. 2.125 The general south to north direction of the glacier flow is visible by the pattern of numerous lakes in northern Norway (approximately *70°20'N* and *27°29'E*). Scene is 40 km wide (Image credit: ©Google earth 2013)

Fig. 2.126 Lakes adapted from the rock formations created by glacial erosion during the Ice Ages. The scene is east of Keller Lake in the Northwest Territories of Canada *(63°59'56.41"N, 121°10'55.46"W)*. The largest lake is 7.5 km in length (Image credit: ©Google earth 2013)

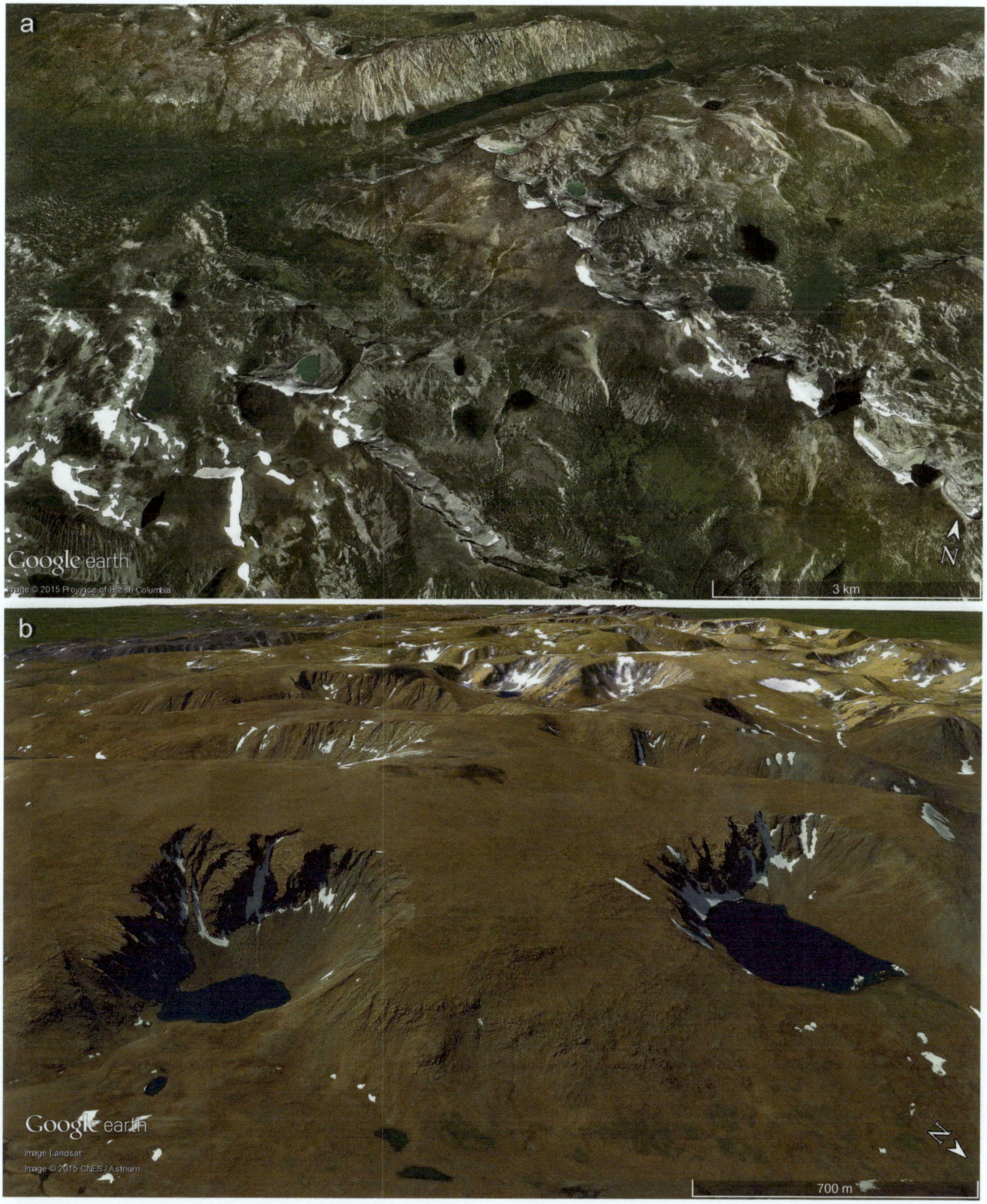

Fig. 2.127 (**a**) Approximately 30 lakes at the site of former glaciers near the summits of the Canadian Rockies *(51°57'N, 125°23'W* around 2000 m asl) lie in cirques; steep-walled semi-circular basins formed in mountains by the erosive activity of glaciers and often contain tarns (small lake) formed as a glacier melts. Width of scene is >10 km. (**b**) Cirque lakes from the northern Ural Mountains of western Russia *(66°40'N* and *64°13'E)*. Distance of lake centers is 2 km. (**c**) Lakes in cirques of Snowdonia Wales, UK (approximately *53°04'N, 4°05'W)*. Longest lake measures 1.7 km. (**d**) Glacial lakes in cirques of central Scotland *(57°41'N, 4°58'W*, approximately 800 m asl). Largest lake is 640 m across (Images credits: ©Google earth 2013, 2015)

Fig. 2.127 (continued)

Fig. 2.128 Lake Michigan one of the "Great Lakes" of USA demonstrates the perfect shape of a glacier tongue (approximately 500 km in length from the north) left from the last Ice Age (Image credit: ©Google earth 2013)

Fig. 2.129 (**a**) An oblique view demonstrates how Lake Garda (50 km long) in northern Italy *(45°33'N* and *10°38'E)* has retreated from the European Alps as one of the largest piedmont glacier tongues of the Ice Ages with a series terminal moraines flanking its front. (**b**) An oblique satellite image demonstrates how Lake Pukaki on the South Island of New Zealand has come down from the New Zealand Alps to the east. The Tasman Glacier carved this lake out in the Ice Ages and lies in the distance in front of the highest summit Mt. Cook (called "Aoraki" in Maori at 3754 m asl). The lake level is 496 m asl, and ends *(44°06'S* and *170°09'E)* with terminal moraines that form a natural dam (Images credits: ©Google earth 2015)

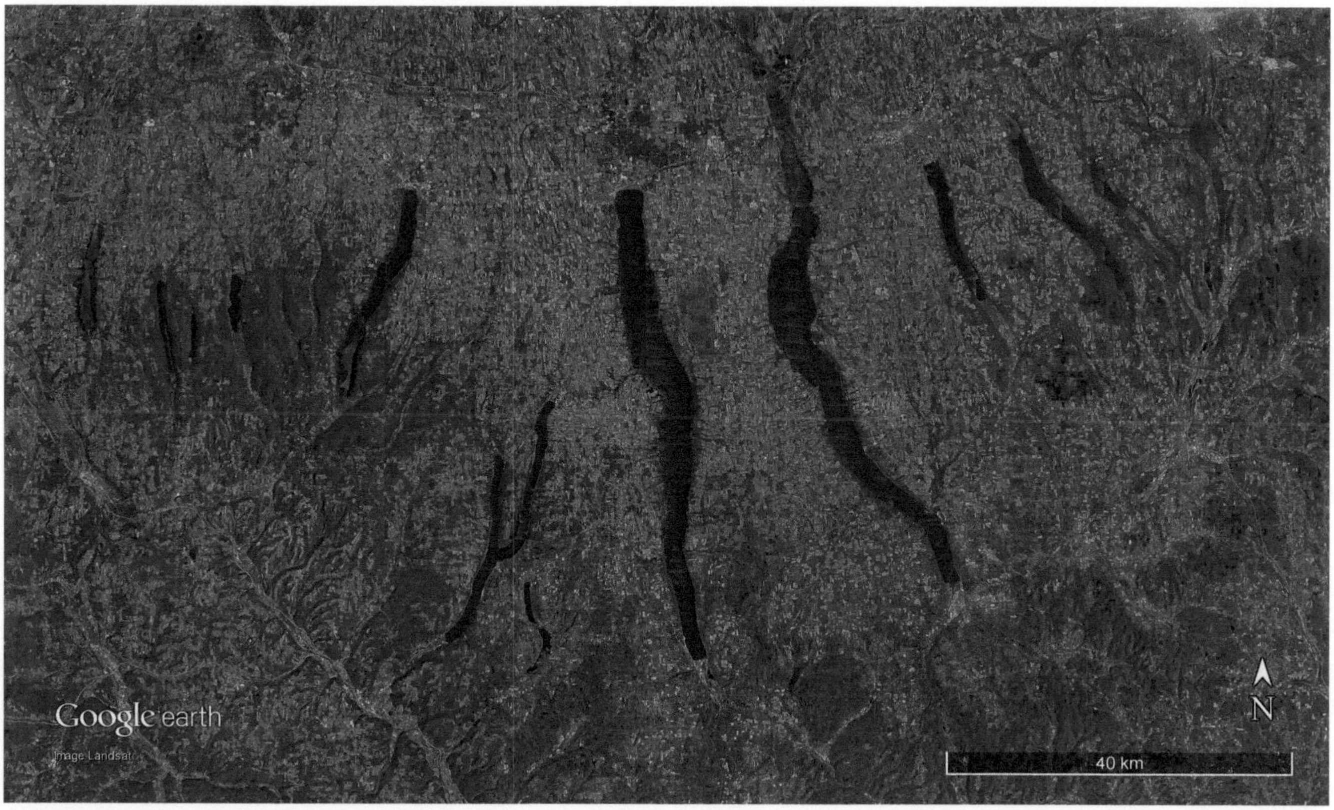

Fig. 2.130 Ten lakes known as "Finger Lakes" represent an extremely wide former glacial-lobe in the north of New York State, USA *(42°43'N, 76°50'W)*. The longest of these lakes was most probably fashioned by subglacial meltwater under hydrostatic pressure and reaches 55 km in length (Image credit: ©Google earth 2013)

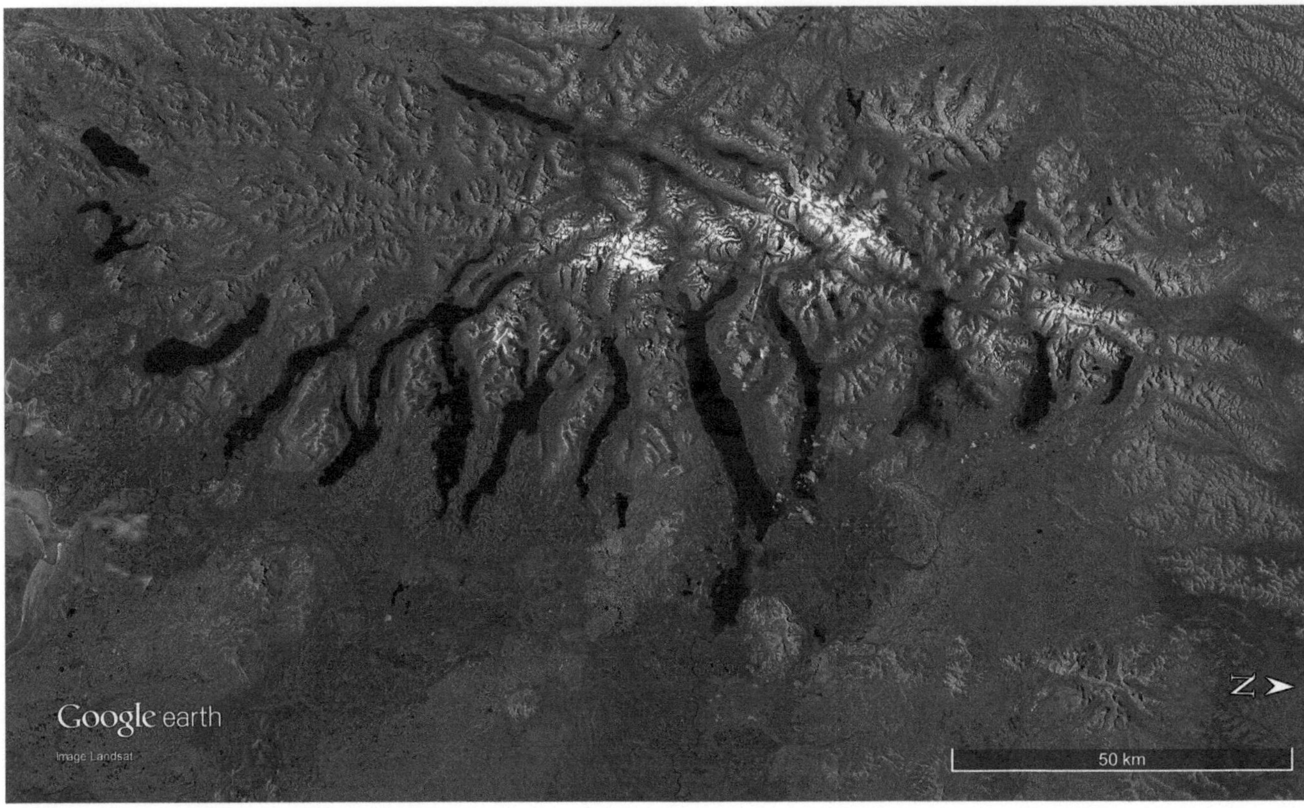

Fig. 2.131 This landscape of glacier lakes in southern Alaska (approximately *59°46'N, 158°54'W*, north is right) is evidence of strong Ice Age glacial erosion along existing valleys, scouring most below sea level. A straight N-S- running fault line has also been carved by glacial erosion, further intensified in this region by high snow precipitation (as nourishment for glaciers) coming from the nearby Pacific Ocean. Lake Nuyakuk is the largest lake, with a length of 32 km and although the lake rests only 98 m above sea level, depth of the lake exceeds 300 m. The highest summits of the Wood River Mountains (part of the Kuskokwim Mountain Ranges) in the image reach 1310 m asl. Scene is 200 km wide (Image credit: ©Google earth 2014)

Fig. 2.132 Glacial lakes in the east Peruvian Andes (approximately *15°15'S* and *70°56'W*, at 4600–4700 m asl) have been dammed by a series of lateral and terminal moraines during late-glacial times as a result of glacier retreat and re-advance. Scene is 8 km wide (Image credit: ©Google earth 2015)

Fig. 2.133 An oblique satellite image viewing the south-western side of volcanic mountains in the Peruvian Andes *(13°59'S, 70°2'W)*. The 20 km long valley (north of Lake Titicaca) is an old glacier bed, evidenced by the sharp-crested lateral moraines. The lowermost lake is 4320 m asl and the highest summit in the background reaches 5739 m asl (Image credit: ©Google earth 2015)

Fig. 2.134 Around 60 years ago, Lake Tasman (approximately *43°41'S, 170°11'E*, and 772 m asl) started to form as a consequence of the Tasman Glacier (New Zealand's largest glacier) receding. The 7 km lake is now over 200 m deep due to stagnant ice that initially scoured the basin, has melted. Thick debris covering the Tasman Glacier protects it from melting quickly, and icebergs calve from time to time at the glacier front with overflow into Lake Pukaki (Image credit: ©Google earth 2015)

Fig. 2.135 The Paron Lake in the Peruvian Andes *(9°0'S, 77°40'W*, lying at 4165 m asl) has been blocked by moraines of a glacier advancing from a side valley. Lake is 3.3 km in length (Image credit: ©Google earth 2015)

Fig. 2.136 A glacier retreating recently in southern Chile (approximately *49°48'S, 73°06'W)* has left two new lakes dammed by terminal moraines in a forest belt (around 700 m asl). Scene is 7 km wide (Image credit: ©Google earth 2015)

Fig. 2.137 The fresh retreat of a melting tongue from the massive Bering Glacier (a piedmont glacier from Wrangell-St. Elias Mountains) in the Panhandle of Alaska *(60°08'N, 143°29'W)* has created a proglacial lake 22 km across that overflows into the sea (Image credit: ©Google earth 2015)

Fig. 2.138 In the Peruvian Andes (approximately *11°57'S, 76°04'W* at 4963 m asl) two merged glaciers calve into a 500 m wide proglacial lake formed in recent years as the glaciers retreat, even at these high altitudes. The lake pattern is a consequence of iceberg calving (Image credit: ©Google earth 2015)

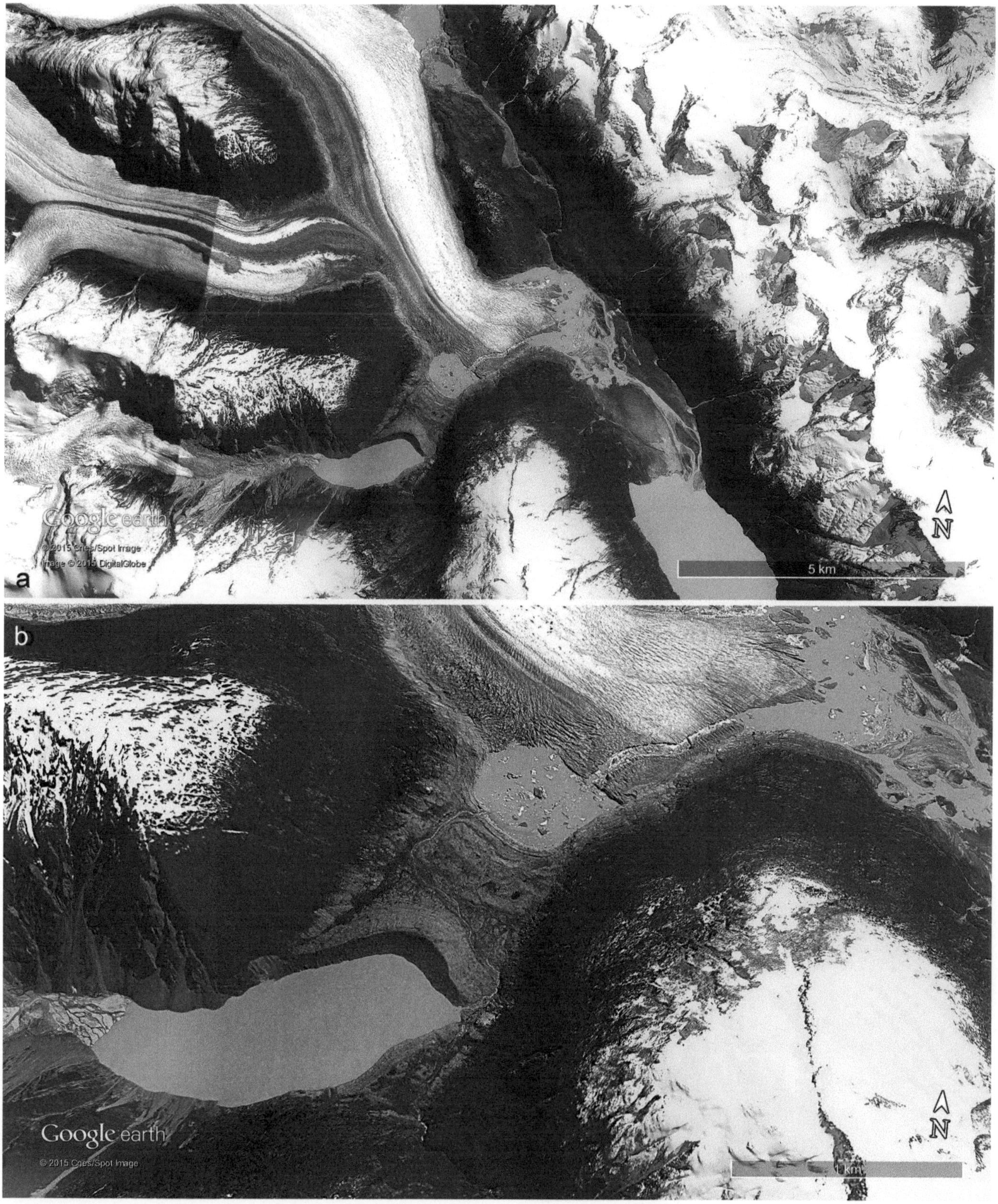

Fig. 2.139 (**a**) The Ventisquero (meaning glacier) Colonia blocks meltwater lakes from side-glaciers in southern Chile *(47°14'50.35"S, 73°14'38.78"W)*. Scene is 15 km wide. (**b**) Detail of the contained lakes in a side valley from (a) in a 6 km wide scene (Images credits: ©Google earth 2015)

Fig. 2.140 (**a**) Meltwater from the main glacier of Alpefjord in eastern Greenland *(72°07'N* and *26°12'W)*, is actually blocked by a side-valley-glacier from the south. Scene 77 km wide. (**b**) Detail of (**a**) in 35 km wide scene (Images credits: ©Google earth 2015)

Fig. 2.141 (**a**) In east Greenland (approximately *73°40'N, 26°18'W*) are two major W-E-running valleys that demonstrate multiple blockages of side-valley-glaciers. Scene is 75 km wide. (**b**) Detail of the southern valley in Fig. (**a**) of 24 km width. (**c**) Detail of the northern valley in (**a**) of 30 km width (Images credits: ©Google earth 2013)

Fig. 2.141 (continued)

Fig. 2.142 On Nunavut island in north Canada *(71°21'47.78"N, 72°21'25.25"W)* side glaciers have blocked the main valley because of their quick reaction to a short-time positive mass balance (the difference between accumulation and melting). Scene is 20 km (Image credit: ©Google earth 2013)

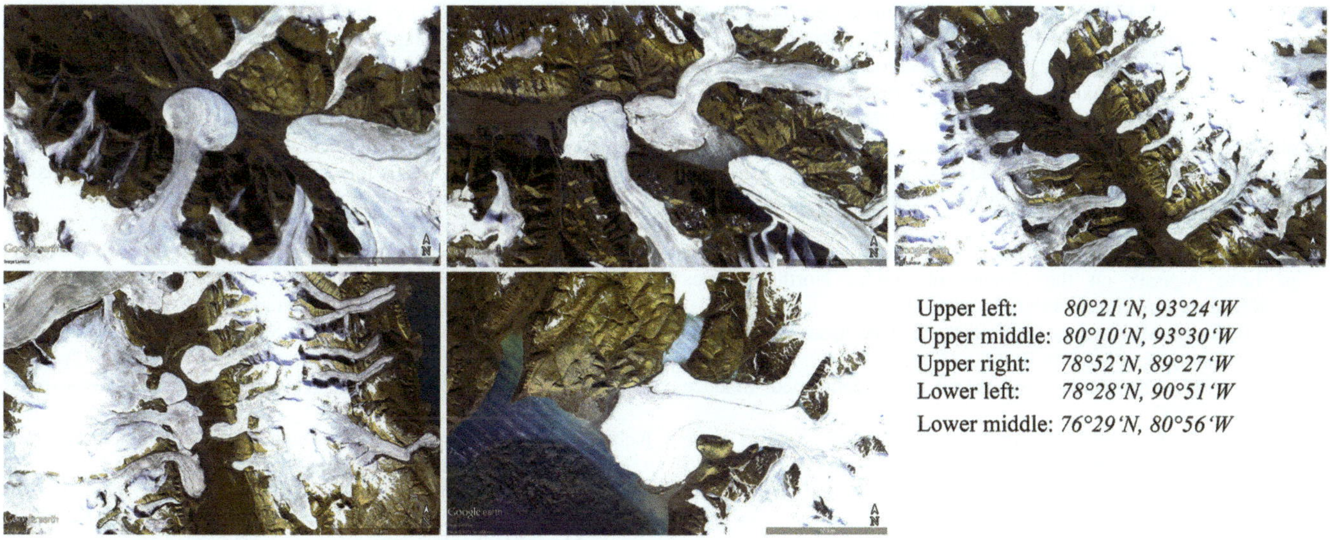

Fig. 2.143 These examples from Ellesmere Island (Northern Canada) show multiple blockages of major valleys caused by side-valley-glaciers (Image credit: ©Google earth 2015)

References

Anima RJ (1990) Annotated bibliography on research conducted in coastal lagoons and estuaries of the Pacific Coast of the United States. USGS open file report, pp 90–347

Bacon CR, Lanphere MA (2006) Eruptive history and geochronology of Mount Mazama and the Crater Lake region, Oregon. Geol Soc Am Bull 118:1331–1359

Barber DC, Dyke A, Hillaire-Marcel C et al (1999) Forcing of the cold event of 8,200 years ago by catastrophic drainage of Laurentide lakes. Nature 400:344–348

Barnett PJ (1985) Glacial retreat and lake levels, north-central Lake Erie Lake Basin, Ontario. Quaternary Evolution of the Great Lakes. Geol Assoc Can Spec Pap 30:185–194

Beach T, Luzzadder-Beach S, Dunning N et al (2008) Human and natural impacts on fluvial and karst depressions of the Maya Lowlands. Geomorphology 101:308–331

Bergner AGN, Strecker MR, Trauth MH et al (2009) Tectonic and climatic control on evolution of rift lakes in the Central Kenya Rift, East Africa. Quat Sci Rev 28:2804–2816

Bird ECF (2010) Encyclopedia of the World's coastal landforms, vol 1. Springer, Dordrecht, p 485

Boamah D, Koeberl C (2003) Geology and geochemistry of shallow drill cores from the Bosumtwi impact structure, Ghana. Meteorit Planet Sci 38:1137–1159

Botts L, Krushelnicki B (1995) The Great lakes: an environmental atlas and resource book. United States Environmental Protection Agency and Government of Canada, 3rd edn, Available via: http://www.epa.gov/grtlakes/atlas/index.html. Accessed 2 Sept 2015

Brenner M, Rosenmeier M, Hodell D, Curtis J (2002) Paleolimnology of the Maya Lowlands. Anc Mesoam 13:141–157

Burton MR, Dawyer GM, Granieri D (2013) Deep carbon emissions from volcanoes. Rev Mineral Geochem 75:323–354

Cohen A, Soreghan M, Scholz CA (1993) Estimating the age of formation of lakes: an example from Lake Tanganyika, East African rift system. Geology 21:511–514

Constantine J, Dunne T (2008) Meander cutoff and the controls on the production of oxbow lakes. Geology 36(1):23–26

Cruz JV, Franca Z (2006) Hydrogeochemistry of thermal and mineral water springs of the Azores archipelago (Portugal). J Volcanol Geotherm Res 151(4):382–398

Davydova NN, Arslanov KA, Khomutova VI et al (1996) Late- and postglacial history of lakes of the Karelian Isthmus. Hydrobiologia 322(1–3):199–204

Delmelle P, Bernard A (1999) Volcanic lakes. In: Sigurdsson H (ed) Encyclopedia of volcanoes. Academic, San Diego

Delvaux D, Moyer R, Stapel G et al (1997) Paleostress reconstructions and geodynamics of the Baikal region, Central Asia, Part 2, Cenozoic rifting. Tectonophysics 282:1–38

Dunning NP, Scarborough V, Valdez F Jr et al (1999) Temple mountains, sacred lakes, and fertile fields: ancient Maya Landscapes in Northwestern Belize. Antiquity 73:650–660

Ferrière L, Koeberl C, Ivanov BA et al (2008) Shock metamorphism of bosumtwi impact crater rocks, shock attenuation, and uplift formation. Science 322(5908):1678–1681

Fisher TG (2003) Chronology of glacial Lake Agassiz meltwater routed to the Gulf of Mexico. Quat Res 59(2):271–276

Fisher TG (2004) River Warren boulders, Minnesota, USA: catastrophic paleoflow indicators in the southern spillway of glacial Lake Agassiz. Boreas 33(4):349–358

Fisher TG, Smith DG, Andrews JT (2002) Preboreal oscillation caused by a glacial Lake Agassiz flood. Quat Sci Rev 21:873–878

Folco L, DiMartino M, El Barkooky A et al (2010) The Kamil crater in Egypt. Science 329(5993):804

Fredriksson K (1973) Lonar lake, India: an impact crater in basalt. Science 180(4088):862–864

Gaidos E, Lanoil B, Thorsteinsson T et al (2004) A viable microbial community in a subglacial volcanic crater lake, Iceland. Astrobiology 4(3):32–44

Gibson JJ, Prowse TD, Peters DL (2006) Partitioning impacts of climate and regulation on water level variability in Great Slave Lake. J Hydrol 329(1):196

Grady W, Litteljohn BM, Damstra ES (2007) The Great Lakes: the natural history of a changing region, David Suzuki foundation series. Greystone Books, Vancouver

Hart DE (2009) Morphodynamics of non-estuarine rivermouth lagoons on high-energy coasts. J Coast Res SI 56:1355–1359

Hart DE, Bryan KR (2008) New Zealand coastal system boundaries, connections and management. N Z Geogr 64(2):129–143
Hodych JP, Dunning GR (1992) Did the Manicouagan impact trigger end-of-Triassic mass extinction? Geology 20:51–54
Holcombe TL, Warren JS, Taylor LA et al (1997) Lakefloor geomorphology of Western lake Erie. J Great Lakes Res 23(2):190–201
Holcombe TL, Taylor LA, Reid DF et al (2003) Revised Lake Erie postglacial lake level history based on new detailed bathymetry. J Great Lakes Res 29(4):681–704
Hostetler SW, Bartlein PJ, Clark PU et al (2000) Simulated influences of Lake Agassiz on the climate of central North America 11,000 years ago. Nature 405(6784):334–337
Jones WB, Bacon M, Hastings DA (1981) The Lake Bosumtwi impact crater, Ghana. Geol Soc Am Bull 92:342–349
Kerr RA (1995) Chesapeake Bay impact crater confirmed. Science 269(5231):1672
Kirk RM, Lauder GA (2000) Significant coastal Lagoon systems in the South Island, New Zealand – coastal processes and lagoon mouth closure. Science for Conservation 146, Department of Conservation, Wellington
Kjerfve B (ed) (1994) Coastal lagoon processes. Elsevier oceanography series, vol 60, Elsevier, Amsterdam
Klimasauskas E, Bacon CR, Alexander J (2002) Mount Mazama and Crater lake: growth and destruction of a cascade volcano. U.S. Geological Survey Fact Sheet, 092–02
Kusakabe M (ed) (1994) Geochemistry of Crater lakes. Geochem J 28(3):137–306
Larson G, Schaetzl R (2001) Origin and evolution of the Great Lakes. J Great Lakes Res 27(4):518–546
Lee TM (1996) Hydrogeologic controls on the groundwater interactions with an acidic lake in karst terrain, Lake Barco, Florida. Water Resour Res 32(4):831–844
Leverington DW, Teller JT (2003) Paleotopographic reconstructions of the eastern outlets of glacial Lake Agassiz. Can J Earth Sci 40(9):1259–1278
Mann JA, Cherry RN (1970) Large springs of Florida's "Sun Coast" Citrus and Hernando counties. Leaflet No. 9, Florida Department of Natural Resources, Bureau of Geology
Manville V, White JDL, Houghton BF et al (1999) Paleohydrology and sedimentology of a post–1.8 ka breakout flood from intracaldera Lake Taupo, North Island, New Zealand. Geol Soc Am Bull 111(10):1435–1447
Master S, Reimold WU (2000) The impact cratering record of Africa: an updated inventory of proven, probable, possible, and discredited impact structures on the African continent. In: Catastrophic events and mass extinctions: impacts and beyond. LPI contribution no. 1053, Lunar and Planetary Institute, Houston, pp 133–134
Miller JA (1986) Hydrogeologic framework of the Floridan aquifer system in Florida and in parts of Georgia, Alabama, and South Carolina. USGS Prof paper 1403-B
Miller JA (1997) Hydrogeology of Florida. In: Randazzo AF, Jones DS (eds) The geology of Florida. University Press of Florida, Gainesville, pp 69–88
Montenegro-Guillén S (2003) Lake Cocibolca/Nicaragua. Lake basin management initiative: experience and lessons learned brief. LBMI regional workshop for Europe, Central Asia and the Americas. Saint Michael's College, Colchester, Vermont, pp 1–29
Motz LH (1998) Vertical leakage and vertically averaged vertical conductance for Karst Lakes in Florida. Water Resour Res 34(2):159–167
Murton JB, Bateman MD, Dallimore SR et al (2010) Identification of Younger Dryas outburst flood path from lake Agassiz to the Arctic ocean. Nature 464(7289):740–743
Osinski GR, Spray JG, Lee P (2001) Impact-induced hydrothermal activity within the Haughton impact structure, arctic Canada: generation of a transient, warm, wet oasis. Meteorit Planet Sci 36:731–745
Pasternack GB, Varekamp JC (1997) Volcanic lake systematics I. Physical constraints. Bull Volcanol 58(7):526–538
Perez NM, Hernandez PA, Padilla G et al (2011) Global CO2 emission from volcanic lakes. Geology 39:235–238
Perkins S (2002) Once upon a lake. Sci News 162(18):283–284
Pirajno F, Hawke P, Glikson AY, Haines PW, Uysal T (2003) Shoemaker impact structure, Western Australia. Aust J Earth Sci 50:775–796
Rao CAN, Alfred JRB (2002) Bibliography of the Indian estuaries, lagoons and backwaters. Zoological Survey of India, Kalkutta
Ribeiro DC, Martins G, Nogueira R et al (2008) Phosphorus fractionation in volcanic lake sediments (Azores – Portugal). Chemosphere 70:1256–1263
Robertson PB, Grieve RAF (1975) Impact structures in Canada: their recognition and characteristics. J R Astron Soc Can 69:1–2
Rosenmeier M, Hodell D, Brenner M, Curtis J, Guilderson T (2002) A 4000-year lacustrine record of environmental change in the southern Maya Lowlands, Petén, Guatemala. Quat Res 57:183–190
Rouwet D, Tassi F, Mora-Amador R et al (2014) Past, present and future of volcanic lake monitoring. J Volcanol Geotherm Res 272:78–97
Rouwet D, Christenson B, Tassi F et al (eds) (2015) Volcanic lakes, Advances in volcanology. Springer, Dordrecht
Rowe GL, Ohsawa S, Takano B et al (1992) Using crater lake chemistry to predict volcanic activity at Potis volcano, Costa Rica. Bull Volcanol 54:494–503
Schallenberg M, Schallenberg L (2012) Eutrophication of coastal Lagoons: a literature review. Prepared for Environment Southland. Hydrosphere Research Ltd, Dalmore
Scheffers A, Scheffers S, Kelletat D (2012) The coastlines of the world with google earth – understanding our environment. Springer, Dordrecht
Schiefer S, Klinkenberg B (2004) The distribution and morphometry of lakes and reservoirs in British Columbia: a provincial inventory. Can Geogr/Le Géographe Canadien 48(3):345–355
Scott S (1990) Late Holocene fluctuations of Mono Lake, eastern California. Palaeogeor Palaeoclimatol Palaeoecol 78:333–381
Scott DL, Etheridge MA, Rosendahl BR (1992) Oblique-slip deformation in extensional terrains: a case study of the lakes Tanganyika and Malawi rift zones. Tectonics 11(5):998–1009
Sichrowsky U, Schabetsberger R, Sonntag B et al (2014) Limnological characterization of volcanic crater lakes on Uvea Island (Wallis and Futuna, South Pacific). Pac Sci 68(3):1–26
Siddiqi SZ (2008) Limnological profile of high-impact meteor Crater lake Lonar, Buldana, Maharashtra, India, an extreme hyprealkaline, saline habitat. In: Sengupta M, Dalwani R (eds) Proceedings of Taal 2007, The 12th world lake conference, pp 1597–1613
Sinclair WC, Stewart JW, Knutilla RL, Gilboy AE, Miller RL (1985) Types, features and occurrence of sinkholes in the karst of westcentral Florida. USGS Water-Resources Investigations Report 4126
Spray JG, Kelley SP, Rowley DB (1998) Evidence for a late Triassic multiple impact event on Earth. Nature 392:171–173
Talbot MR, Delibrias G (1977) Holocene variations in the level of Lake Bosumtwi, Ghana. Nature 268:722–724
Thorleifson LH (1996) Review of Lake Agassiz history. Sedimentology, geomorphology, and history of the Central Lake Agassiz Basin, Geological Association of Canada Field Trip Guidebook for GAC/MAC Joint Annual Meeting, pp 55–84
Tihansky AB (1999) Sinkholes, west-central Florida – a link between surface water and ground water. In: Galloway D, Jones DR,

Ingebritsen SE (eds) Land subsidence in the United States, vol 1182, USGS, circular., pp 121–141
Tihansky AB, Knochenmus LA (2001) Karst features and hydrogeology in West-central Florida – a field perspective. In: Kuniansky EL (ed) USGS Karst interest group proceedings, water-resources investigations report 01–4011:198–211
Valero-Garcés B, Morellon M, Moreno A et al (2014) Lacustrine carbonates of Iberian Karst lakes: sources, processes and depositional environments. Sediment Geol 299:1–29
Varekamp JC, Rowe GL Jr (eds) (2000) Crater lakes. J Volcanol Geotherm Res 97(1–4):1–508
Waples JT, Eadie B, Squires M et al (2008) The Laurentian Great lakes. North American Continental Margins. In: Hales B et al (eds) American continental margins: a synthesis and planning workshop, Washington, DC, pp 73–81
Wetzel RG (2001) Limnology: lake and river ecosystems, 3rd edn. Academic, New York

Swamps, Bogs and Other Wetlands

3

Abstract

In geological time scales, lakes are young features on the Earth's surface (merely decades up to many thousands of years). The existence of lakes often depends on climatic conditions that can guarantee the amount of surface water will exceed evaporation in the same region. In contrast, wetlands are able to develop in very shallow relief (and even on plains) where incoming water becomes stagnant. One important reason to the present day existence of high numbers of swamps or bogs or other kinds of wetlands, is a lake may pass through this phase of being a wetland through its early development and may pass it again when loosing water. The principal difference between lakes (with open water) and wetlands is the presence of plants (*Hydrophytes*), particularly along the wide fringes of the shores, but also as floating islands breaking loose from the fringe, or growing from the shallow bottom to the surface. Wetlands can quickly change appearance where open water may be totally covered by floating plants and water of some depth may still exist below, or growing from the bottom, a carpet of reed and grass with no visible water volumes. If decomposition is slow as a result of unsuitable climate conditions (e.g. high precipitation and low temperatures), wet vegetation mats may also grow from flat low-lying areas forming domed surfaces demonstrated in the dominant moss bogs (raised bogs) of circum-polar latitudes. Marshes or mangrove woodlands describe plants in wetlands along ocean shores (algae, grass, herbs, even trees) that have adapted to survive the salt- water conditions including tidal currents and some wave action.

Special categories of lakes are those that are variable—they decrease and increase in size primarily due to periodic or episodic humid climatic conditions sometimes leading into longer climate shifts (years to decades)—while other lake forms get water through periodic flooding. Although variable lakes resemble more wetlands and floodplains than open lake forms, if they are permanent wetlands then special ecosystems evolve, with many of them unique and biologically diverse.

Wetlands can represent tidal marshes (salt or brackish water without trees), swamps (with shrubs or trees similar to flooded savannah) or non-tidal (swamps or bogs). Based on hydrology principles, wetlands can be categorized as riverine (associated with streams), lacustrine (associated with lakes) and palustrine (isolated).

Hydrophytes (plants in water) can be divided into four groups:

- submerged water plants as sea grasses and eelgrass;
- floating water plants with small roots such as lilies, lily pad and duckweed;
- emergent plants as in marshes, where roots are completely submerged; and finally
- swamp trees and shrubs including mangroves or the cypress species *Taxodium*.

Additionally, three groups of algae occur in wetlands: as microscopic plankton (Cyanophytae, Cyanobacteria), the primary food production in wetlands, and free-floating macroscopic duckweeds (e.g. *Spirodela polyrrhiza* or *Wolffia*

A.M. Scheffers, D.H. Kelletat, *Lakes of the World with Google Earth*, Coastal Research Library 16,
DOI 10.1007/978-3-319-29617-3_3

globasa). The second group of algae are floating, often forming large mats (e.g. *Paspalum repens* or *Echinochloa polystrachya* in the Amazon floodlands, or more commonly the water hyacinth *Eichhornia crassipes*). A third group is fixed to the ground, evident along rocky shorelines (brown algae including *Fucus, Pelvetia, Macrocystis, Nereocystis, Laminaria, Durvillea,* and many others).

Worldwide more than 2000 sites are protected under the RAMSAR Convention on Wetlands of International Importance, particularly important to the protection of waterfowl habitat biodiversity. Large RAMSAR sites include the Pantanal within the countries of Bolivia, Brazil and Paraguay, the Okavango Delta in Botswana, the Niger Inland Delta, and the Nile swamps of Sudan (called Sudd in south Sudan). Table 3.1 lists the countries with a minimum of 20 protected RAMSAR sites (2014).

Discriminating between a lake and a swamp is not well defined because of differences associated with the "floating" aspect. The availability of light and nutrients is the reason for

Table 3.1 Number of RAMSAR-sites ordered by country (Only 20 and more mentioned) (State of 2014, compiled from UN- and RAMSAR Internet sites)

Country	No. of RAMSAR-sites (status as of 2014)
United Kingdom	170
Mexico	142
Spain	74
Sweden	66
Australia	65
Norway	63
Netherlands	53
Italy	52
Algeria	50
Finland	49
China	46
Japan	46
Ireland	45
Denmark	43
Tunesia	40
Canada	37
USA	36
Russia	35
Germany	34
Ukraine	33
Portugal	31
Hungary	29
India	26
Iran	24
Morocco	24
Austria	23
Argentina	21
South Africa	21

plant growth in stagnant water and both requirements are best guaranteed in shallow water and close to the shorelines. In a benthic situation, i.e. in deeper water apart from the shorelines, floating plants and in particular planktonic algae occur. This can be interpreted by the green colour, diminishing in intensity from the shores to the centre of swamps.

Many swamp environments have floating islands made up of aquatic plants, mud and peat. Reaching up to several metres in thickness, floating is guaranteed on a buoyant mat of plant roots or other organic detritus. At the start, sedge and reeds grow along the shorelines, however in deeper water these aquatic plants may then lose ground contact and begin to float. Strong winds can dislocate whole sections from the shore causing these floating islands to migrate over the lakes and from these movements often become round in shape that may eventually reattach to another section of the shoreline. Typically shallow and located in humid climates, the great lakes of the world such as Lake Kyoga in Uganda, Lake Malawi in eastern Africa, Lake Upemba in the Democratic Republic of the Congo and Lake Chad in Mali can also have floating reed islands. In addition, on Lake Titicaca there are artificial floating islands with huts, made from bundled reeds by the Uros people to live in and used in historic times to protect themselves from aggressive neighbours.

Aside from large expanses of swamp environments are thousands of smaller regions with wetlands of different origin including (peat) bogs and numerous lakes progressively being diminished by sediment infill. As the lake shallows, plants may eventually seed and attach to the bottom while in the deeper central sections open water still may exist for some time. Excluding periodic rainfall, wetlands can materialize from fluctuating levels of ground water that quite often never reach the state of open lakes and may have a particularly wide distribution along shallow coastlines wetlands, often brackish being aside open ocean coasts. Wetlands created by fluctuating groundwater that replace former lagoons can appear along the landward fringe of mangroves and may also form into long narrow strips such as swales between beach ridges.

A sequence of Google earth images displays swamps and other wetlands from several continents. The largest wetland area of the world is Pantanal in South America (approximately 230,000 km^2, Fig. 3.1) fed by Rio Paraguay with an inclination of merely 30 m over a 600 km length. During the November to March wet season, rivers from north to northwest distribute water unable to drain from the Pantanal plains fast enough resulting in a wide floodplain. This region was dry up until at least 10,000 years ago and during the last millennia up to 80 % of the area flooded where many animals were forced to seek shelter on higher river dams and drier places (Cerrado), now appearing as savannahs with shrubs and trees (see Fig. 3.2) More than 2000 diverse plants have been identified in the Pantanal wetlands and the region can

Fig. 3.1 A labyrinth of old river courses and swamps in the Pantanal, southwest Brazil *(20°30'S, 57°50'W)*. Pantanal is the largest wetland of the world and alternates in size from 140,000 to 200,000 km^2. This scene is 20 km wide (Image credit: ©Google earth 2015)

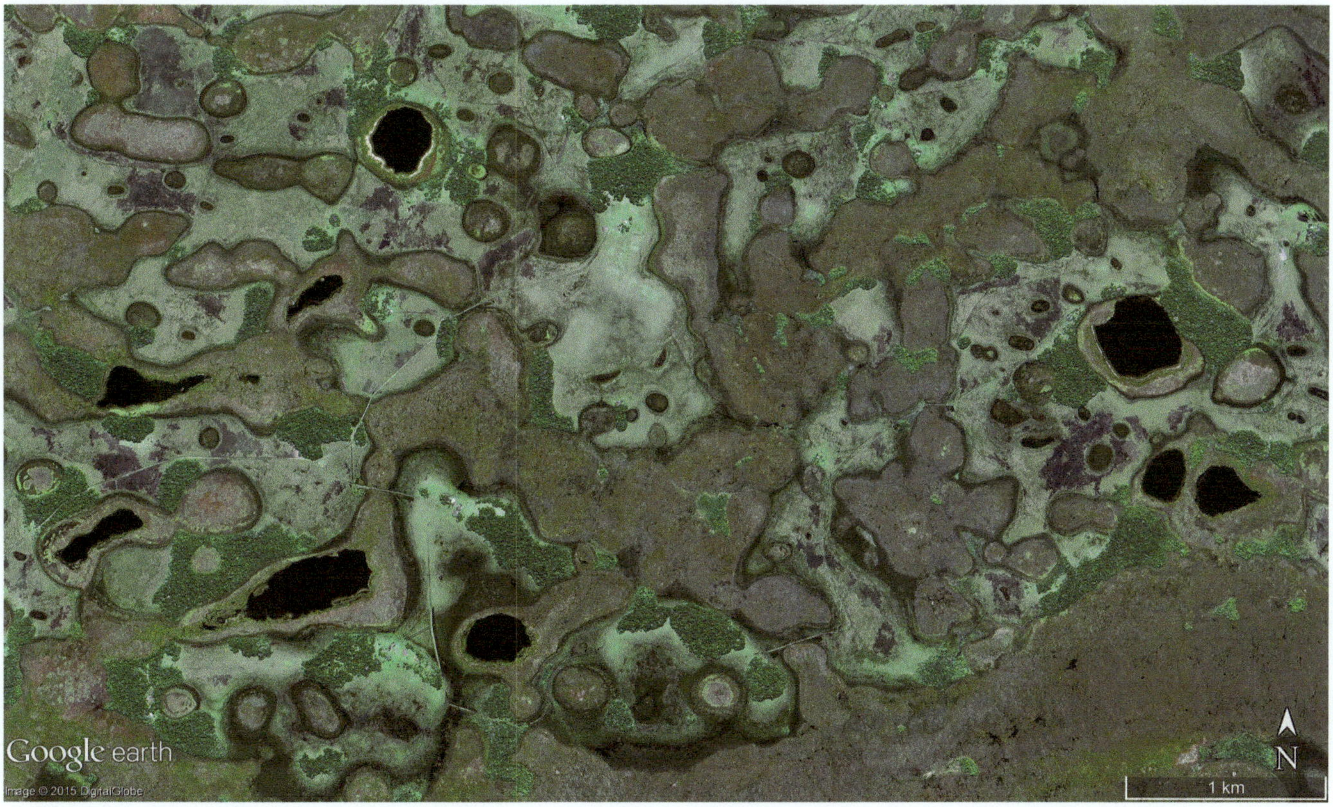

Fig. 3.2 A 30 km^2 section of the southern Pantanal wetlands in southern Paraguay (approximately *27°05'S* and *57°24'W*). The character of the swamp and in particular the vegetation patterns are clearly discernable even though the relief is less than 3 m in height. Scene is 7 km across (Image credit: ©Google earth 2015)

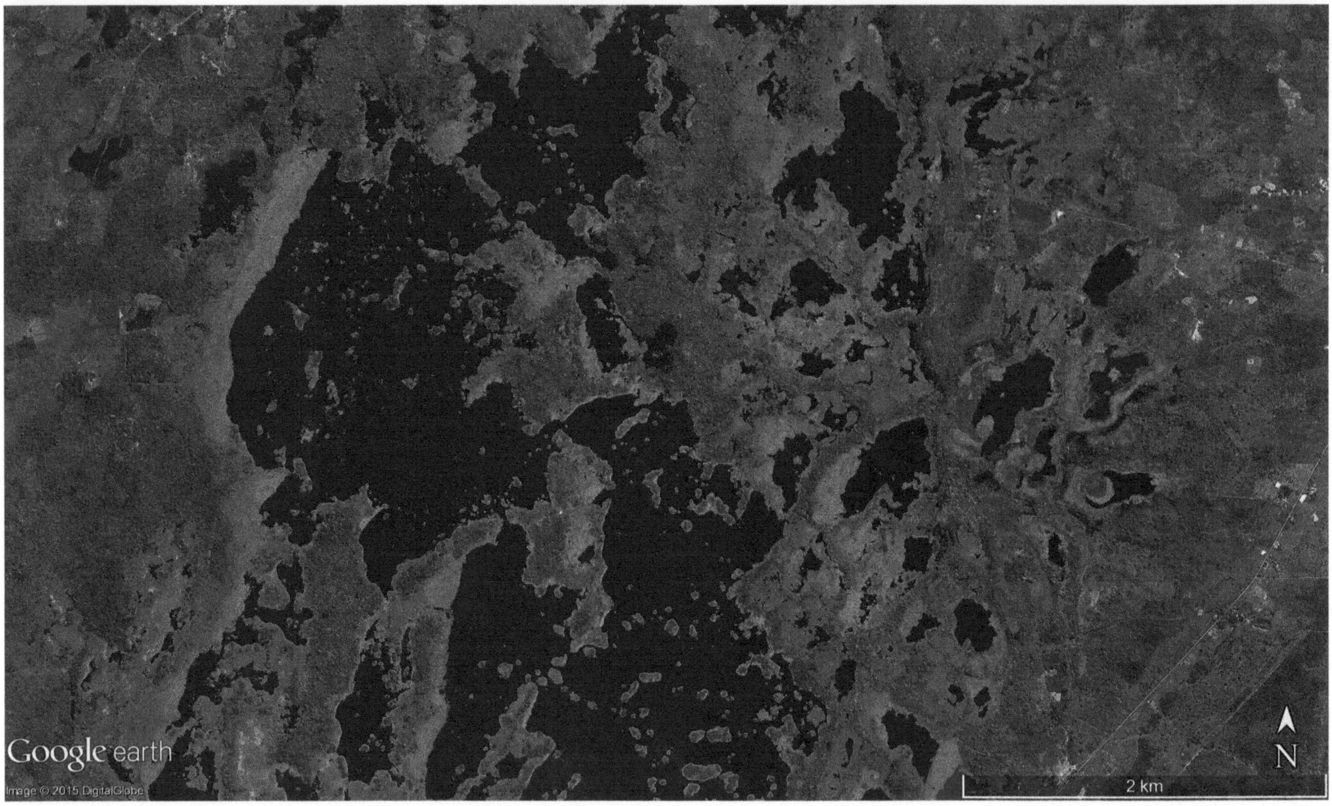

Fig. 3.3 Lakes and swamps in north-central Brazil *(3°37'S, 44°46'W)* lie within a tropical rain forest and demonstrate flooding independent from river beds. Scene is 7.5 km wide (Image credit: ©Google earth 2015)

be divided into six widespread ecosystems: constantly open waters (rivers and lakes), floating plant islands (*Echinochloa stagnina*, *Salvinia auriculata* or the water lettuce *Pistia stratoides*), rooted water plants (such as *Cyperus giganteus*, *Scirpus validus* and *Typha dominguensis*), river-flooded grasslands, rain-flooded grasslands, and wooded grasslands.

Similar extensive permanent wetlands exist (excepting their sizes fluctuate periodically or episodically) in other regions of humid tropics including Brazil (Fig. 3.3), Bolivia (Fig. 3.4), the Congo Basin of Central Africa (Fig. 3.5), in northernmost Argentina (Fig. 3.6.), the Upper Nile Swamps (Sudd) in south Sudan (Fig. 3.7a, b), and the inland delta of Niger River in West Africa (Fig. 3.8). Large wetland areas also exist in extra-tropical latitudes, e.g. in central and NW Russia or Canada and Alaska (Fig. 3.9).

A comparable wetland process to that of the Pantanal of South America has developed in the Sudd of South Sudan where waters from the wet season are unable to drain fast enough resulting in a wetland extending 500 km to the north-south with a maximum 200 km east-west-extension—dependent on the water discharge from the Ethiopian Highlands and Lake Victoria—covering an area between 30,000 and 130,000 km^2 (57,000 km^2 is protected by RAMSAR). Floating vegetation islands are formed in the centre of the wetlands by *Echinochloa stagnina*, *Vossia cuspidata*, and *Cyperus papyrus* (ancient Egyptians made paper from papyrus), and along the periphery rooted plants like *Phragmites communis*, *Echinochloa pyramidal*is and *Oryza barthii* occur (amongst many others). On higher ground, although shrubs and trees grow resembling savannah the grasses (high in the humid season and low in the dry season) dominate.

The inland delta of Okavango River in Botswana southwest Africa (Figs. 3.10, 3.11, and 3.12) is a vast wetland of approximately 20,000 km^2, fed from high rainfall in Angola to the north that evaporates in the Kalahari semi-desert to the south and along the southern part of the wetland is fringed by an arm of the East African Rift Valley (i.e. a tectonic element). The northern section of the Okavango Delta contains water constantly filled by the Okavango River that flows southward into permanent swamps of around 2–5 m in depth. A periodic change of wetland and dry savannah conditions occur around the fringe of the delta where the southern part of the plain is interrupted by several hills (max. relative height of only 3 m becoming islands during flood) and old dunes from the "Sandveld".

Approximately 1300 plant species identified in Okavango Delta have formed different ecosystems according to the duration and depth of annual flooding. Along the water chan-

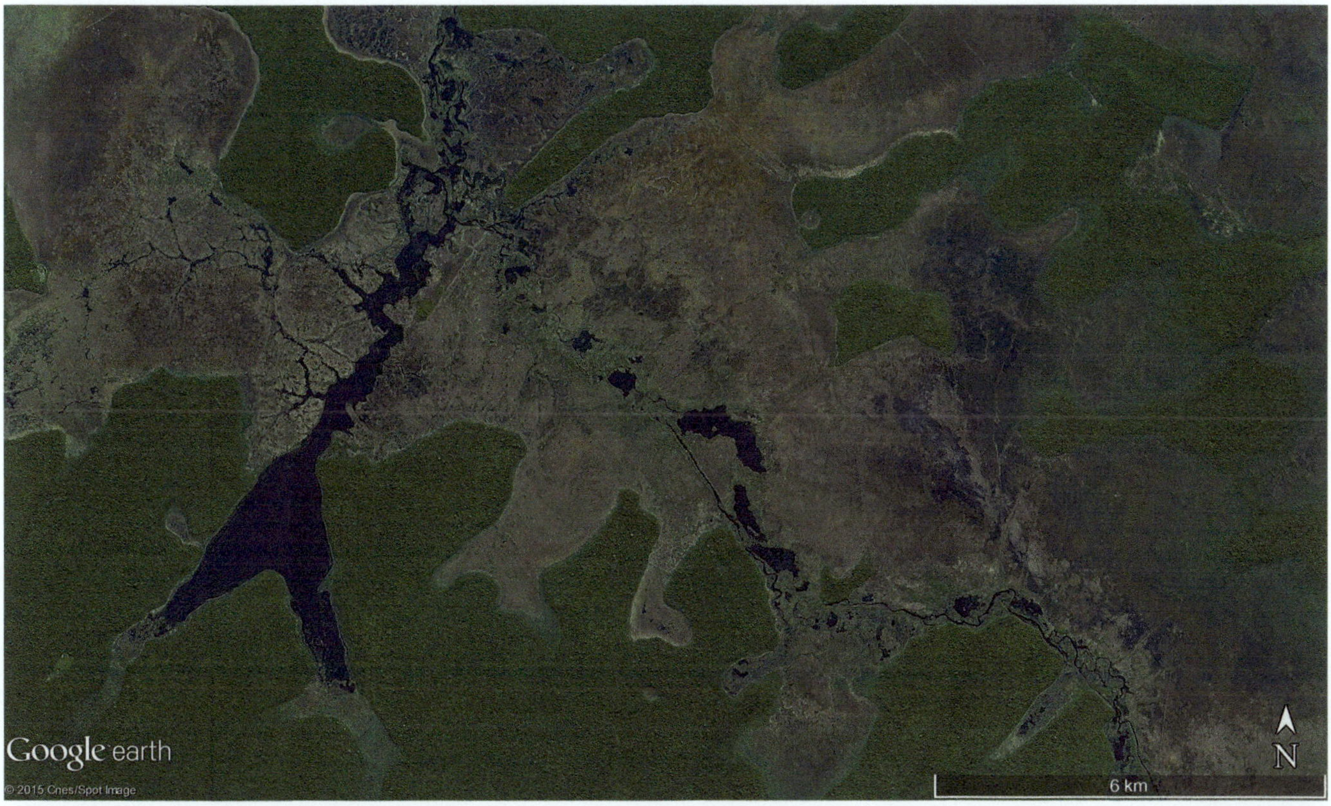

Fig. 3.4 This site in NE Bolivia at *13°58'S, 63°28'W* (20 km wide scene) shows Bolson te Oro with a cover of floating plants within a rain forest environment (Image credit: ©Google earth 2015)

Fig. 3.5 Swamps with open waters in a 25 km scene at Upemba Lake in Katanga District of the Democratic Republic of Congo at around *8°47'S, 26°08'E* (Image credit: ©Google earth 2015)

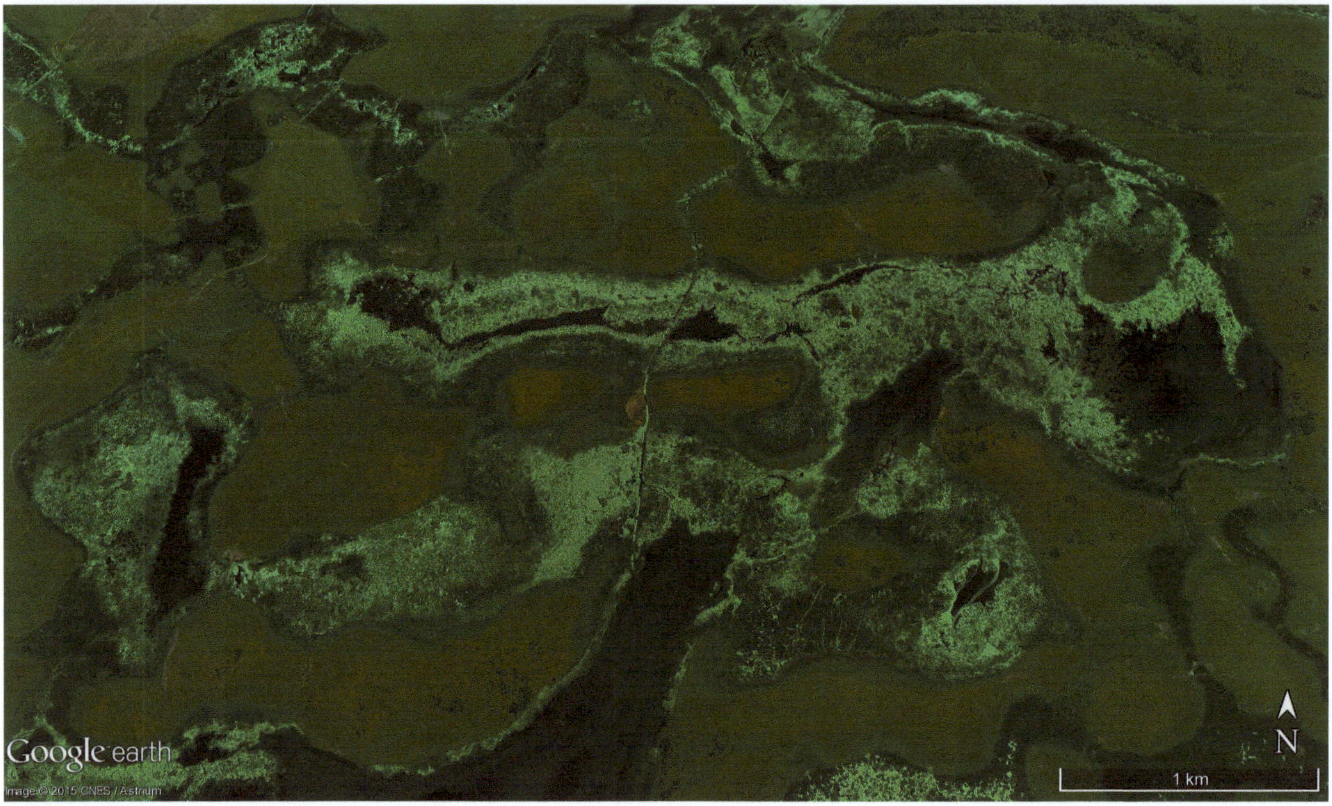

Fig. 3.6 A network of old river beds in northern Argentina at about *36°14'S* and *56°58'W* has transformed into long stripes of freshwater swamps in a 6 km wide scene (Image credit: ©Google earth 2013)

nels are primarily papyrus species (*Cyperus papyrus*) and in open swamps the floating vegetation islands dominate. On elevations shrubs and trees (*Phoenix, Ficus, Acacia*) occur and the higher the ground are longer periods of evaporation and salt concentration in the soil that normally excludes trees in the central parts of islands.

In contrast to large wetlands that diversify into a myriad of ecosystems, the small marsh and swamp wetlands may produce one ecosystem with a strict order of organisms (in particular plants) according to water depth and distance to shorelines. Nevertheless even small wetland patches can include a high variation in flora, often identified in satellite images by their different surface structure and colour (Figs. 3.13, 3.14, 3.15, and 3.16).

Conditions for diverse ecosystems and richness of species are limited in basins isolated by the topography often without river input and only replenished by little discharge from the surrounding slopes and groundwater. Here water depth and shoreline distance are the main factors as evident in the examples of Figs. 3.17, 3.18, 3.19, 3.20, 3.21, 3.22, 3.23, 3.24, 3.25, 3.26, 3.27, and 3.28.

In arid and semi-arid climates reliant on local conditions (i.e. isolated from the water discharge of unconnected humid zones) the salinity of a swamp's waters can change within a short period of time. Torrential rain from thunderstorms may recharge the swamp water from days to several weeks. The increased volume of fresh water inevitably evaporates with a subsequent rise in salt concentration that restricts the variety of flora and fauna, and encourages endemism with the development of species unique to this geographic location. Good examples in Australia are found in salt lakes with saltpans attached (Fig. 3.29).

Often, strange aspects in many swamps of any size are floating vegetation islands and the particular round or oval island shapes formed by collisions with other floating plant assemblages or shorelines. Water and nutrient availability may also lead to round patterns of vegetation as demonstrated in the Yukatan pans of Mexico (Figs. 3.30, 3.31, and 3.32).

Although saline conditions predominate open ocean shorelines, salt content is often substantially less than that found in salt lakes and saltpans of arid climates. Here, the primary landscape is salt marshes by way of halophytic grass communities spread along open lowland coasts of middle and higher latitudes (Fig. 3.33a, b). The grass species dominates higher levels with less inundation (influenced by episodic timing and flooding depths) and species grow where every tide inundates the area.

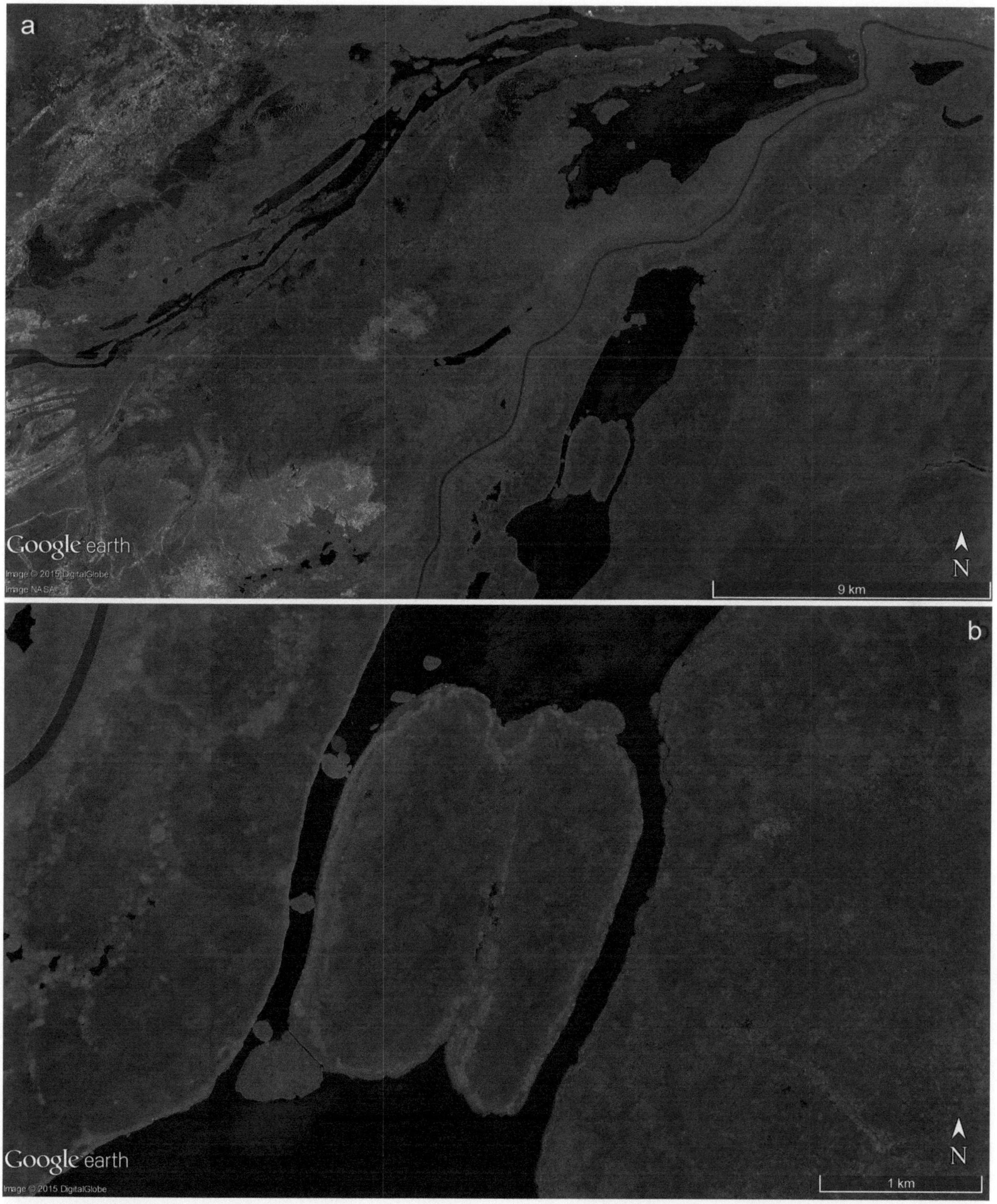

Fig. 3.7 (**a**) Swamps of the Upper Nile valley (in northern part of South Sudan) in a 40 km wide scene at about *9°26'N* and *30°23'E*. These swamps of the White Nile, called "Sudd", may cover an area of up to 130,000 km^2 if flooded high, or only 30,000 km^2 in the dry season. 57,000 km^2 have been protected as a RAMSAR site. The flooding level may change up to 1.5 m depending on discharges from the White Nile and from Lake Victoria. Both sources are responsible for the environmental conditions, because in general, 550–650 mm/ year of precipitation is much lower than evapo-transpiration in these latitudes. Within the swamps, five different ecosystems have to be differentiated: waters of lakes and rivers, areas of floating plants (e.g. *Echinochloa stagnina, Vossia cuspidate*, and *Cyperus parpyrus)*, grassland flooded periodically by rivers, where along the banks fringes of *Phragmites communis* may root in water depth of around 1 m, rain-flooded grassland aside of rivers and lakes, and wooded grassland on slightly higher ground. (**b**) Detail of the Sudd landscape (upper Nile swamps, Sudan), 6.5 km wide scene (enlargement of (**a**), Images credits: ©Google earth 2015)

Fig. 3.8 Part of the inland delta of River Niger in Mali, West Africa, as a swampy region at the southern fringe of the Sahara Desert. Scene is 130 km across and centre is at about *15°02'N* and *4°17'W*. The central part is formed like a very flat fan, documented by the diverging river arms (Image credit: ©Google earth 2013)

Fig. 3.9 Swamps accompanying a wide valley in the "Taiga" (Russian for forest) of NE Russia at about *67°35'N* and *53°06'E* in a 2.4 km wide scene (Image credit: ©Google earth 2015)

Fig. 3.10 Okavango Delta (at nearly 1000 m asl) in Botswana (southern Africa) in August, 2007. Scene is 290 km wide and the periodical wetlands are about 20,000 km^2 large. Like the inland delta of the Niger, the Okavango delta is formed like a very flat fan, from which the periodic river arms diverge to the west and the east. The delta and flooding area is limited in the SE by a branch of the East African Rift valley. Centre of the scene is at about *19°20'S* and *22°49'E* (Image credit: ©Google earth 2013)

Fig. 3.11 Detail 7 km wide of the eastern Okavango Delta at *19°10'20.80"S* and *23°13'08.36"E*, taken on Dec. 14th, 2010. It shows meander bends of a formerly active river system (Image credit: ©Google earth 2015)

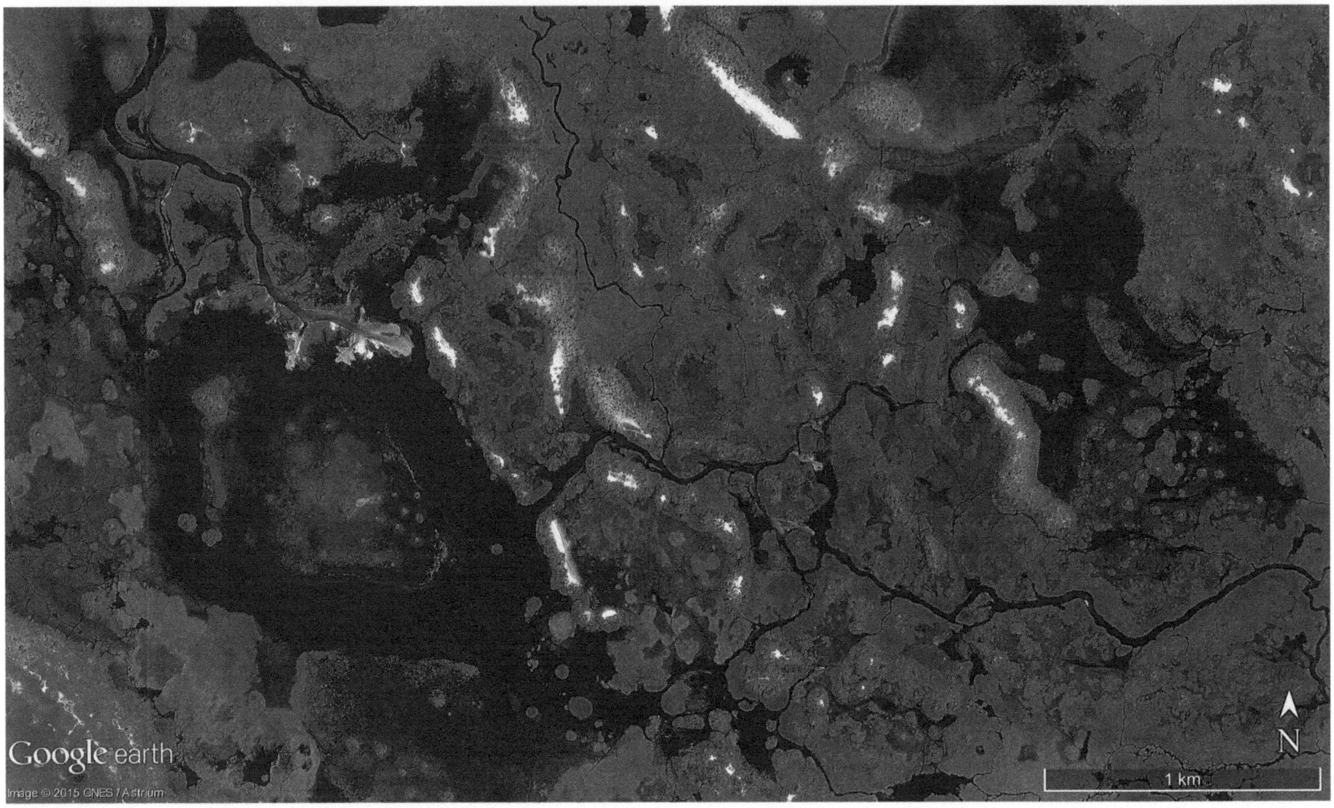

Fig. 3.12 Detail 5 km wide in western Okavango Delta at *19°19'40.41"S* and *22°42'27.47"E*, taken Sept. 30th, 2013 White spots are salt concentrations along up to 1 m higher ground (Image credit: ©Google earth 2015)

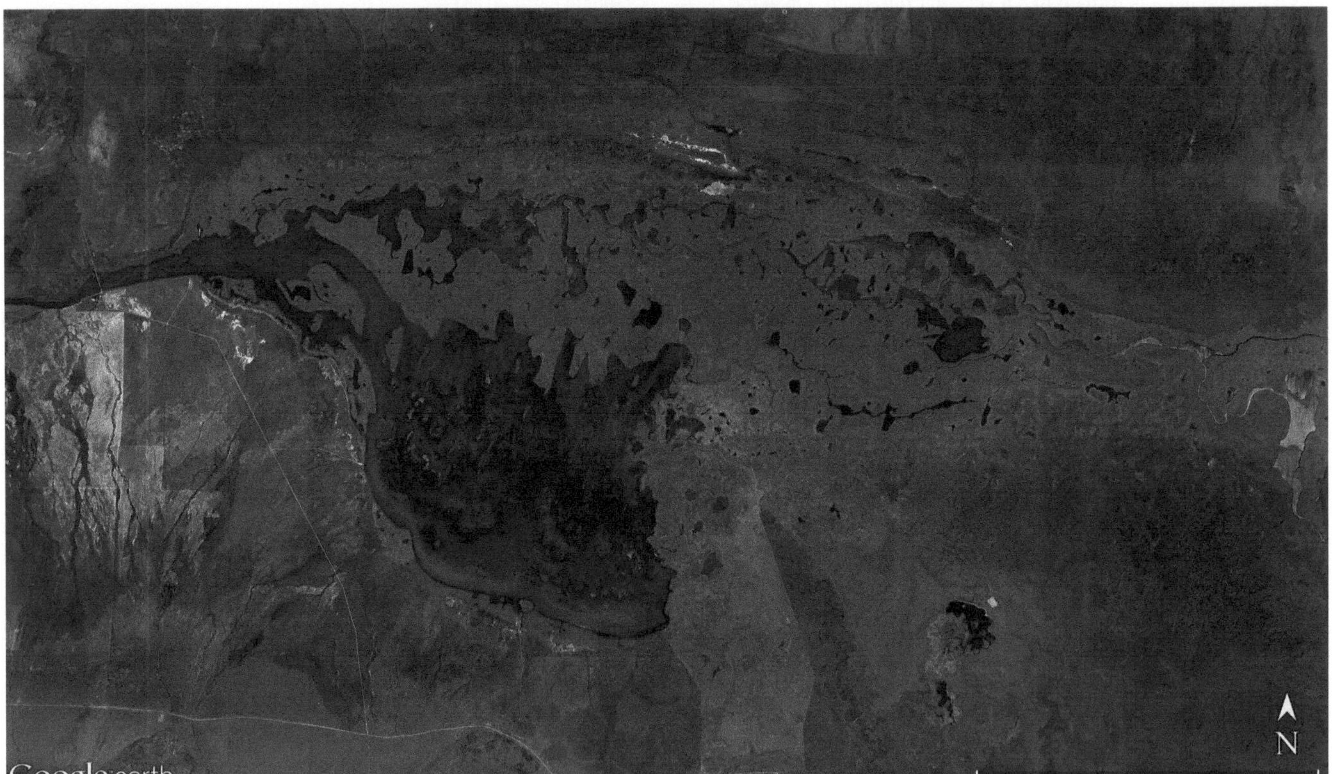

Fig. 3.13 A swamp with floating vegetation developed in Lower Red Rock Lake in southernmost Montana, USA. The scene is 11 km wide at *44°38'N, 111°50'W* (Image credit: ©Google earth 2013)

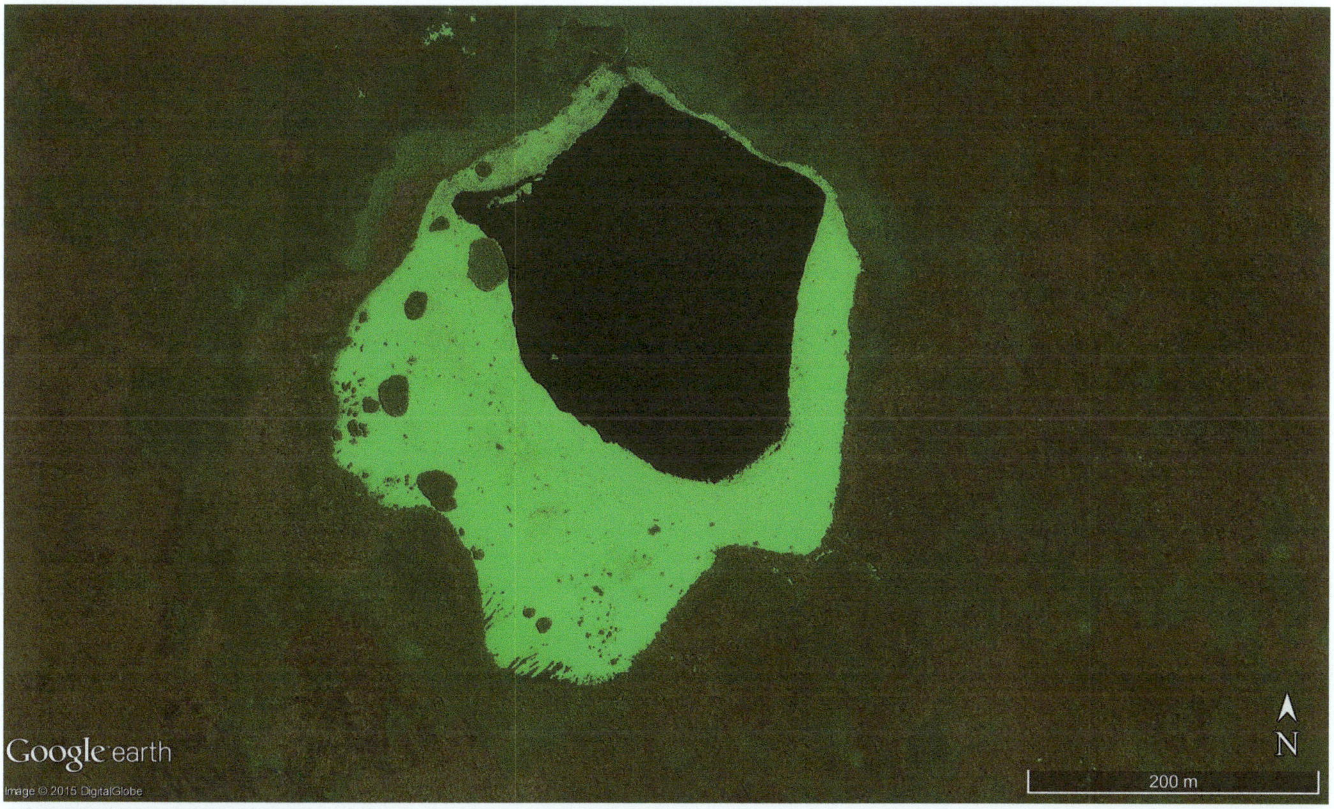

Fig. 3.14 A lake transforming into a swamp in a 800 m wide scene in the Province of Tabasco, southern Mexico, taken on April, 24th, 2010, at *18°07'N, 92°37'W* (Image credit: ©Google earth 2015)

Fig. 3.15 A small lake in NW Yukatan, Mexico, 600 m across *(18°18'15.93"N, 92°17'05.17"W)* with round floating vegetation islands (Image credit: ©Google earth 2015)

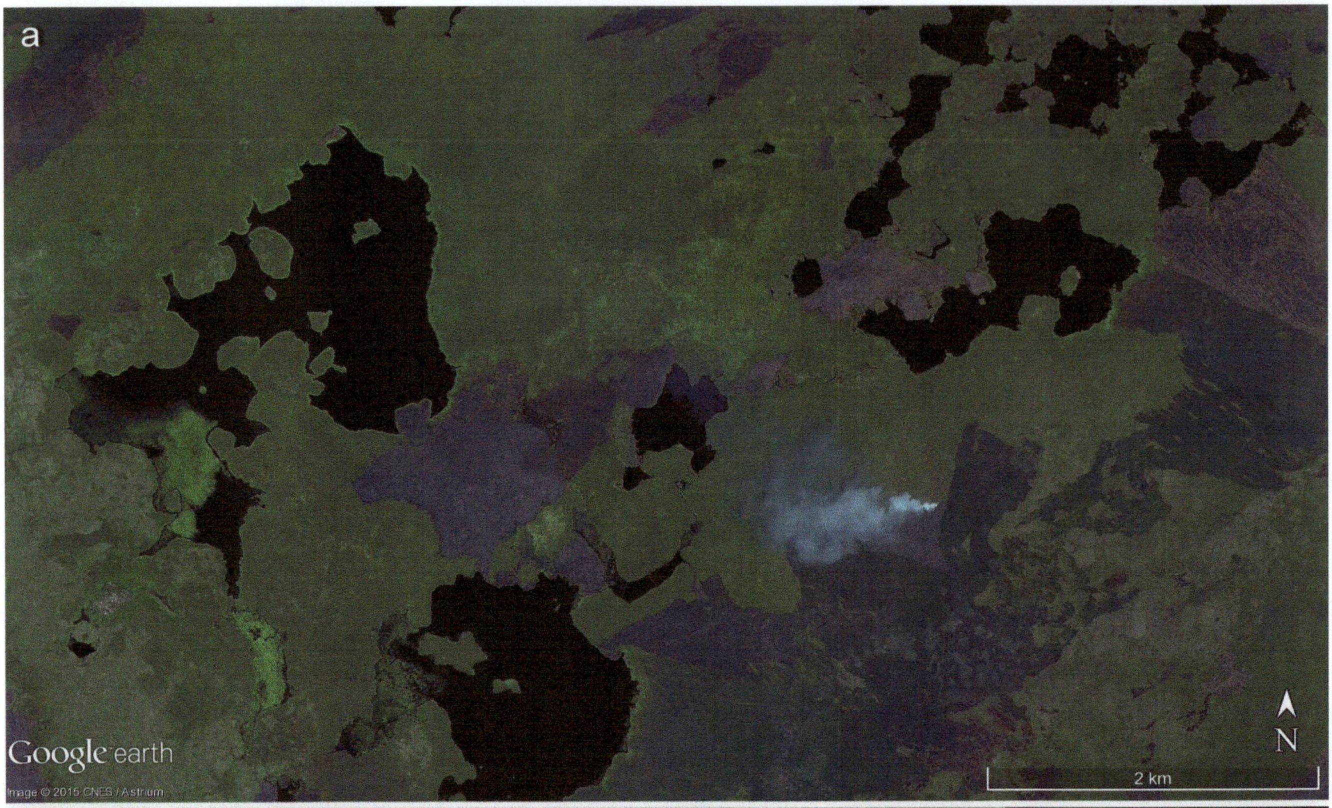

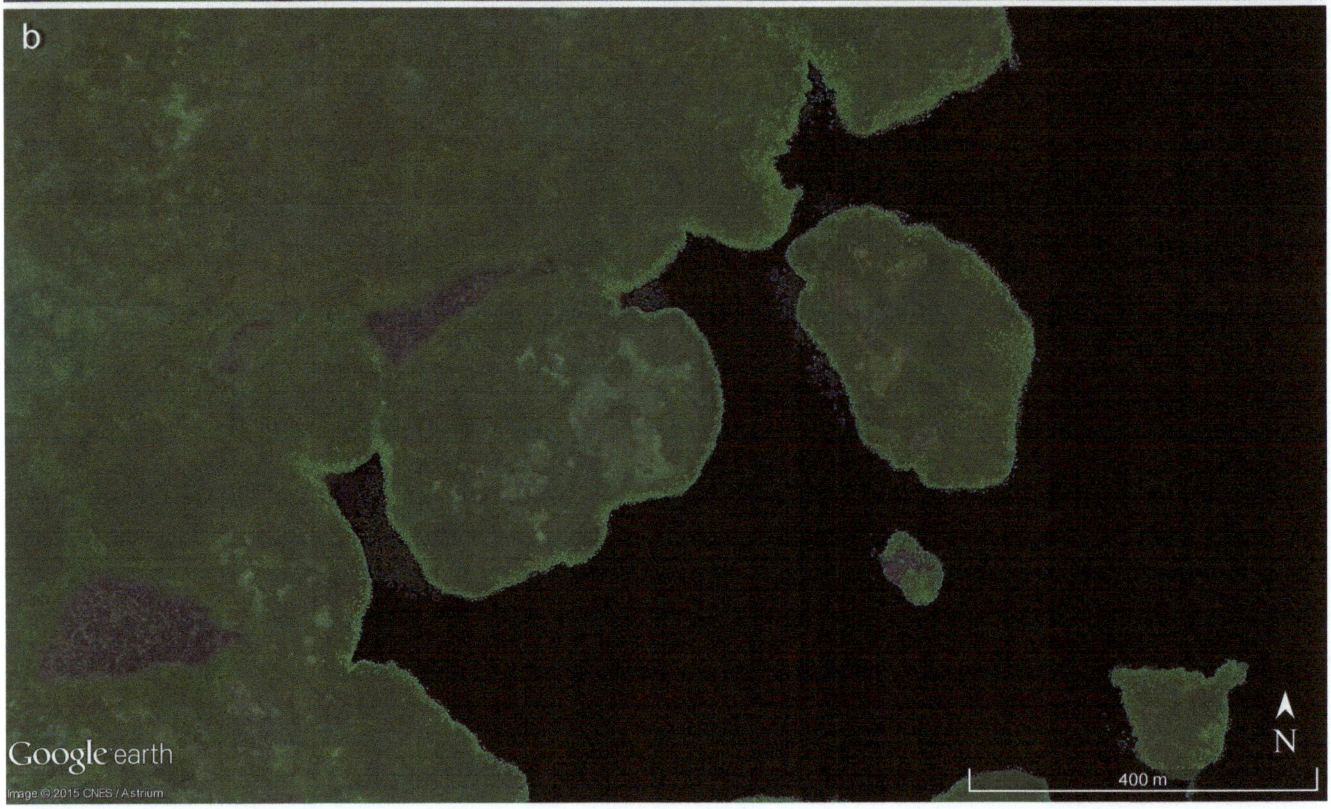

Fig. 3.16 (**a**) Eastern Madagascar fresh water swamps: 9 km scene at *17°34'S* and *48°21'E*. The round contours along the open water document the existence of floating vegetation islands. (**b**) Detail of (a), 1.4 km wide at *17°33'27.56"S* and *48°19'24.62"E*. The vegetation along the open water clearly is different from that on the swampy flats or in the centre of floating islands. (**c**) Detail of (a), 1.2 km wide at *17°34'04.86"S* and *48°19'19.98"E*, showing the fresh extension of water hyacinth (*Eichhornia crassipes*). (**d**) Detail of (a) with many round and floating vegetation islands. Scene is 660 m at *17°34'03.83"S* and *48°20'55.45"E* (Images credits: ©Google earth 2015)

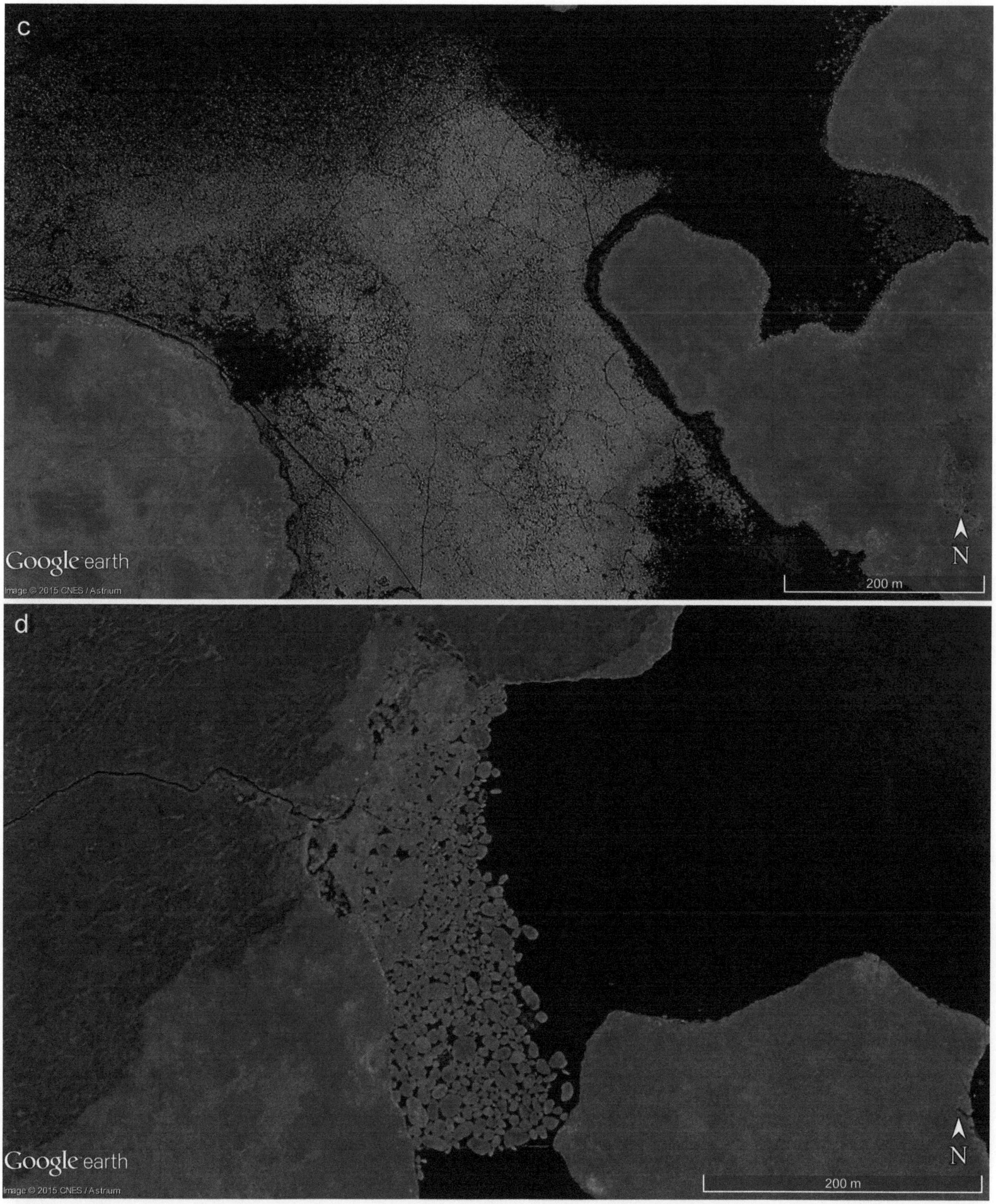

Fig. 3.16 (continued)

In contrast to the middle and higher latitude grasses of *Spartina* and *Salicornia*, mangrove species (trees) is the dominant vegetation flourishing in the intertidal zone of tropical and sub-tropical coastlines. Diverse forms of mangrove root systems primarily attach to muddy substrates because mangrove communities do not occur on sandy beaches due to higher wave energy preventing the anchoring of a new plant, and growing on rocks is possible but rare because of the unsuitable conditions of roots anchoring to rock substrate.

Mangroves can often be identified on satellite images by a thick band of higher vegetation along the shoreline or observed in the same direction to the coastline as separate bands of plant communities (Figs. 3.34, 3.35, 3.36, and 3.37).

Peat bogs are a unique type of inland wetland of the cool and cold latitudes, developed when intensive plant growth occurs in the short summer periods and the accumulation of rotten plant remains due to low temperatures influencing the slow decomposition of vegetation. The water content in the peat reaches 90 % and explains internal pressure processes during freezing. The results are pressure ridges of peat with wider depressions lying between the ridges, filled by water in summer and shaped into narrow, elongated ponds—and according to appearance are named "string bogs" in cold latitudes (in German "Strangmoore" and in Swedish "Aapa-moore", see Figs. 3.38a–d). In cool and humid climates, peat growth may exceed the level of groundwater and form bogs with a flat domelike surface where vegetation must survive under low nutrient conditions as only rainwater feeds the high parts of the raised bogs (Fig. 3.39a, b).

A special wetland type is the commonly named "river of grass" consisting of numerous more or less parallel and shallow-flowing water channels amid bands of high grass. This aspect is best seen in southern Florida (Fig. 3.40). Fed by groundwater from karst springs and the Kissimmee River, the landscape is constantly saturated and excess surface water flows down from the limestone terrains into the Ocean (Gulf of Mexico). As a result of storm surges, mangrove seedlings (in particular *Rhizophora mangle*) originating from the brackish and salt water sections of the open coastline may drift up to over 30 km inland to areas containing only fresh water, and found to be growing well.

Fig. 3.17 Western Mongolia at *48°02′N* and *92°33′E*, a 130 km scene with Khar-Us-Nuur at 1175 m asl. Despite the arid environment, Khar-Us-Nuur is a freshwater lake up to 1578 km² wide but only 2.2 m deep. It is the remnant of a larger lake, which began to disappear around 5000 years BP by a climate change. The northeastern and shallower part of the lake has been totally transformed into a swamp (Image credit: ©Google earth 2013)

Fig. 3.18 The freshwater Achit Lake (about 10 m depth) in east Mongolia *(49°31'N, 90°31'E* at 1436 m asl). The 240 km^2 lake is a RAMSAR site and has been receding for many years because of a long-lasting drought in the region. Scene is 80 km wide. (Image credit: ©Google earth 2015)

Fig. 3.19 The Sadyrbay Lake *(50°37'22.07"N, 70°21'36.25"E* at 325 m asl) in the Kazakhstan steppe (a flat large area of unforested grassland) was in fact a riverbed (flowing west to east) and partly enlarged by deflation (wind removal of rock, sand, etc.) into wide pans. Scene is 27 km wide (Image credit: ©Google earth 2015)

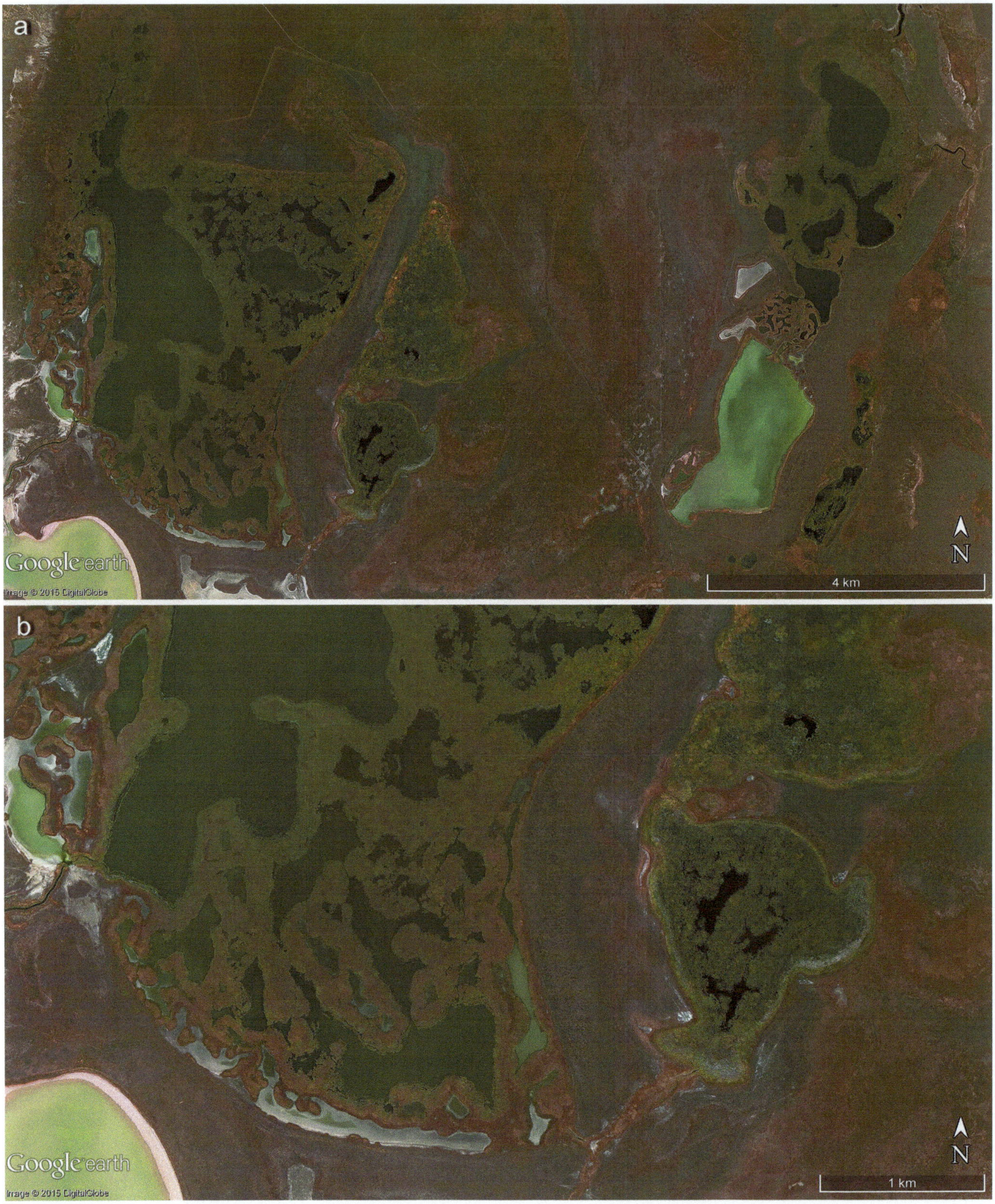

Fig. 3.20 (**a**) Detail of floating vegetation patterns on Kazakhstan lakes and pans (approximately *51°39'N, 64°27'E)* Scene is 15 km wide. (**b**) Magnified view of the floating vegetation patterns in (a). Scene is 6 km wide. (**c**) Magnified detail of (a). The low salinity of the lake (due to its greater depth) is suitable for vegetation, whereas in the adjacent flat pans the water has evaporated into saltpans Scene is 2.3 km (Images credits: ©Google earth 2015)

Fig. 3.20 (continued)

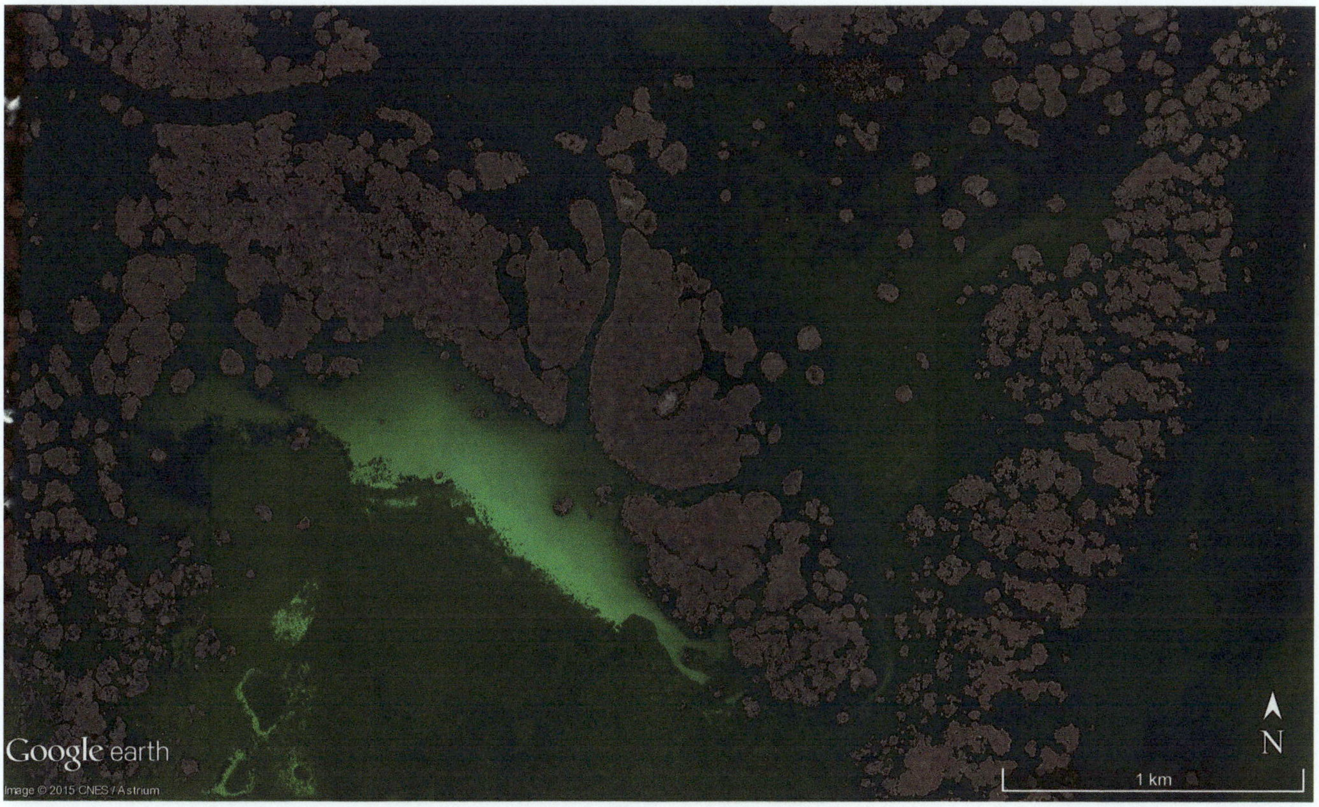

Fig. 3.21 In the northwest section of Lake Titicaca, Peru (*15°24'04.54"S, 69°52'44.72"W*, at 3819 m asl) rooted reeds locally called Totora (*Schoenoplectus californicus*, or sub-species *Sch. acutus.*) grow in shallow water and are harvested by the traditional Uru people for boat building and to construct floating island settlements. Originally the purpose of the island settlements was for defense and could be moved when under threat. Scene is 4 km wide (Image credit: ©Google earth 2015)

Fig. 3.22 Water hyacinth covers a swamp nestled between old dune belts in western Madagascar *(18°24'00.43"S, 44°05'41.60"E)*. Scene is 2 km wide (Image credit: ©Google earth 2015)

Fig. 3.23 A freshwater lake with the common reed *Phragmites* in the eastern Peruvian Andes *(14°53'S, 70°19'W* at 3800 m asl) close to Lake Titicaca. Scene is 530 m wide at (Image credit: ©Google earth 2015)

Fig. 3.24 Blue-green algae (*Cyanophyceae*) growth determines the colour of 4 km long Lake Towydkin in Canada *(51°53'N, 123°20'W)*, where water depth is only a few metres as a maximum. (Image credit: ©Google earth 2015)

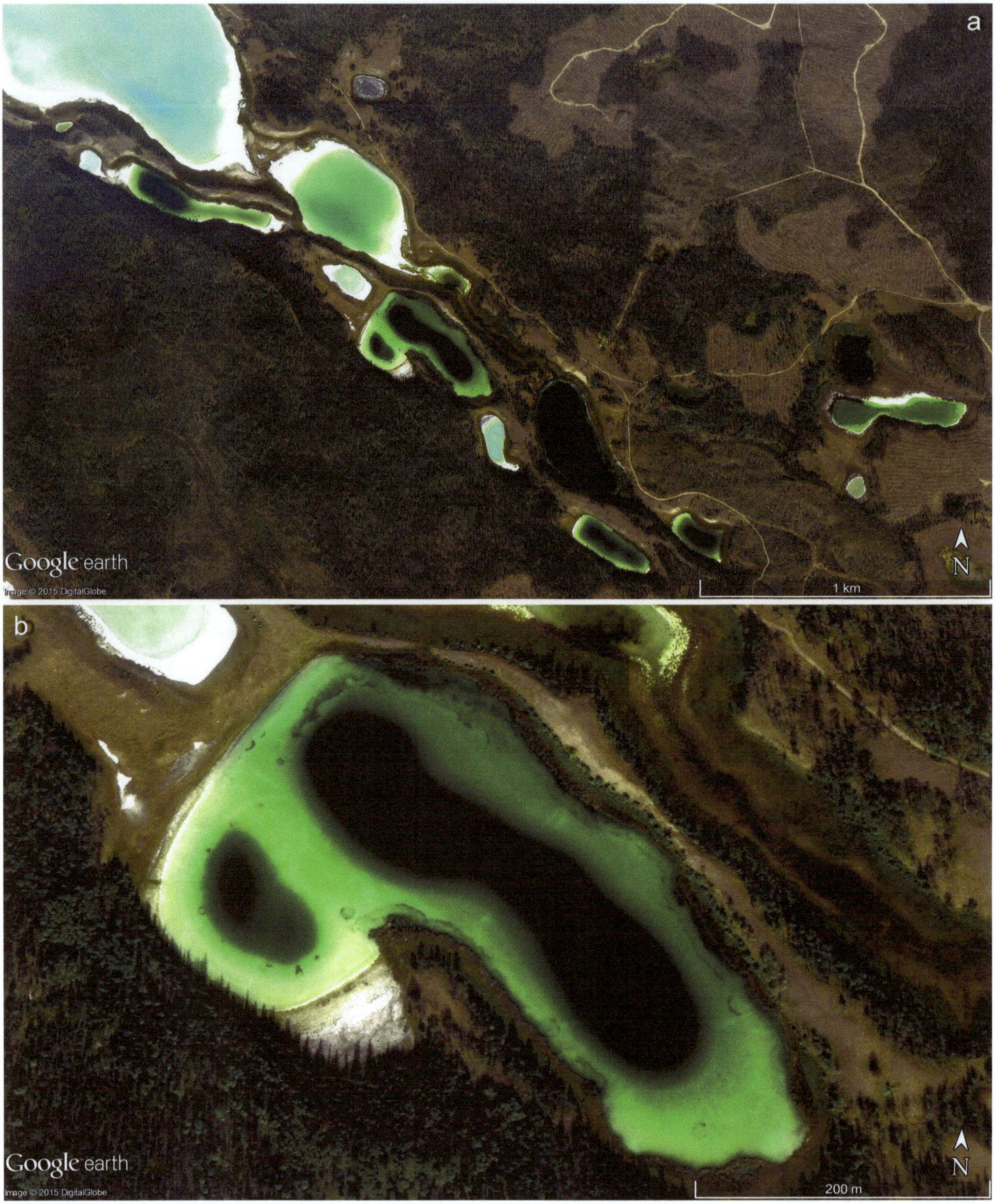

Fig. 3.25 (**a**) Naturally occurring algal mats fringe lakes (Alberta Lake is pictured to the northwest with Pollard Lake bottom south) in Canada (*51°20'N* and *121°37'W*). Scene is 3.5 km wide (Image credit: ©Google earth 2013). (**b**) Detail of Fig. 3.24 (a) in a 600 m wide scene (Images credits: ©Google earth 2015)

Fig. 3.26 (**a**) Freshwater swamps in semi-arid western Queensland, Australia *(14°31'S, 141°36'E)*. Scene is 1.2 km wide. (**b**) Freshwater swamps in the northern Western Australia, the largest is 320 m across, at *12°29'S, 131°31'E* (Images credits: ©Google earth 2015)

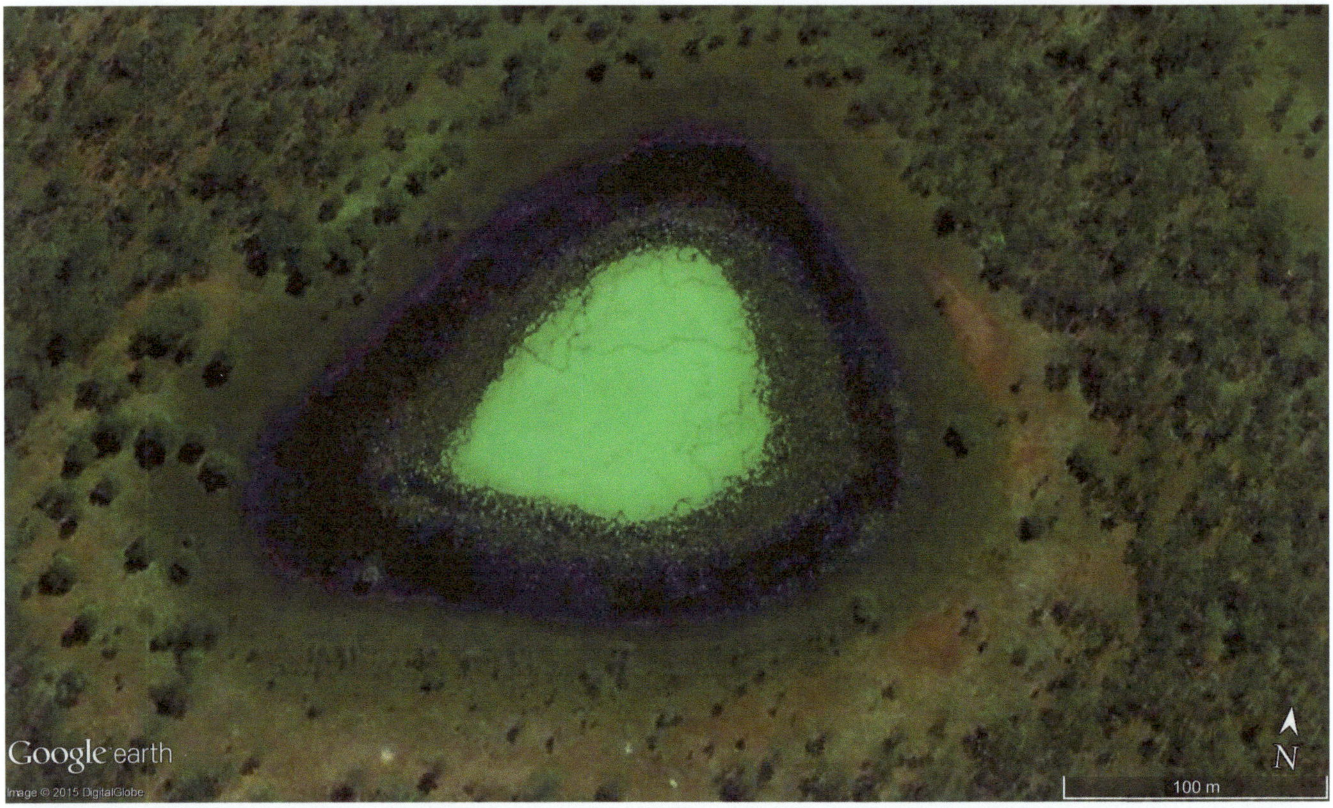

Fig. 3.27 Freshwater swamp in western Queensland of Australia *(14°53'S, 141°53'E)*. The 130 m wide light green area illuminates the tracks of wild buffalos in water hyacinth cover (Image credit: ©Google earth 2015)

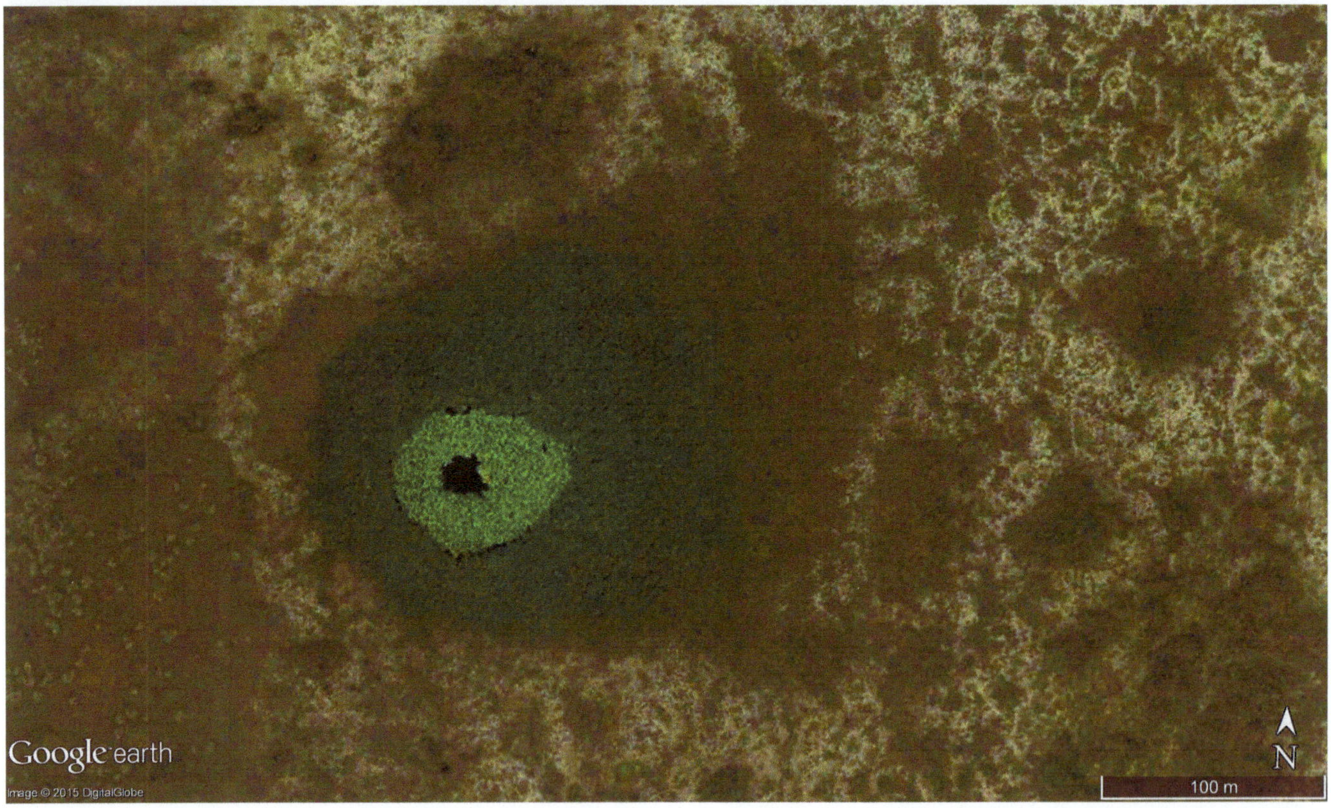

Fig. 3.28 A swamp filling a former sinkhole resulting from the collapse of limestone bedrock (cenote) on the Yukatan peninsula of Mexico, west of Cancun *(21°15'N, 87°11'W)*. The light green area reaches 70 m across (Image credit: ©Google earth 2015)

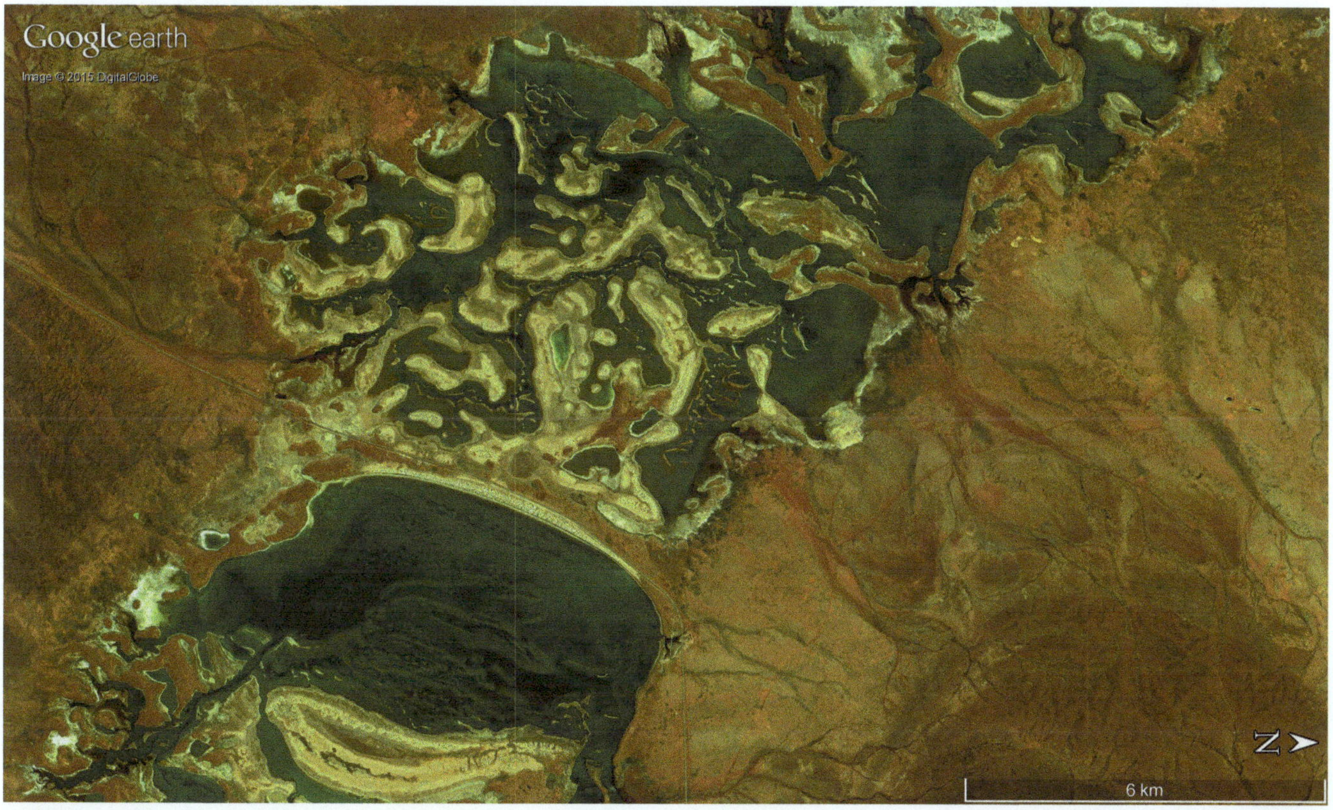

Fig. 3.29 The area of Lake Anneen in West Australia (approximately *26°56'S, 118°1'E)* fluctuates significantly reaching a maximum of around 120 km^2, although smaller eastern sections of Lake Anneen constantly hold water. Scene here is 20 km wide (Image credit: ©Google earth 2015)

Fig. 3.30 In a small lake near Mexico City *(19°16'N, 99°04'W)* rounded islands of floating vegetation are shaped by continually bumping into each other and its shorelines. Scene is 2 km wide (Image credit: ©Google earth 2013)

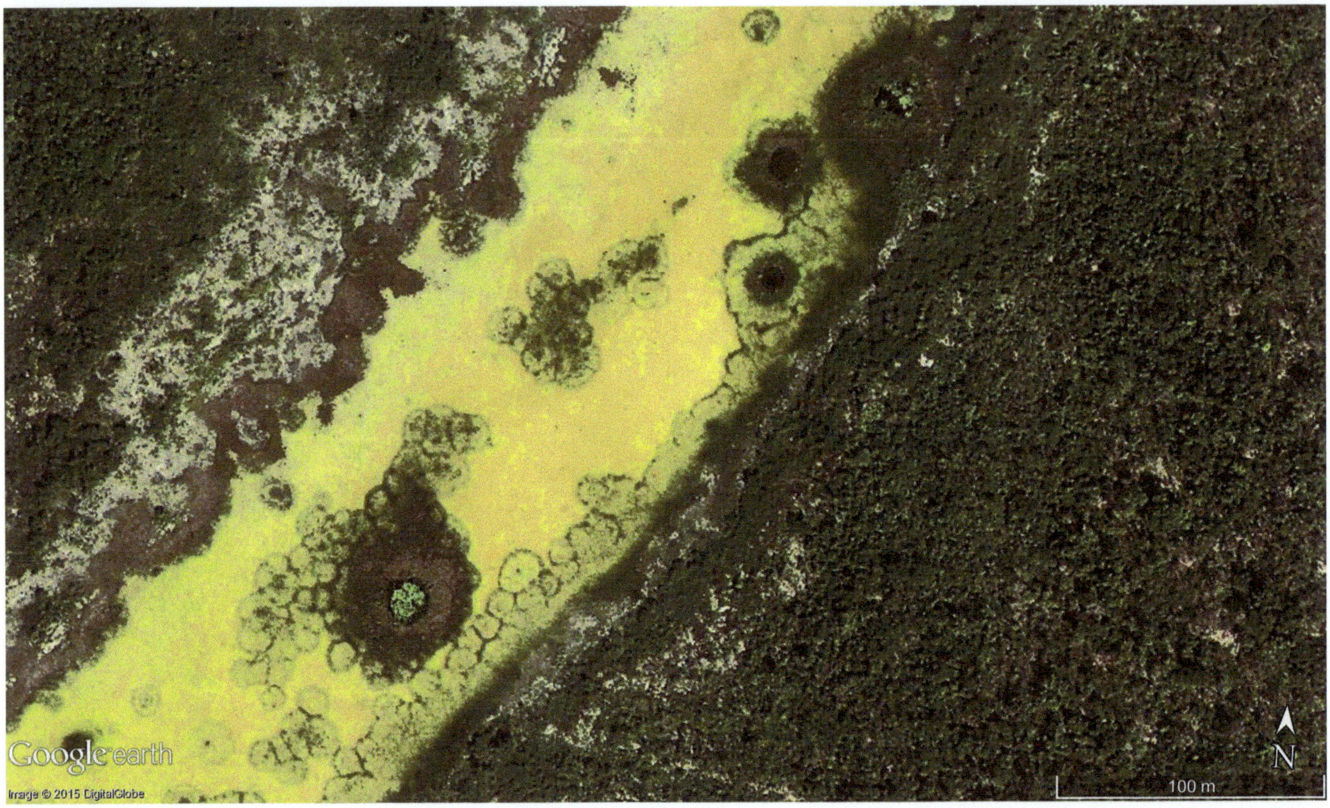

Fig. 3.31 A swamp with circular vegetation patterns in central Yukatan, Mexico *(21°04'N, 87°22'W)*. Diameter of the circled vegetation is around 15 m and the scene is 420 m wide (Image credit: ©Google earth 2015)

Fig. 3.32 In southern Angola *(16°18'S, 19°44'E)* dry river beds demonstrate various circular vegetation patterns that are only fed by groundwater in the dry season. Width of scene is 4 km and the largest diameter of circles reaches 120 m. (Image credit: ©Google earth 2015)

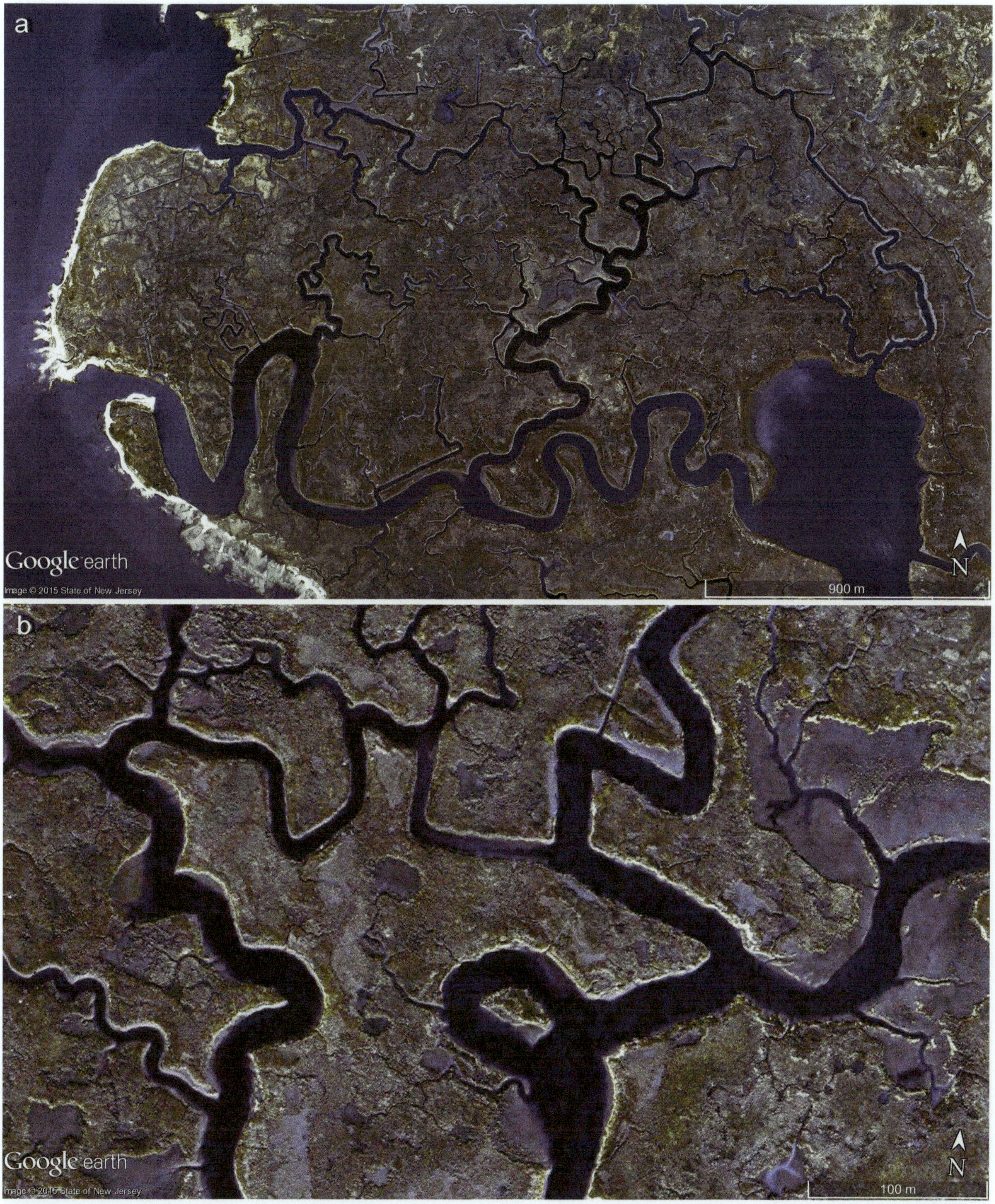

Fig. 3.33 (**a**) A salt marsh of chordgrass (*Spartina* sp.) on the New Jersey east coast of USA (Image taken Dec. 31st, 2002 at *39°12'N, 75°09'W*). Scene is 3 km wide. (**b**) A salt marsh (*Spartina* sp.) from the New Jersey east coast of USA (Image of Dec., 31st., 2002 at *39°12'34.06"N* and *75°08'44.82"W*). Scene is 500 m wide (Images credits: ©Google earth 2015)

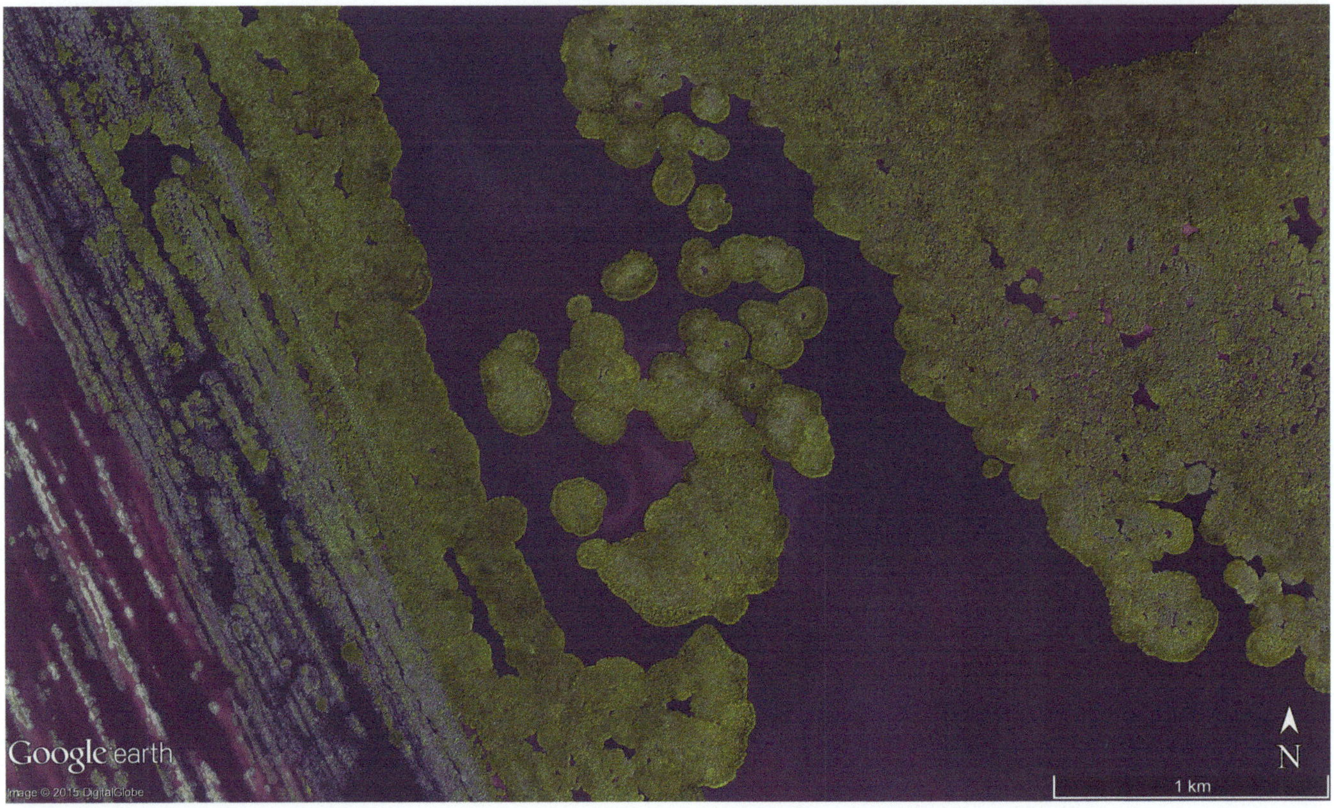

Fig. 3.34 A mangrove swamp with circular floating islands of diameters ranging from 100–200 m is positioned landward from a beach ridge sequence in west Mexico *(22°02'58.31"N, 105°32'24.54"W)*. Scene is 5 km wide (Image credit: ©Google earth 2015)

Fig. 3.35 Mangrove bushes (*Avicennia sp.*) surround a pond of brackish water along the Kona coast northwest of Hawaii Island *(19°48'N and 156°01'W)*. Scene is only 270 m wide (Image credit: ©Google earth 2015)

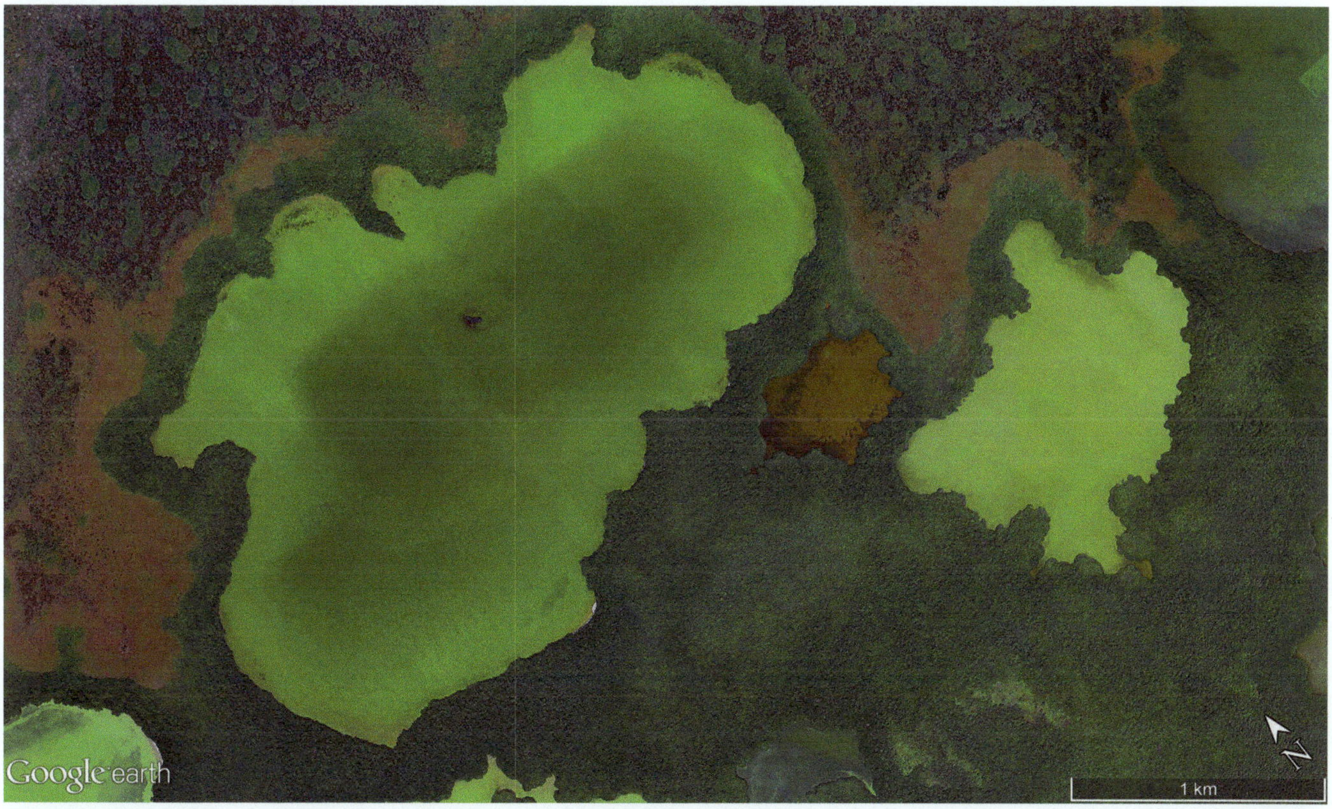

Fig. 3.36 Large shallow ponds in southern Florida *(25°12'13.11"N, 80°46'20.87"W)* are fringed (dark green borders) by the mangrove species *Rhizophora mangle* a 3–4 m high bush. Scene is 6 km wide (Image credit: ©Google earth 2013)

Fig. 3.37 Stands of mangrove forest swamps of southern Florida (approximately *25°55'14.19"N, 81°29'41.54"W)* with open ponds stained brown from the plant compound tannin, a perfect camouflage and home for alligators. Scene is 900 m wide (Image credit: ©Google earth 2013)

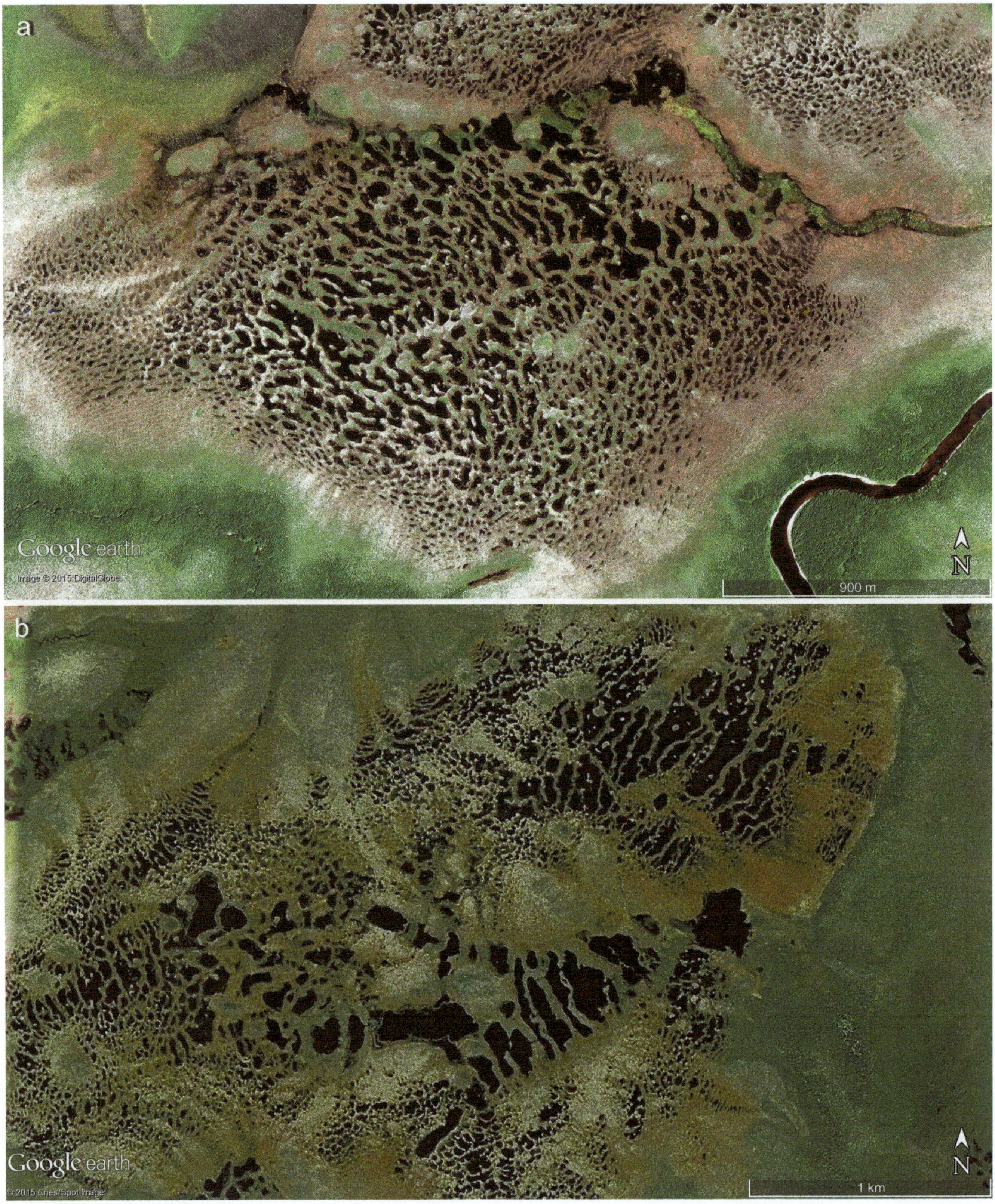

Fig. 3.38 (**a**) Open patches of water have developed when the pore water freezes and compresses in string bogs (In Sweden "Aapa-moore" and in Germany "Strangmoore") from central Canada *(51°09'46.19"N, 82°03'W)*. Scene is 3.5 km wide. (**b**) A setting of string bogs in Central Canada *(51°07'12.94"N, 82°00'23.23"W)*. Scene is 4 km wide. (**c**) A magnified view of a string bog (centre at *51°06'06.01"N* and *82°05'13.32"W)* details the single "strings" are made of many narrow pressure ridges. Scene is 1 km wide. (**d**) Narrow pressure ridges are set amid broader peat strings in central Canada southwest of Hudson Bay *(50°56'20.20"N, 82°14'35.73"W)*. Scene is 1.2 km wide (Images credits: ©Google earth 2015)

Fig. 3.38 (continued)

Fig. 3.39 (**a**) The colourful peat bog Laguna Pedro Aguirre, in the Andes of southern Chile *(45°02'S, 72°07'W)*. In contrast to swamps, here the peat grows from the ground up and may exceed the height of the former pond or lake resulting in raised bogs. Lake is 1 km by 1.5 km in length. (**b**) The internal pattern of the peat bog in Tierra del Fuego of Argentina *(54°53'41.21"S, 65°34'30.34"W)* is a result of frost pressure processes. Scene is 3.4 km. (Images credits: ©Google earth 2015)

Fig. 3.40 The Shark River Slough in southern Florida *(26°05'N, 80°43'W)* is a peculiar type of swamp commonly known as "river of grass" and consists of numerous freshwater courses flowing from Lake Okeechobee and the Kissimmee River, including smaller karst springs to the south, that eventually seep into the Gulf of Mexico. The swampy region can expand to 10,000 km^2 in area with widths extending 90 km and an overall length from Lake Okeechobee to the Gulf of Mexico nearing 160 km. Primarily, most of the habitat is saw grass marsh (*Cladium mariscus*) with sloughs (swamp-like soft muddy ground) and Cypress swamps (*Taxodium distichum*). Tree species of Southern Live Oaks (*Quercus virgiana*), Royal Palms (*Roytonia*) and palmettos (*Serenoa repens*) grow on hammocks, and pine lands with *Pinus elliotti* in the southeast also occur. Width of scene is 18 km. (Image credit: ©Google earth 2015)

References

Allanson BR, Hart RC, O'Keeffe JH et al (1990) Inland waters of Southern Africa. An ecological perspective. Kluwer, Boston

Beadle LC (1981) The inland waters of tropical Africa. An introduction to tropical limnology, 2nd edn. Longman, London

Bedford BL (1996) The need to define hydrologic equivalence at the landscape scale for freshwater wetland mitigation. Ecol Appl 6(1):57–68

Bridgewater P (2008) A new context for the Ramsar Convention: wetlands in a changing world. Reciel 17(1):100–106

Evangelisti D, D'Amelia D, Di Lallo G et al (2013) Relationship between salinity and bacterioplankton in three relic coastal ponds (Macchiatonda Wetland, Italy). J Water Resour Prot 5(9):859–866

Fraser LH, Keddy PA (eds) (2005) The World's largest wetlands: ecology and conservation. Cambridge University Press, Cambridge UK

Glaser PH (1992) Raised bogs in eastern North America: regional controls for species richness and floristic assemblages. J Ecol 80:535–554

Hails AJ (1996) Wetlands, biodiversity and the Ramsar Convention: the role of the Convention on wetlands conservation and wise use of biodiversity. Ramsar Convention Bureau, Gland

Hart TM, Davis SE (2011) Wetland development in a previously mined landscape of East Texas, USA. Wetl Ecol Manag 19:317–329

Hughes FMR (ed) (2003) The Flooded Forest: guidance for policy makers and river managers in Europe on the restoration of floodplain forests. Department of Geography, University of Cambridge, Cambridge, UK

Johnson WC, Millett BV, Gilmanov T et al (2005) Vulnerability of northern prairie wetlands to climate Change. Bio Sci 10:863–872

Jordan TE, Andrews MP, Szuch RP et al (2007) Comparing functional assessments of wetlands to measurements of soil characteristics and nitrogen processing. Wetlands 27(3):479–497

Keddy PA (2010) Wetland ecology: principles and conservation, 2nd edn. Cambridge University Press, New York

Lehner B, Döll P (2004) Development and validation of a global database of lakes, reservoirs and wetlands. J Hydrol 296:1–22

MacDonald GM, Beilmann BW, Kremenetski KV et al (2006) Rapid early development of circum-arctic peatlands and atmospheric CH4 and CO2 variations. Science 314(5797):285–288

MacKenzie WH, Moran JR (2004) Wetlands of British Columbia: a guide to identification, Land management handbook 52. Ministry of Forests, Government Publication Services, Victoria
Maltby E, Barker T (eds) (2009) The wetlands handbook. Wiley-Blackwell, Oxford
Mitsch WJ, Gosselink JG (2007) Wetlands, 4th edn. John Wiley and Sons, New York
Mitsch WJ, Gosselink JG, Zhang L et al (2009) Wetland ecosystems. John Wiley and Sons, New York
Nelson ML, Rhoades CC, Dwire KA (2011) Influences of bedrock geology on water chemistry of slope wetlands and headwaters streams in the southern Rocky Mountains. Wetlands 31:251–261
Roulet NT (2000) Peatlands, carbon storage, greenhouse gases, and the Kyoto protocol: prospects and significance for Canada. Wetlands 20(4):605–615
Smith MJ, Schreiber ESG, Kohout M et al (2007) Wetlands as landscape units: spatial patterns in salinity and water chemistry. Wetl Ecol Manag 15(2):95–103
Solomeshch AI (2005) The west Siberian lowland. In: Fraser LH, Keddy PA (eds) The World's largest wetlands: ecology and conservation. Cambridge University Press, Cambridge UK, pp 11–62

Saline Lakes and Saltpans

4

Abstract

There are several reasons why lakes are (or become) saline. Evidence lies in lagoons along open oceans where seawater may percolate through sandy barriers and lakes become saline or brackish, occasionally transforming their character. Further inland, saline lakes accumulate salt content by freshwater inflow and river discharge that carries minimal amounts of salt, and as a consequence salt concentrations increase in a basin (and a lake) through evaporation. Particularly if a basin is endorheic (without outflow), salinity can range from a few to over 300 % per mil with a tendency to increase concentration as the basin ages, allowing for the many different elements found in salt or brine. With the exception of the extraordinary ancient Dead Sea (the deepest hypersaline lake in the world exceeding 300 m in depth), concentrations of salinity in deep lakes is generally lower than in shallow lakes, as vertical circulation needs more time in deeper waters. In lakes comprising of various depths, salinity can vary from section to section clearly demonstrated in the Caspian Sea where the north is shallow and saline, the south is deep and almost fresh in character. Certain saline lakes no longer appear as lakes; a stable salt crust (over a brine) may cover its surface several metres thick such as the Salar de Uyuni in southwest Bolivia, the world's largest salt flat at 10,600 km^2. With climate changing over time to more arid conditions, a salt lake often transforms into saltpans with a salt crust intercalated with silt and clay, and becomes the uppermost strata of a sediment archive. Saline lakes are specific ecosystems and as salinity increases the number of species decreases toward negligible. In addition, several lakes became saline (or more saline) by anthropogenic influence over water input; the Aral Sea catastrophe has receded up to 90 % since humans diverted the two rivers that feed it for irrigation purposes. Saltpans in particular exhibit many additional aspects clearly discernable from satellite imagery: intensive colours (alternating with seasons), significant variations between shallow and deeper waters associated with evaporation deposits, and internal patterns left on salt crust such as networks or polygons formed by salt expansion and shrinkage. The combination of Halobacteria and Cyanophytes containing beta-carotene an organic compound, results in the red to orange colours of saline lakes and saltpans, and particularly the relationship between the genus *Dunaiella salina* and *Halobacterium cutirubrum* regulates the intensity of the colour.

4.1 Saline Lakes

Saline lakes are an important feature to the landscape of many regions around the world (Table 4.1). The genesis of saline lakes does not differ very much to that of freshwater lakes; one exception is their position in endorheic basins (no outflow) where most (but not all, see the Dead Sea, Figs. 4.1 and 4.2) of the saline lakes are shallow, several below 10 m depths. While most of today's salt lake basins are small and shallow and many exhibit playa characteristics of flat dried-up areas from which water evaporates, there also exist enormous saline lacustrine basins. Modern

A.M. Scheffers, D.H. Kelletat, *Lakes of the World with Google Earth*, Coastal Research Library 16,
DOI 10.1007/978-3-319-29617-3_4

Table 4.1 Largest saline lakes of the world (exceeding 5000 km²) (compiled from various sources)

	Area	Volume	Mean depth	Maximum depth	Salinity
Name	km^2	km^3	m	m	%
Caspian	422,000	79,000	187	1072	1–1.2
(Aral)	(66,000)	(1064)	(16)	(69)	(0.5–1.0)
Balkhash	22,000	122		27	0.05–0.7
Eyre North	7000	23	3	6	10
Issykköl	6300	1730		720	0.5–0.6
Urmia	5000	25	5	16	23

Fig. 4.1 The southern section of the Dead Sea in the Jordan Rift Valley *(31°22'N, 35° 8E)* is used to mine minerals from its extreme salt content. The saline conditions make for a harsh environment in which life cannot survive, hence its name. The Dead Sea is a salt lake 55 km long and 18 km at its widest point, bordering Jordan to the east with Israel and the West Bank to the west. In 2015 the present shoreline claims the lowest land elevation on Earth and lies 429 m below sea level. With a maximum depth of 330 m and a mean depth of 145 m, the Dead Sea is the deepest hypersaline lake in the world and with a salinity of 33.7 % is 8.6 times higher than ocean salinity. The Dead Sea is steadily shrinking, and presently covers an approximate area of 600 km² down from 1050 km² in 1930. Primarily, the water taken for irrigation from its main tributary the Jordan River and the arid climate (50–100 mm/year of precipitation) rapidly lowered the water level of the Dead Sea from −398 to −429 m in only 33 years. The overall levels of the Dead Sea during the last millennia represent past climatic conditions, in particular the relation of precipitation and run off from the Jordan discharge area with high temperatures and high evaporation rates (Image credit: ©Google earth 2013)

salt lakes exhibit a wide diversity of hydrology, morphology, chemistry, and sedimentary processes, and despite elevated salinities, saline lakes serve a variety of uses in mineral industries. The salt and ionic composition of salt lakes demonstrate a multitude of varying chemistry types represented in the brines, often within the same geographic region, and investigation of the stratigraphic records into these chemical signatures of ancient saline lakes (palaeolimnology) is in its infancy.

In terms of limnology, the definition salinity represents the total sum of all ions. A conventional saline value is 3 mg per litre (3 mg/l or 3 ppm); and is near the salinity at which humans first begin to taste salt, below concentrations where biota typical of higher salinities do not grow and above salt

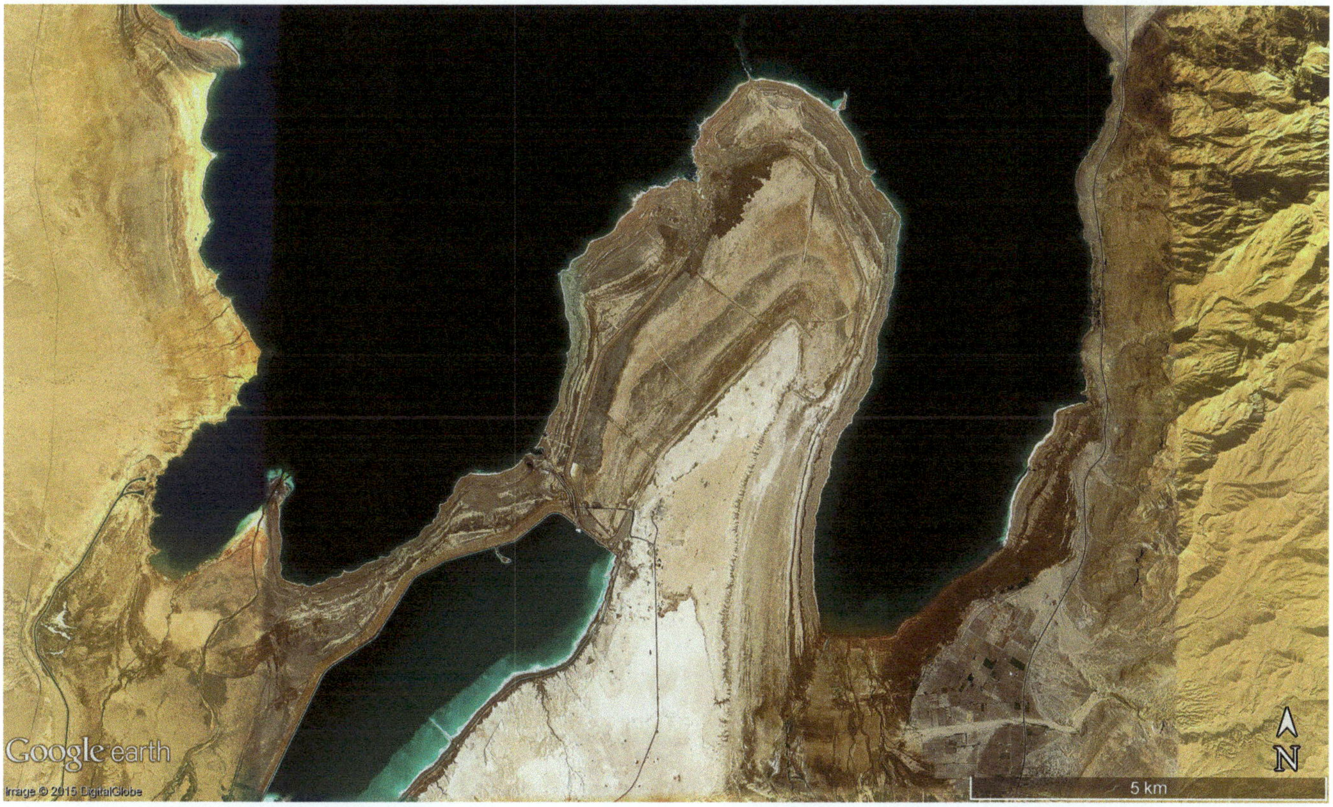

Fig. 4.2 Elevated parallel shorelines along the southern part of the Dead Sea (Lisan peninsula, from the centre of Fig. 4.1 at about *31°20'N* and *35°29'E*, width of image is 20 km) (Image credit: ©Google earth 2015)

concentrations where freshwater biota begin to quickly disappear or stunted. Generally salt lakes are more varied in their physico-chemical features than are freshwater lakes where salinity in salt lakes may vary on a seasonal or long-time basis from less than 50 to over 300 mg/l. Only few inland salt lakes attain salinities from underground or local salt sources (Figs. 4.3, 4.4, 4.5, and 4.6).

Salt lakes permanently, intermittently or episodically contain water where water levels can fluctuate widely or be constant on a seasonal or secular basis, often paralleling the concentrations of salinity. Salt lakes range from deep to shallow, small to extremely large, round to dendritic in shape (Figs. 4.7, 4.8, 4.9, 4.10, 4.11, 4.12, 4.13, 4.14, 4.15, 4.16, 4.17, 4.18, 4.19, 4.20, 4.21 and 4.22) and 70 % of all inland salt water is contained in the Caspian Sea. Many of the world's largest lakes are saline, including Aral Lake, Balkhash Lake, Lake Eyre, Lake Issykköl, Lake Urmia (Fig. 4.3), Lake Qinghai, the Great Salt Lake (Fig. 4.5), Van Lake and the Dead Sea. The most concentrated saline lake in the world is the Don Juan Pond in the McMurdo Dry Valleys of Antarctica, reaching levels in excess of 44 %. This prevents the pond from freezing even though temperatures reach below −50 °C. Outside of the Antarctica, Lake Assal in Djibouti is the most saline lake reaching a salinity level of 34.8 % and the best-known examples of hypersaline lakes are the Dead Sea of Israel and Jordan of 33.7 % salinity and the Great Salt Lake in Utah USA with a variable salinity ranging from 5 to 27 %.

Saline lakes boast a vast array of colours: from green through to brown and pink to all shades of red (Figs. 4.23, 4.24, 4.25, 4.26, 4.27 and 4.28). Salt crusts may decorate the outer fringe of saline lakes (Figs. 4.29 and 4.30) and in rare cases, mineral deposits can develop into minor formations in the shallowest parts of the lake and from the bottom deposits build toward the surface. Examples are given with the commonly named "Spotted" or "Dotted Lakes" of the Okanagan Valley in British Columbia, western Canada (compare Figs. 4.31, 4.32 and 4.33). These lakes often become saltpans toward the end of summer and belong to the heritage of the First Nations, the indigenous societies still use the minerals for healing and the sites remain sacred (to all). Because of high evaporation and the semi-arid climate in this region of Canada, circular forms are created within the lakes, constructed of a central depression with an elevated rim these formations are evidence of high concentrations of sulphates (from calcium, sodium and magnesium). The localized conditions and time of the year can also influence these salt lake deposits to exhibit different colours from lake to lake, and from various basins within a lake

Fig. 4.3 Lake Urmia, northwest Iran (approximately *37°35'N, 45°26'E* at 1268 m asl) was one of the largest salt lakes in the world, however the southern section has become a saltpan because the small 1.5 km wide opening in the dam road will not allow the waters from the deep northern section to circulate efficiently. Lake Urmia is 120 km long in this aspect (Image credit: ©Google earth 2013)

Fig. 4.4 Zhari Namco is a shrinking salt lake covered with ice from winter frost and in northern Tibet *(30°48'N, 85°47'E* at 4686 m asl). The lagoon rests at 4616 m asl and is 3.3 km wide (Image credit: ©Google earth 2015)

Fig. 4.5 The Great Salt Lake in 2011, Utah, USA *(41°10'N, 112°34'W)* with a length of 125 km, width to 50 km, area of 4400 km² and water volume approximately 19 km³. Today the maximum depth of 10 m with a mean depth of only 4.3 m aids in making the lake hyper-saline and because of the high salinity in the northern region (up to 28 %) and the southern region (around 15 %), no fish species can survive in the lake except brine shrimps and brine flies. The Great Salt Lake is the remnant of ancient Lake Bonneville from late-glacial times (once 10 times larger) and minor changes to its water-surface elevation results in large changes in the surface area of the lake. In 1847 the Mormons arrived at the lake when the surface area averaged 4420 km² and in 1963 the lake diminished to a historic low of 2450 km², the difference in elevation being only 1.77 m. In years with high run-off (e.g. 1983–1987) the lake spills into the western desert, and compared to 1963, during 1986–1987 the lake reached 2.6 m higher and extended its surface area to approximately 8520 km². The consequence was inundation of wide low-lying landscapes with losses of agricultural land, costly salt leachates from nearby flooded salt basins, and along railway lines created blockages resulting in dammed water with the dislodgment of highways (Image credit: ©Google earth 2013)

4.2 Saltpans and Related Features

As discussed in the chapter on salt lakes, saltpans mostly occur in arid climates and are usually shallow endorheic basins. Many are bordered by evaporative fringes—where even a very small climate variation (too little rain of one season) may transform a shallow salt lake into a saltpan—and because this process can be reversible, the relationship between saline lakes and saltpans (often called "Salar" from the Spanish language) remains strong. In addition, Sabkha (or Sebkha) formations exist as a flat region between the desert and sea and are formed primarily through the evaporation of seawater seeping upward from a shallow water table (and from windblown sea spray drying out), characterized by a surface crust of evaporates. The term is from the Arabian language used in Algeria or Libya, whereas in Tunisia, these features are called "Schott".

The transformation of lakes into pans has accelerated during the last decades not only from the natural processes of higher temperatures and increased evaporation, but also by human-induced climate change now adding further increases to temperature and evaporation levels, and particularly anthropogenic impacts resulting from bad irrigation practices or the lowering of ground water reservoirs by deep wells. Examples are the large saline lakes Lake Aral and the Great Salt Lake given in the introductory chapter.

The following satellite images demonstrate the extremely wide catalogue of saltpans related to size, forms and colours of evaporates (aside from salt, evaporates contain many other minerals often valuable to mining). Parallel to lakes as "eyes of the continents" are saltpans with their strange forms and decorative vivid colours. Many saltpans are small and miniscule, however new satellite image techniques of recent years make these extraordinary formations observable with excellent imagery.

Fig. 4.6 The Salton Sea in southern California, USA (approximately *33°18'N, 115°47'W)* reaches a length of 59 km and covers an area of 891 km^2, with a maximum depth of 12 m and a mean depth of 7.6 m, its salinity is at 3.3 %. Situated in a tectonic depression (southern part of California's Imperial Valley) the lake emerged from a dam breakage at the Colorado River in 1905 and after 2 years of flooding reached its current size (varying slightly by evaporation and fluctuating river discharge). The lake surface lies 69 m below sea level and a cryptodepression (deep sections of the lake basin below sea level) is at a similar level to the deepest part of Death Valley (Image credit: ©Google earth 2013)

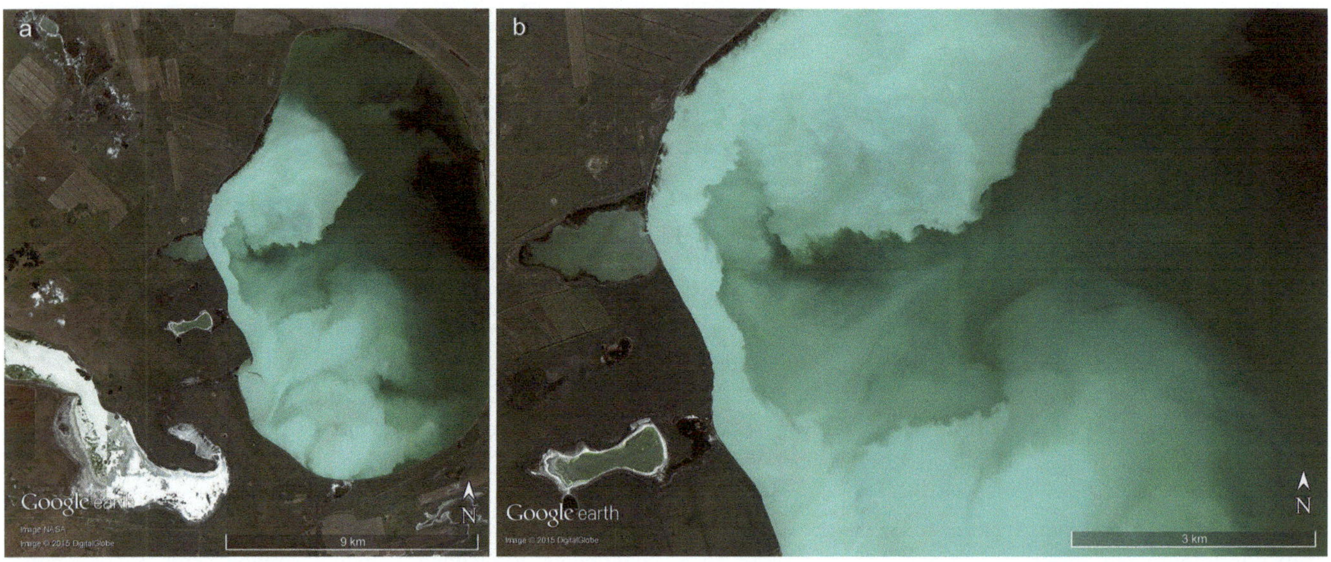

Fig. 4.7 (**a**) The shallow salt lake Bol'shoye Topol'noye in Kazakhstan *(53°18'11.76"N, 77°58'35.69"E)* is 16 km long, lies in a N-S-direction and is mixed by westerly winds. (**b**) Enlarged detail of (**a**) in a 12 km scene (Images credits: ©Google earth 2015, picture token June 16th, 2007)

Fig. 4.8 The eastern section of the main Etosha Pan in Botswana *(18°48'06.50"S, 16°47'29.42"E* at 1085 m asl) exhibits drift of hyper-saline waters caused by wind stress. Scene is 13.5 km wide (Image credit: ©Google earth 2015, picture token August 16, 2011)

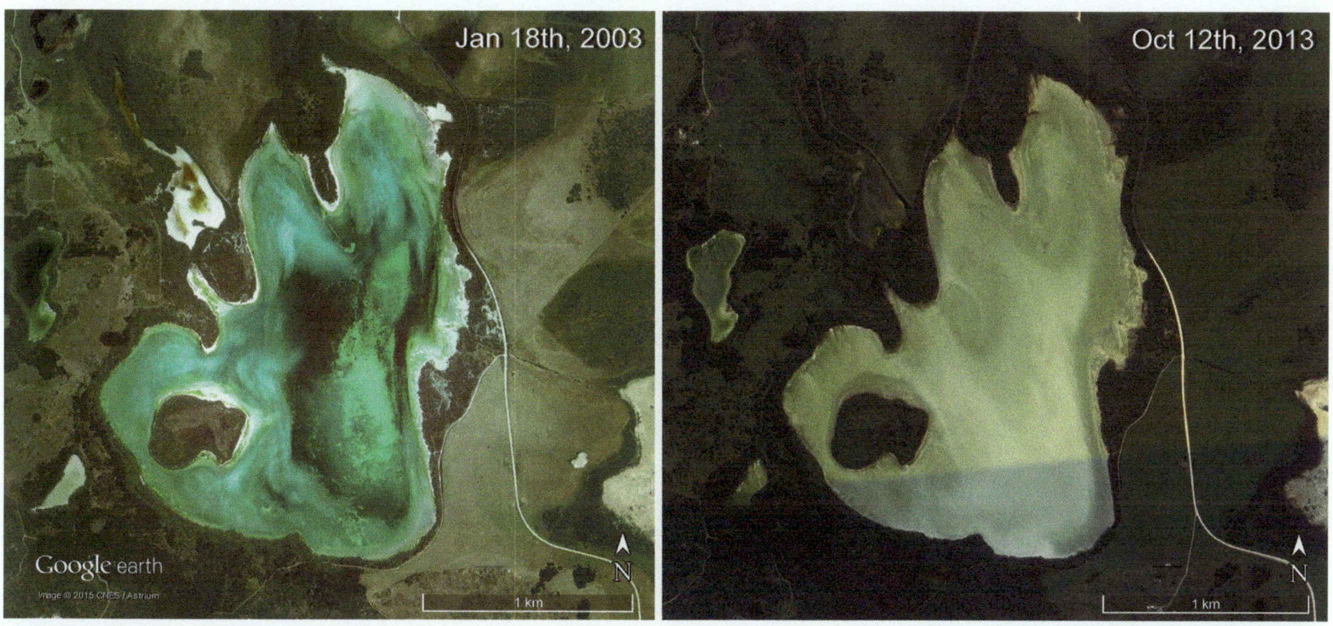

Fig. 4.9 A saline lake from the State of Victoria, Australia *(37°12'49.91"S, 139°47'31.41"E)* demonstrates changes in appearance caused by various water levels. Scene is 4.6 km wide (Image credit: ©Google earth 2015)

Fig. 4.10 Four different aspects of Laguna Verde in Bolivia *(22°47'40.09"S, 67°50'16.36"W* at 4329 m asl). The colours change not only through water depth (compare the 2010 low and the 2013 high), but also by the angle of the sunlight. The 2007 level presents a deeper blue colour than the 2004 lower level. Scene is 3.6 km wide (Image credit: ©Google earth 2015)

Fig. 4.11 Set on the land bridge *(41°35'N, 52°40'E)* between the Caspian Sea and the Garabogazköl Basin, small lakes present various stages from hyper-saline lakes to saltpans. Scene is 1.1 km wide (Image credit: ©Google earth 2015)

Fig. 4.12 The Eastern Toshka Lake *(23°07'N, 31°15'E)* is part of several lakes in the western Sahara Desert, filled from 1998 onward by the Sadat Canal (overflow from Nasser Lake), located behind the Nile's Aswan High Dam situated along the southern border of Egypt to Sudan. The Eastern Toshka Lake stretches 25 km in a W-E-direction and while the expansion of all lakes together nears 500 km^2, strong evaporation and less inflow is causing this area to recede (Image credit: ©Google earth 2013)

Fig. 4.13 Evaporation enriches the salt content of a salt lake in southern Bolivia *(22°22'S, 67°06'W* at 4553 m asl) although it is fed by a freshwater spring from the southwest. Lake length is 3 km (Image credit: ©Google earth 2013)

Fig. 4.14 (**a**) Salt lakes are situated in the Andean mountains of South America (Catamarca Province of eastern Argentina approximately *27°35'S, 68°37'W*). The northern lake is Laguna Verde. Scene is 24 km wide. (**b**) Laguna de los Aparejos demonstrates a strong salt encrustation along its northern shoreline and lies south of Laguna Verde in (**a**). Scene is 12 km wide (Images credits: ©Google earth 2015)

Fig. 4.15 (**a**) Two salt lakes (both approximately 1.5 km in length) are located in the semi-desert of southern Kazachstan (approximately *43°14'N, 70°31'E),* northeast of the town Qaratau. (**b**) Detail from the eastern lake of (**a**) presents a salt crust in different colours depending on the pore water and certain bacteria. Scene is 600 m wide (Images credits: ©Google earth 2015)

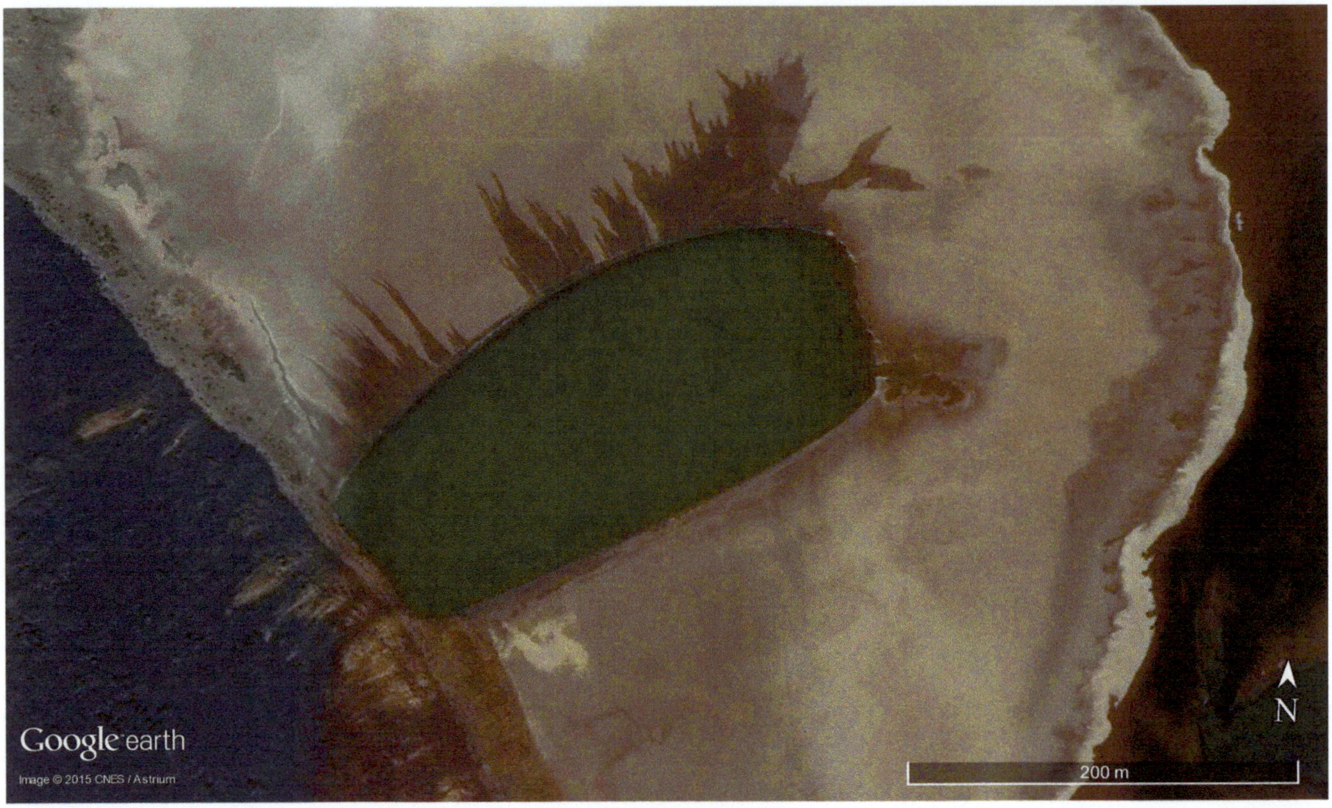

Fig. 4.16 In one of the many small rift valleys of Djibouti in eastern Africa *(11°50'N, 42°22'E)* a 300 m long salt lake (9 m below sea level) is situated within extended saltpans and withstands the harsh desert environment (Image credit: ©Google earth 2015)

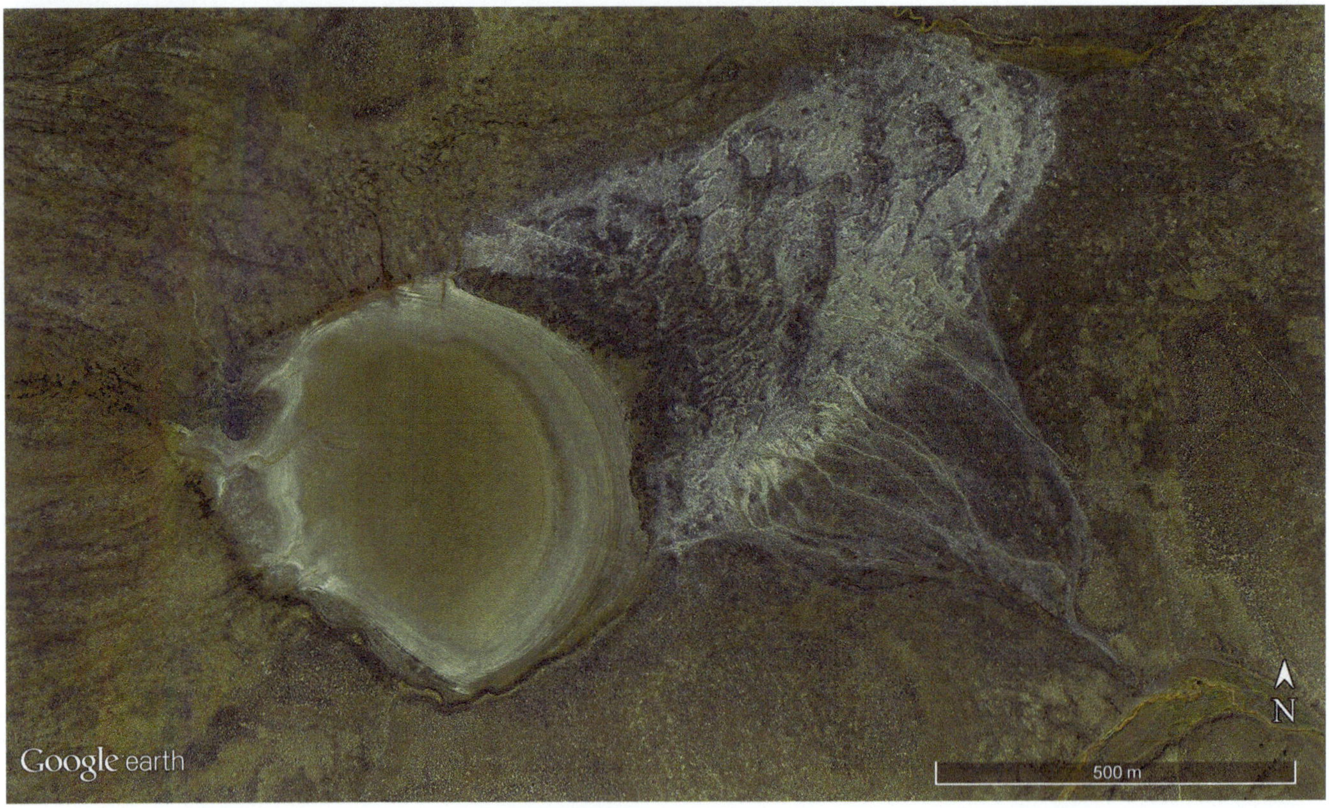

Fig. 4.17 This Oregon salt lake (directly east of Poker Jim Lake at *42°36'N, 119°34'W)* is approximately 600 m across and exhibits a salty region of wind drift from the southwest direction (Image credit: ©Google earth 2013)

Fig. 4.18 Positioned in the Andes near the eastern border between Chile and Argentina in the Atacama region lies the 4 km long salt rich Laguna Brava (approximately *26°19'S, 68°36'W* at 4228 m asl) (Image credit: ©Google earth 2013)

Fig. 4.19 Laguna Escondido *(26°39'S, 68°31'W* at 4364 m asl) is a 3 km long salt lake in the Atacama region of northern Chile close to the Bolivian border. The salt content in the small southern lagoon is higher than Laguna Escondido where northwesterly winds drive salt dust to the southeast (Image credit: ©Google earth 2015)

Fig. 4.20 In the temperate climatic region of Oregon USA, freshwater input is insufficient to avoid evaporation and dilute the high salt content in this mountain lake *(43°14'N, 119°37'W)*. The dark coloured eastern section of the mountain lake is known as the "Mud Lake". Lake diameter is 14.5 km (Image credit: ©Google earth 2013)

Fig. 4.21 Many lakes in Australia show well rounded contours including Lake Gore in the south-eastern section of Western Australia *(33°46'S, 121°30'E)*. Lake Gore extends 3.2 km in N-S-direction and is shallow (max. 2.2 m in 1986), developing into a saltpan during most years (Image credit: ©Google earth 2015)

Fig. 4.22 Lakes in an open environment may take astonishing similar forms such as this ensemble in the western part of New South Wales, Australia *(32°23'S, 142°24'E)*. Lake Menindee is the largest (16 km across and up to 7 m deep) and over the last few decades, the lakes fed from the Darling River system have been connected to serve as a water storage system for the environment causing changes in salinity as a result of these artificial conditions. Scene is 48 km wide (Image credit: ©Google earth 2015)

Fig. 4.23 Laguna de la Azufrera in Antofagasta Province of northern Chile *(25°05'S, 68°31'W* at 4242 m asl) part of the Atacama Desert. The eastern shore of the salt lake presents a delicate harmonic curve encasing a barrier and lagoon. Further landward, older beach contours suggest an extensive shrinking process. The scene is 5 km wide (Image credit: ©Google earth 2015)

Fig. 4.24 A small saline lake 240 m wide contains algal blooms in the northeast of Chile close to the Bolivian border *(26°18'S, 68°35'W* at 4240 m asl). Old shorelines along the eastern perimeter demonstrate the lake is receding (Image credit: ©Google earth 2015)

Fig. 4.25 An algal bloom in a small saline lake of southern Bolivia 20 km north of Laguna Bush *(22°25'S, 67°10'W* at 4702 m asl). The lake is only 400 m long and salt banks lie along its eastern shoreline (Image credit: ©Google earth 2015)

Fig. 4.26 Green algae in the southern section of Lake Preston in Western Australia south of Perth *(32°54'S, 115°40'E)*. Lake Preston is only 1 km wide but more than 20 km long, and during the second-last high sea-level event of the Pleistocene epoch was a former lagoon separated by a dune barrier. (Image credit: ©Google earth 2015)

Fig. 4.27 A salt lake in the northern Province of Santa Cruz in eastern Argentina *(46°55'S, 67°06'W)* holds bacteria containing beta-carotene that generates the reddish colour. Lake is 900 m across (Image credit: ©Google earth 2015)

Fig. 4.28 A lake is fringed by re-advance terminal morainic ridges and filled by a young and active fluvial fan in the Bolivian Andes *(16°20'S, 68°09'W* at 4562 m asl,). At this altitude, high radiation coupled with clear and dry air support development of beta-carotene producing algae producing the intensive red colour. Scene 1.5 km wide (Image credit: ©Google earth 2015)

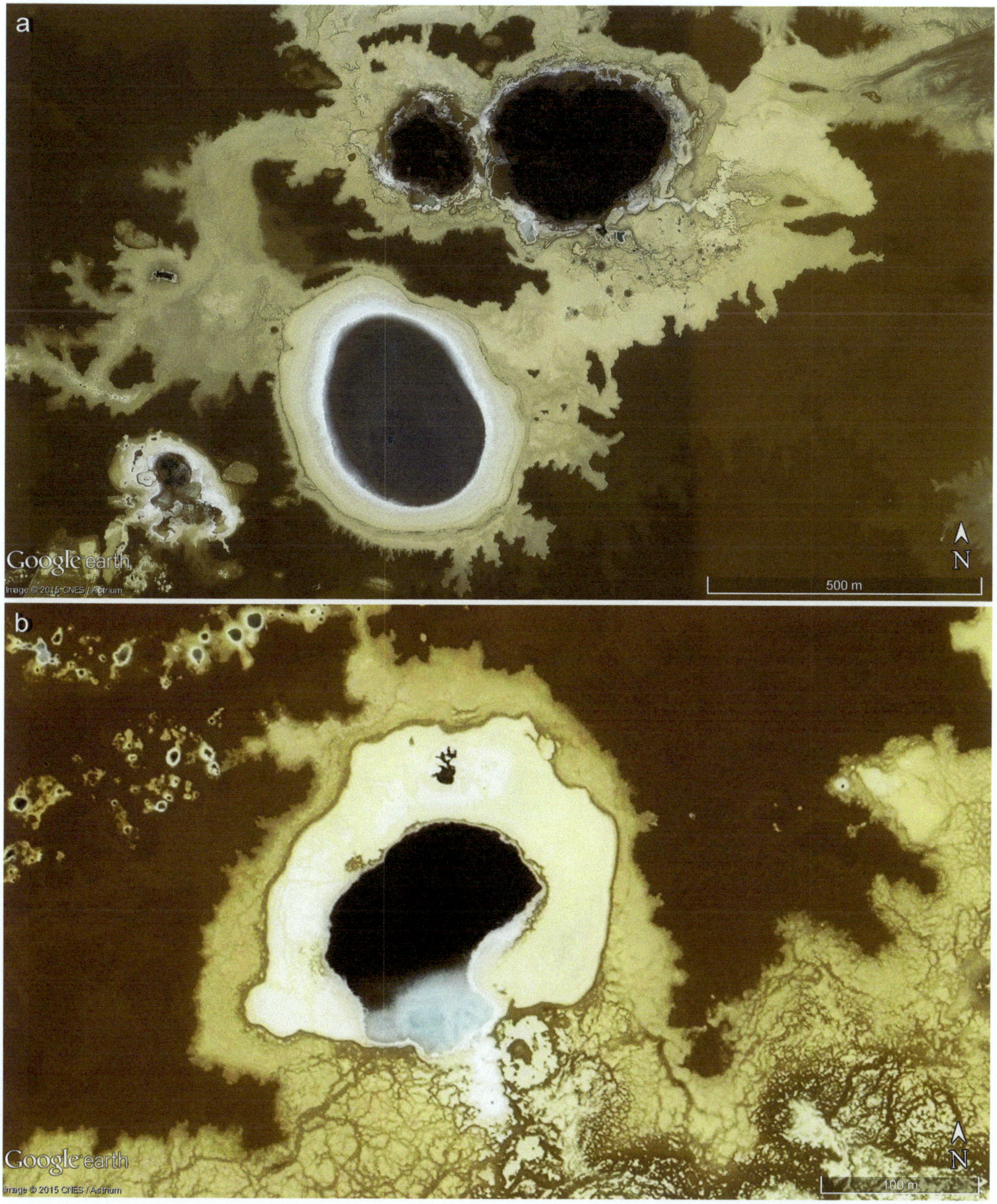

Fig. 4.29 (**a**) Situated 17 m below sea level, salt lakes and saltpans of the Sivah Oasis in Egypt. The lake *(29°10'N, 25°44'E)* is almost 500 m across. (**b**) A tiny section of the Sivah Oasis in Egypt *(29°11'N, 25°43'E)* shows a salt lake and surrounding saltpans resting 17 m below sea level. Scene is 600 m wide (Images credits: ©Google earth 2015)

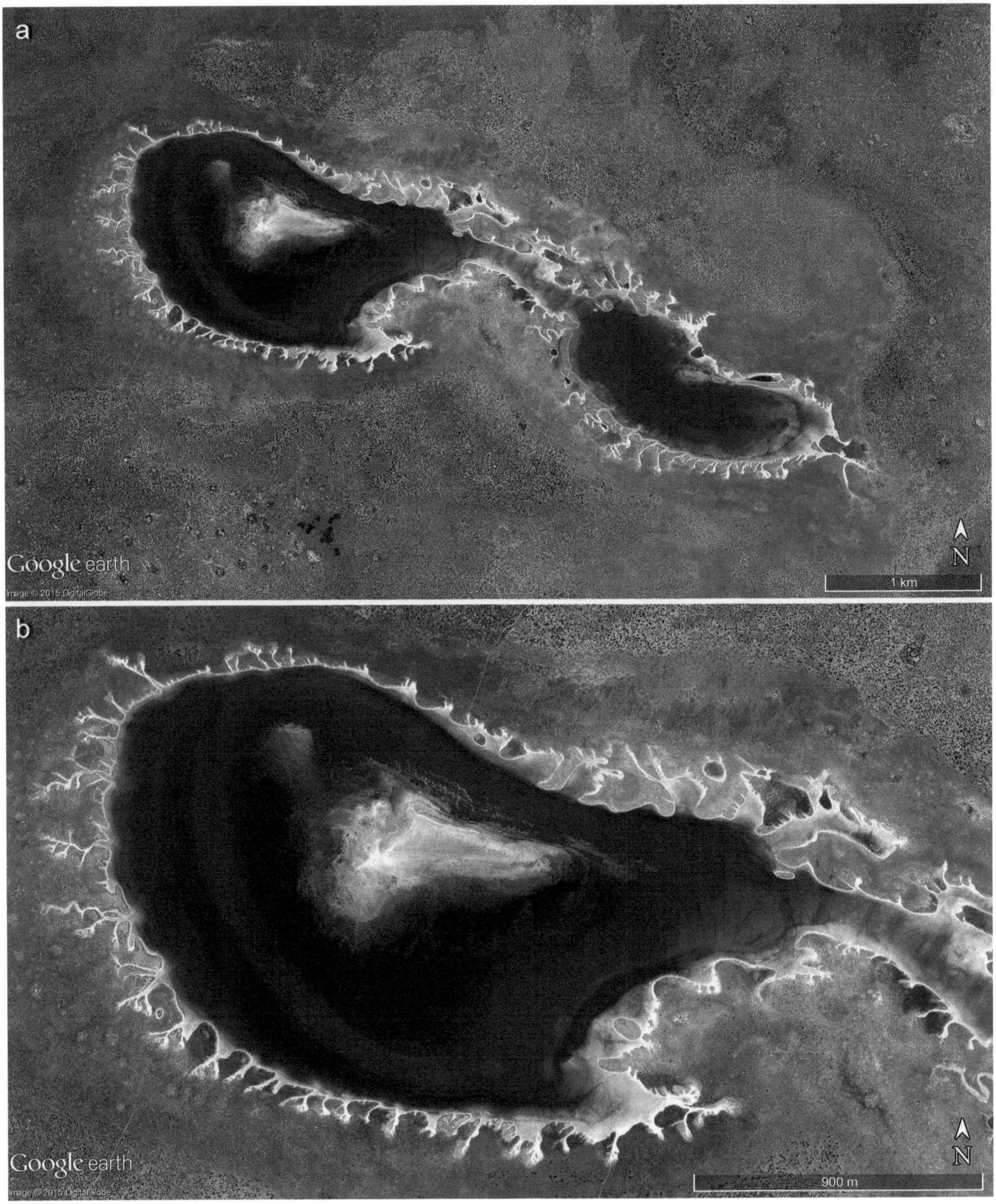

Fig. 4.30 (**a**) Two combined salt lakes (together 5 km long) west of the large Etosha Pan in northern Namibia *(18°34'S, 15°24'E)*. (**b**) The north-westerly part of the two lakes of (a) in detail (Images credits: ©Google earth 2015)

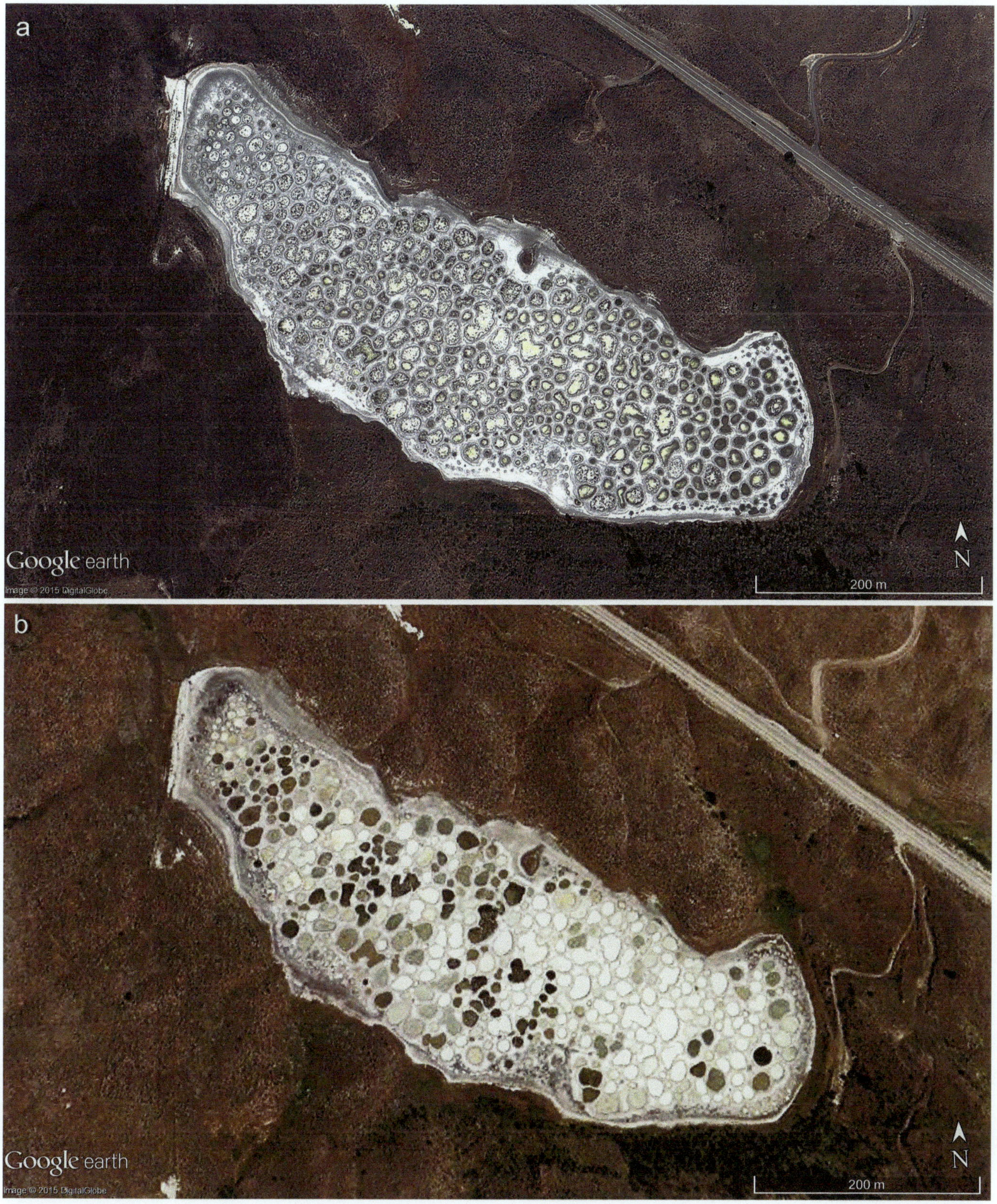

Fig. 4.31 (**a**) One of the commonly named "Spotted Lakes" in British Columbia, Canada *(49°04'40.48"N, 119°33'59.83"W)*. The 700 m long lake exhibits the many different mineral compounds left after water at higher levels evaporated (Image credit: ©Google earth 2015, September 4th, 2010). (b) The same Spotted Lake as in (a) 6 years before (Image credit: ©Google earth 2015, December 31st, 2004)

Fig. 4.32 A 200 m long Spotted Lake in British Columbia, Canada *(48°13'32.92"N, 119°35'38.61"W)* along with other spotted lakes are situated in a glacier-sculptured landscape south of Osoyoos. The lake's internal pattern represents deposits of minerals (Image credit: ©Google earth 2015, September 1st, 2013)

Fig. 4.33 **(a)** A 440 m long Spotted Lake of British Columbia *(48°17'34.20"N, 119°32'29.37"W)* (Image credit: ©Google earth 2015, November 4th, 2011). **(b)** The same Spotted Lake as in (a) after 2 years (Image credit: ©Google earth 2015, September 1st, 2013)

Fig. 4.34 The largest saltpan in South America is Salar de Uyuni in western Bolivia (around *19°56'S, 68°W* at 3662 m asl). Salar de Uyuni is 150 km wide from west to east with an area of 10,500 km^2 and the thickness of its salt crust can extend to several metres. The salt crust lies on a highly concentrated brine of salt with a probable depth to 100 m, containing the elements kalium, magnesium and the rare earth element lithium. This information allows qualitative estimates on the amount of water evaporated and/or to the time of evaporation from former sequential transformation lakes, during Late Pleistocene climates with less evaporation. In this instance, Lake Minchin existed about 30,000–40,000 years ago and later transformed into palaeo-lake Tauca around 10,000 years ago. The older lakes were once much larger and connected to modern Lake Poopo and the saltpan Salar de Coipasa. The flatness of the Altiplano landscape around Salar de Uyuni is demonstrated by the fact that during the wet season Lake Titicaca (that lies 500 km away) may discharge and deliver water into neighbouring Lago Poopo, in turn flooding the Salar de Uyuni. The satellite image presents a view of the saltpan to the northeast; its white surface in the foreground, the smaller Salar de Coipasa (65 km across) at near the same level (3665 m asl), the light blue Lago Poopo to the east-northeast (80 km long at 3698 m asl), Lake Titicaca (3810 m asl) is the black patch to the north and the green area to the top northeast is the Amazon lowlands of Bolivia and Brazil (Image credit: ©Google earth 2013)

Following the impressive saltpans of Salar de Uyuni in Bolivia (Fig. 4.34), Badwater in the Death Valley National Park in California (Fig. 4.35), and the Etosha Pan of northern Namibia in southwest Africa (Fig. 4.36), simple saltpan forms are now discussed (Figs. 4.37, 4.38, 4.39, 4.40, 4.41 and 4.42), followed by more complex forms (Figs. 4.43, 4.44, 4.45, 4.46, 4.47, 4.48, 4.49, 4.50, 4.51, 4.52, 4.53, 4.54, 4.55, 4.56, 4.57, 4.58, and 4.59) with changing colours (Figs. 4.60, 4.61, 4.62, 4.63, 4.64, 4.65, and 4.66). River-like contours of saltpans also exist (Figs. 4.67, 4.68, 4.69, 4.70, 4.71, 4.72, 4.73, 4.74, 4.75, 4.76, 4.77, and 4.78) and several even exhibit contours resembling "insect-antennas" located in small gullies that deliver water seeping from excess groundwater at the end of a rainy period. Internal patterns of salt crusts, mostly in the form of polygons are also documented (Figs. 4.79, 4.80, 4.81, 4.82, 4.83, 4.84, and 4.85).

Fig. 4.35 The large saltpan of "Badwater" in the southern Death Valley National Park in California, USA (approximately *36°14'N, 116°49'W*) rests 86 m below sea level, representing the lowest point in North America. Spanning 8 km across from east to west, the size of the saltpan is almost 70 km^2, yet is a remnant of the much larger Pleistocene Lake Manly that left old strandlines of salt crust up to 1.5 m thick on mud deposits, along the slopes of Death Valley (Image credit: ©Google earth 2015)

Fig. 4.36 Etosha saltpan in northern Namibia (around *18°49'S, 16°20'E* at 1085 m asl.) covers an area over 4000 km^2 and is part of the Kalahari Basin, a remnant of a far greater lake in past cooler climates. Scene is 140 km wide (Image credit: ©Google earth 2013).

Fig. 4.37 A 5.2 km long saltpan fills a void within a linear dune system in Namibia, southwest Africa *(23°48'S, 17°49'E)* (Image credit: ©Google earth 2015)

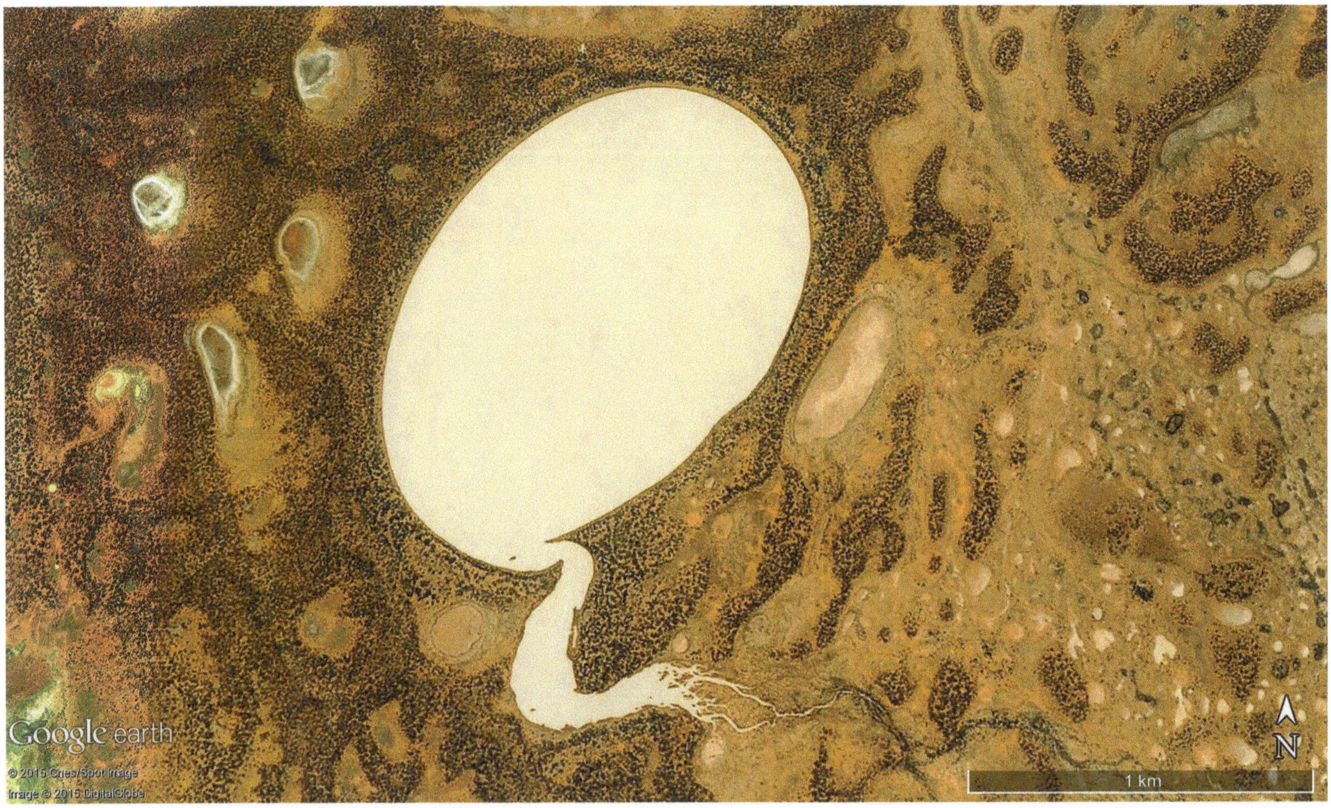

Fig. 4.38 A perfect oval saltpan formation (1.7 km across) from the southern part of the former Lake Carnegie in western Australia *(26°18'04.50"S, 122°25'04.63"E)* (Image credit: ©Google earth 2015)

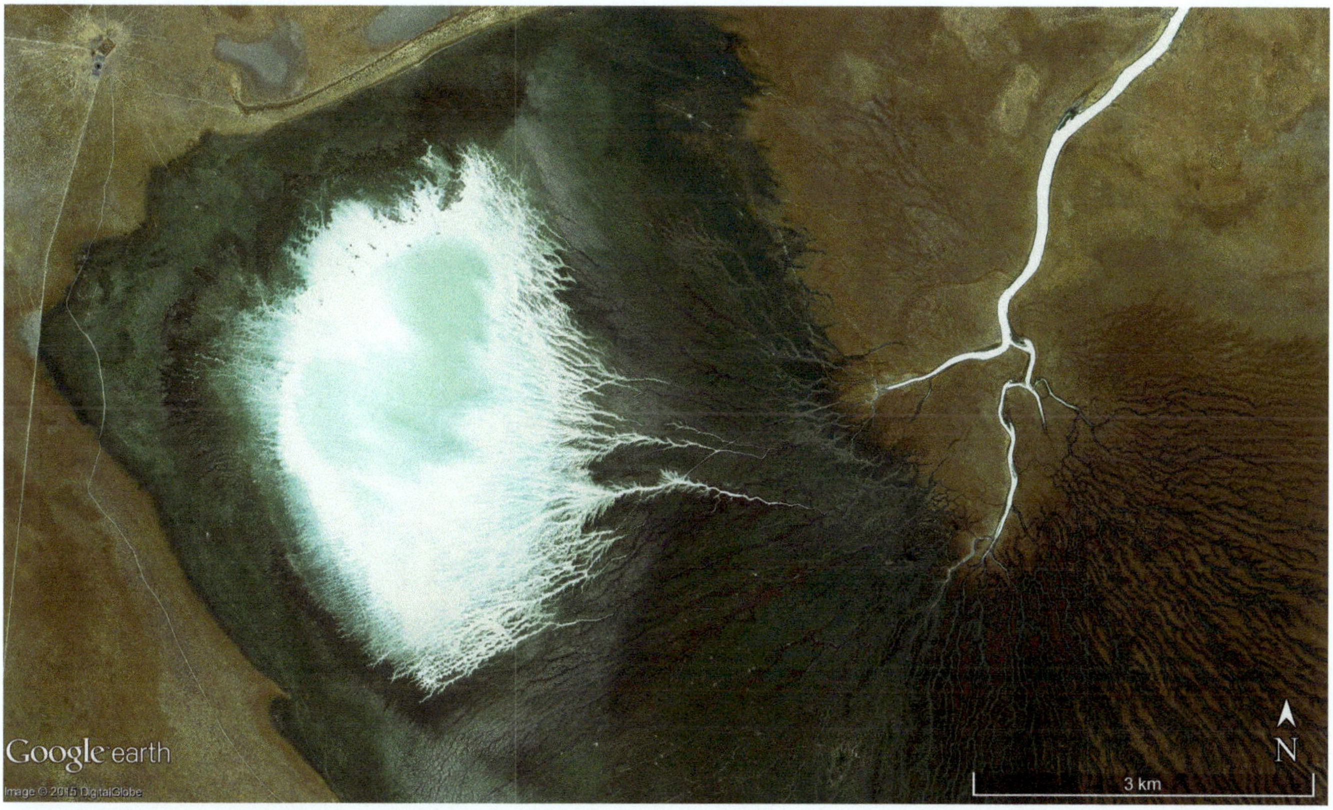

Fig. 4.39 The Corella Lake fed from the northeast in Northern Territory, Australia *(18°36'40.87"S, 135°35'34.76"E)* Scene is 10 km wide (Image credit: ©Google earth 2015)

Fig. 4.40 A 8 km wide saltpan with several working salt mines, called "Laguna Secca" in the Antofagasta region of northern Chile (approximately *24°25'S, 69°07'W* at 2900 m asl) (Image credit: ©Google earth 2015)

Fig. 4.41 The 6 km wide saltpan of Lake Gilles in South Australia *(33°01'S, 136°37'E)* was previously a depression of a dune system moving to the east (Image credit: ©Google earth 2015)

Fig. 4.42 **(a)** One of many saltpans in the steppe of West Kazakhstan is 2.8 km long *(49°02'N, 48°59'E* at 8 m below sea level). **(b)** Another saltpan in southwest Kazakhstan *(48°47'N, 49°42'E* at 13 m below sea level). Scene is 860 m wide (Images credits: ©Google earth 2015)

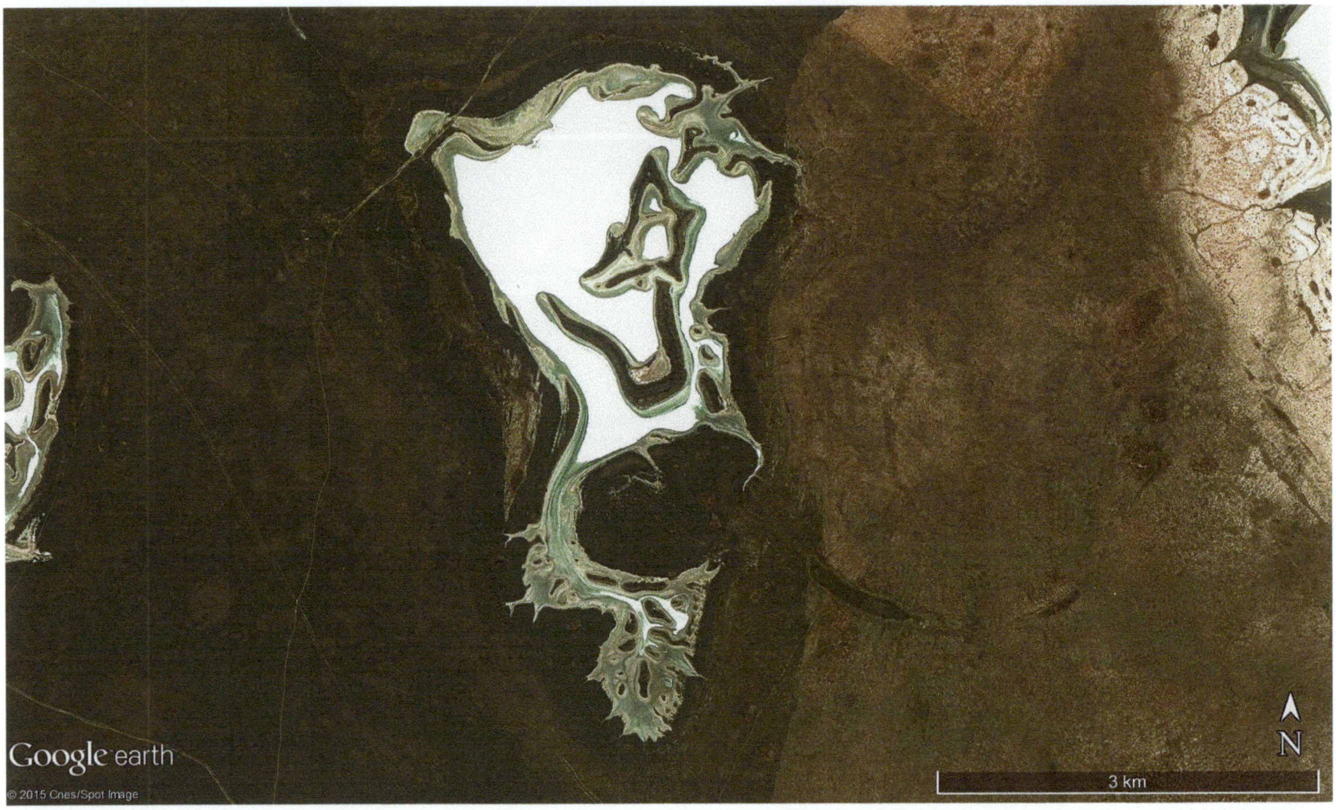

Fig. 4.43 A large saltpan in south Kazakhstan *(49°05'07.99"N, 49°09'54.18"E)*. Scene is 10 km wide (Image credit: ©Google earth 2015)

Fig. 4.44 The 35 km long Lake Rebecca in Western Australia, now a saltpan at around *29°53'S, 122°09'E* (Image credit: ©Google earth 2015)

Fig. 4.45 Laguna Busch in southern Bolivia *(22°38'S, 67°10'W* at 4542 m asl). A series of beach ridges describe previous large eastward expansions of the saltpan. Scene is 4 km wide (Image credit: ©Google earth 2015)

Fig. 4.46 (**a**) Salar de Incahuasi in northwest Argentina *(24°14'S, 67°35'W* at 3473 m asl). The Salar stretches 25 km from north to south. (**b**) A detailed view from the northwest section of Salar de Incahuasi in Argentina (a). Ephemeral springs deliver waters to far regions on the Salar flat and subsequently evaporates into salt crusts. Scene is 7 km wide. (c) Detail from the east flank of Salar de Incahuasi in Argentina (a). A distinct salt border marks a former water level. Scene is 7 km wide (Images credits: ©Google earth 2015)

Fig. 4.46 (continued)

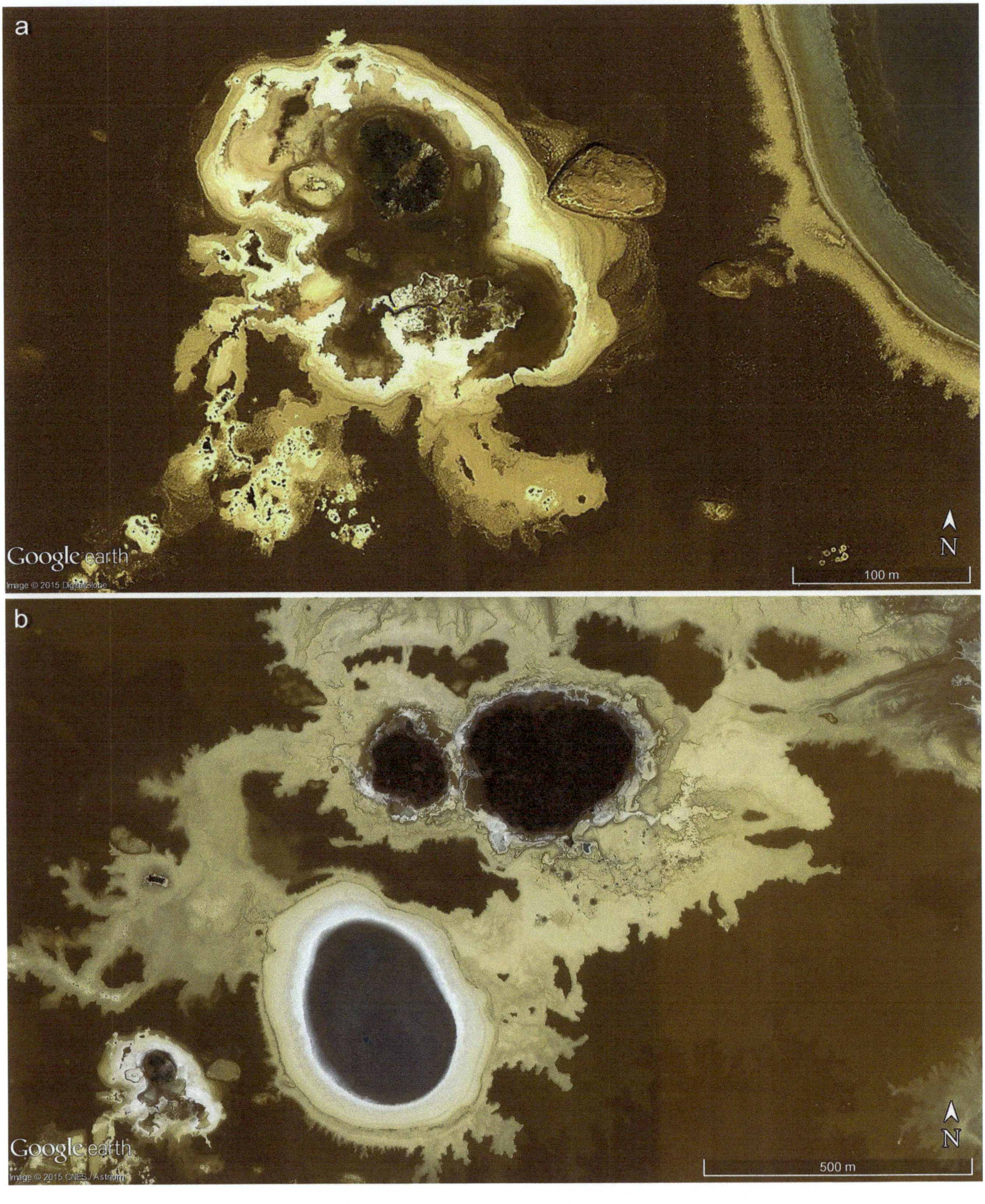

Fig. 4.47 (**a**) A decorative saltpan in the wide Siwa Oasis of Egypt *(29°10'N, 25°44'E* at 17 m below sea level). Fresh water input rises from a groundwater aquifer in this part of the Qatarra Depression near the border to Libya. Scene is 600 m wide. (b) A salt lake and a saltpan pattern in the Siwa Oasis of northwest Egypt *(29°12'38.61"N, 25°40'30.82"E* at 17 m below sea level). Scene is 1.8 km wide (Images credits: ©Google earth 2015)

Fig. 4.48 Fed by freshwater springs a saltpan is part of the Lago Verde region in the Antofagasta district of northern Chile *(21°21'S, 68°22'W* at 3699 m asl) (Image credit: ©Google earth 2015)

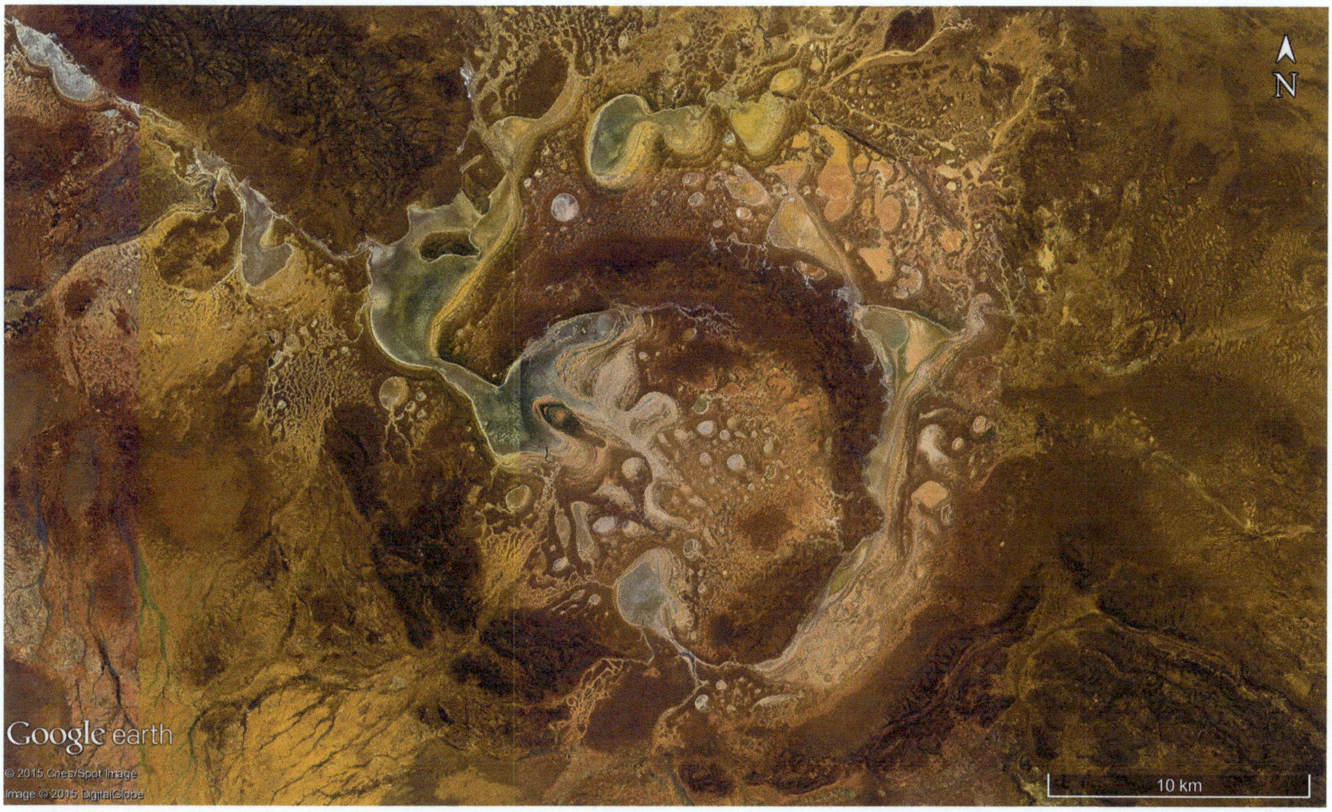

Fig. 4.49 Fed on three flanks by ephemeral creeks during thunderstorms, the origin of the circular round patterns to these saltpans remains unknown (impact structure?). This 22 km wide scene near Lake Carnegie in Western Australia *(25°57'S, 123°08'E)* is 220 km east of the impact structure known as Shoemaker crater (Image credit: ©Google earth 2013)

Fig. 4.50 The southern section of the Seletyteniz saltpans in Kazakhstan *(53°14'N, 72°54'E)* lies within inactive dunes. Scene is 32 km wide (Image credit: ©Google earth 2015)

Fig. 4.51 Saltpans and swamps are situated in the steppe of Kazakhstan at *(53°25'N, 78°10'E)*. Scene is 3 km wide (Image credit: ©Google earth 2015)

Fig. 4.52 Strong winds have removed salt sediments from saltpans in Kazakhstan *(51°43'N, 79°47'E)* to form depressions (salt blowouts), which are situated amongst inactive dune fields. Scene is 36 km wide (Image credit: ©Google earth 2015)

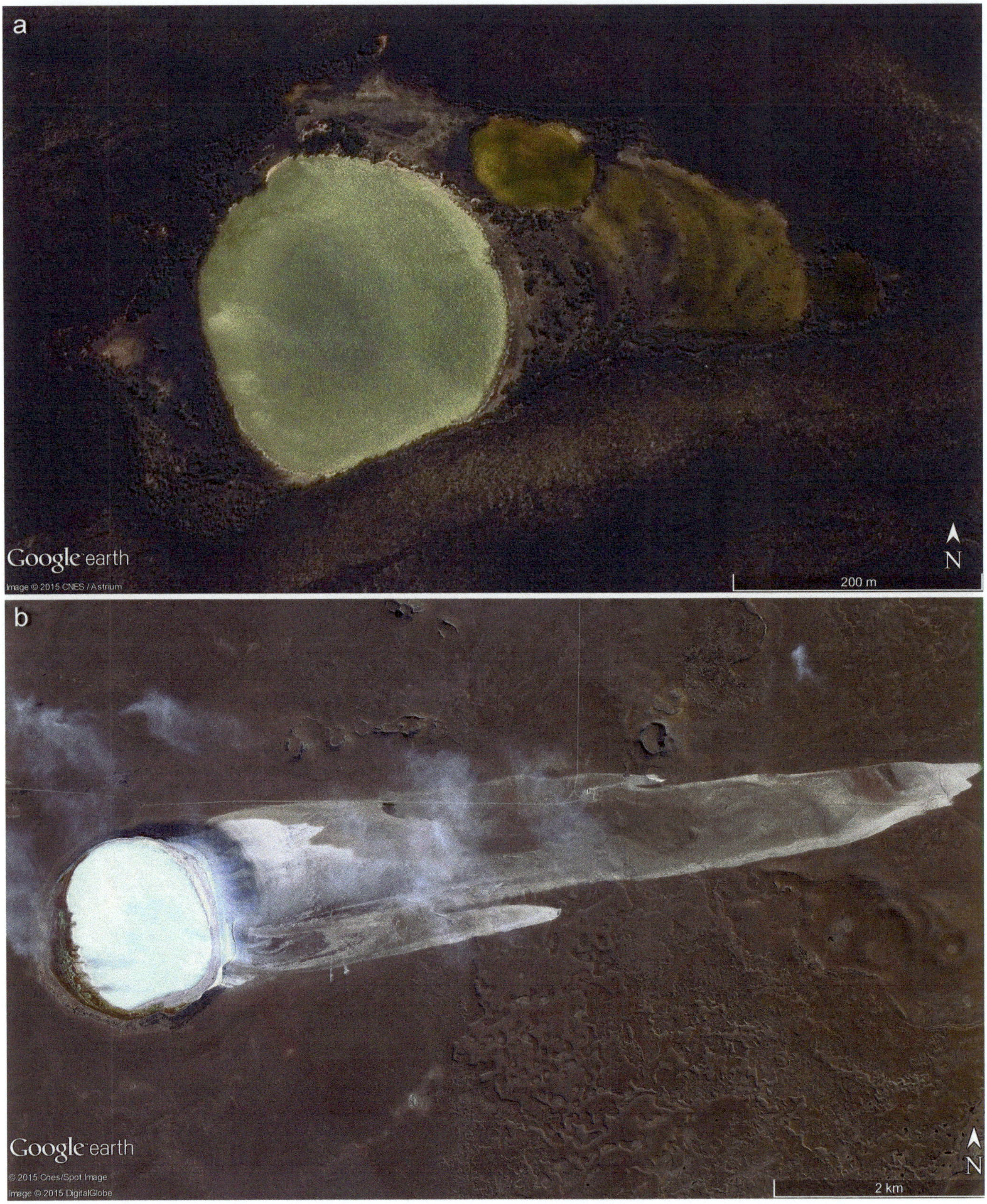

Fig. 4.53 (**a**) A 300 m wide saltpan from a large group of similar formations in the eastern part of South Africa's Cape Province *(33°50'S, 123°31'E)*. The westerly winds have drifted salt into the vegetation eastward and left bare flats behind. The total length of the aeolian feature is 630 m. (**b**) Very strong and steady westerly winds created the elongated salt blowouts from Laguna Bismarck (1.2 km pan diameter) in eastern Patagonia Argentina *(51°52'S, 69°41'W)*. The total feature is 8 km long (Images credits: ©Google earth 2015)

Fig. 4.54 **(a)** Former small rivers are now chains of tiny saltpans that lead to a 7.3 km long depression (Lake Malata) in South Australia *(34°11'S, 135°33'E)*. Lake Malata has a covering of salt and exhibits several beach ridges developed by westerly storm wind processes on a former lake. **(b)** Saltpans replace a former wide river course viewed from the northeast of Lake Malata *(34°11'S, 135°33'E)* in South Australia. Scene is 5.3 km wide (Images credits: ©Google earth 2015)

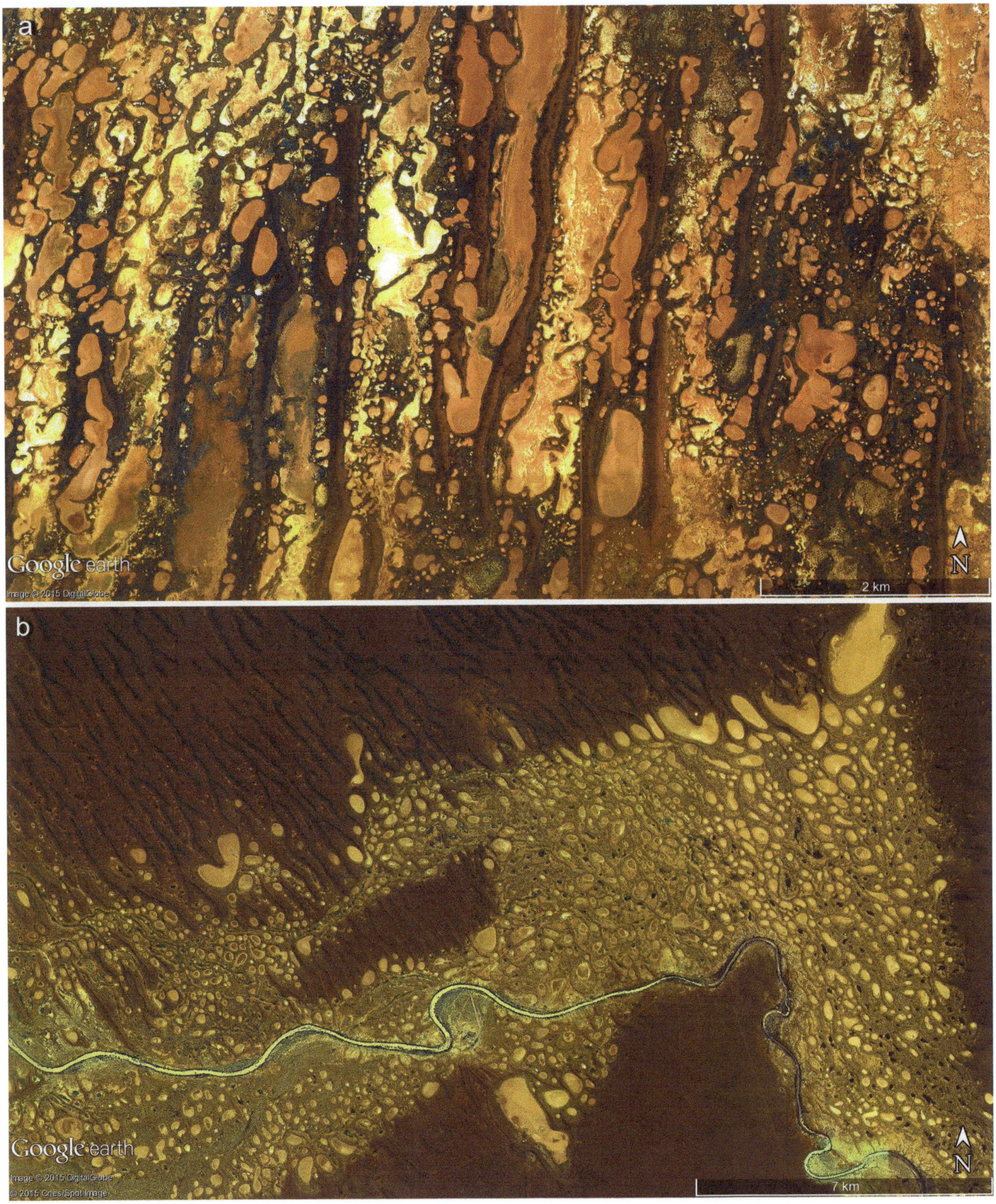

Fig. 4.55 **(a)** In many regions of arid Australia (particularly in the west) are vast fields of elongated saltpans mostly positioned in between relic Pleistocene dunes. Their origin is under debate and could explain depressions formed by vortices in the course of strong flash floods. The elongated saltpan shown is located near Onslow, Western Australia *(22°05'S, 114°46'E)*. Scene is 10 km wide **(b)** The Wooramel riverbed in Western Australia *(25°38'S, 114°38'E)* mostly carries an east to west flowing river that crosses relic dune systems running north-northwest to south-southeast. The arrangement of the salt/silt pans along the riverbed point to an influence of strong floods creating these configurations and

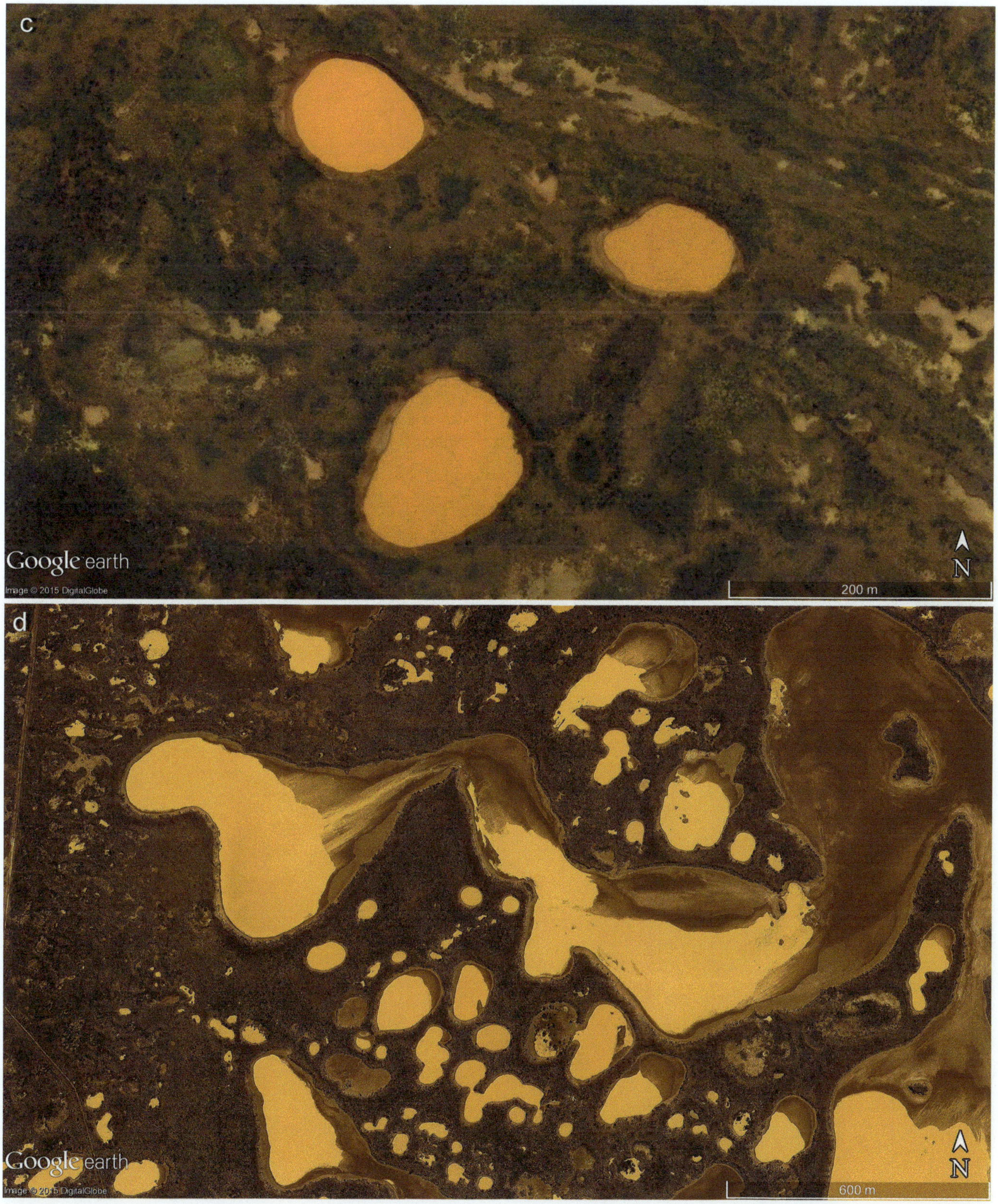

Fig. 4.55 (continued) are comparable to (unconnected) valley patterns in other parts of Western Australia (see a). Scene is 23 km. **(c)** Along the coastline of Western Australia *(21°36'46.11"S, 115°20'00.31"E)* a wide belt of inactive dunes are interrupted by smooth-edged circular basins of many different sizes. Following rainfall, muddy waters fill the basins and influence colours resutant of intense bedrock (sandstone) weathering. The origin of the basins is under debate. Scene is 760 m wide. **(d)** Alternative shapes and colours of mud pans found in old dune fields of Western Australia *(21°49'S, 114°55'E)*. Scene is 1.9 km wide (Images credits: ©Google earth 2015)

Fig. 4.56 Lake Carnegie is an ephemeral lake and saltpan in Western Australia *(26°16'S, 122°40'E)* and during flash floods from strong monsoonal rains, fills an area exceeding 5000 km^2. Scene is 80 km wide (Image credit: ©Google earth 2015)

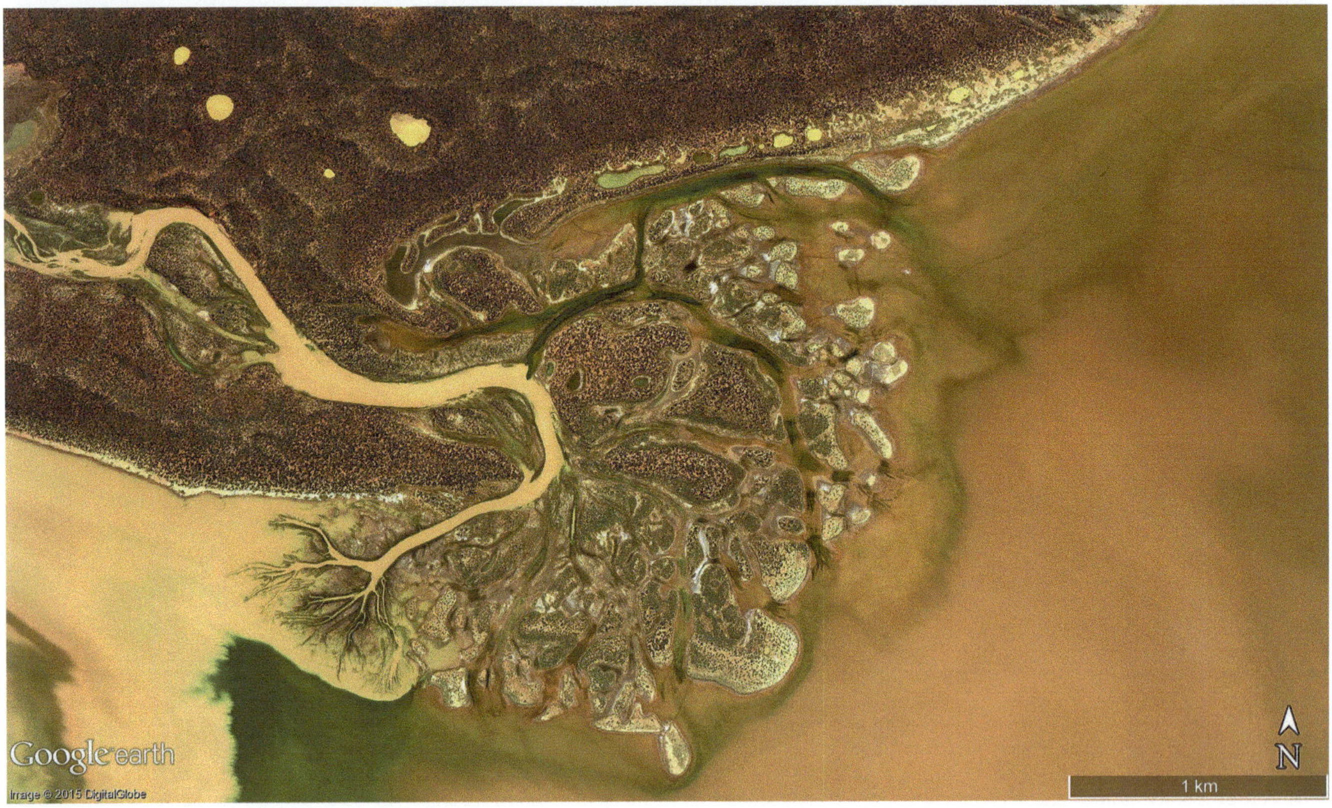

Fig. 4.57 A delta formation evident in the northwest section of Lake Carnegie in Western Australia *(26°05'35.18"S, 122°09'33.20"E)*. Scene is 5.5 km wide (Image credit: ©Google earth 2015)

Fig. 4.58 A detailed western section (compare Fig. 4.56) of the 70 km wide Lake Carnegie (as a giant saltpan) in Western Australia *(26°06'S, 122°08'E)* exhibiting episodic inflow seen by the delta in the northeast of the image. Approximately 5 km wide scene (Image credit: ©Google earth 2015)

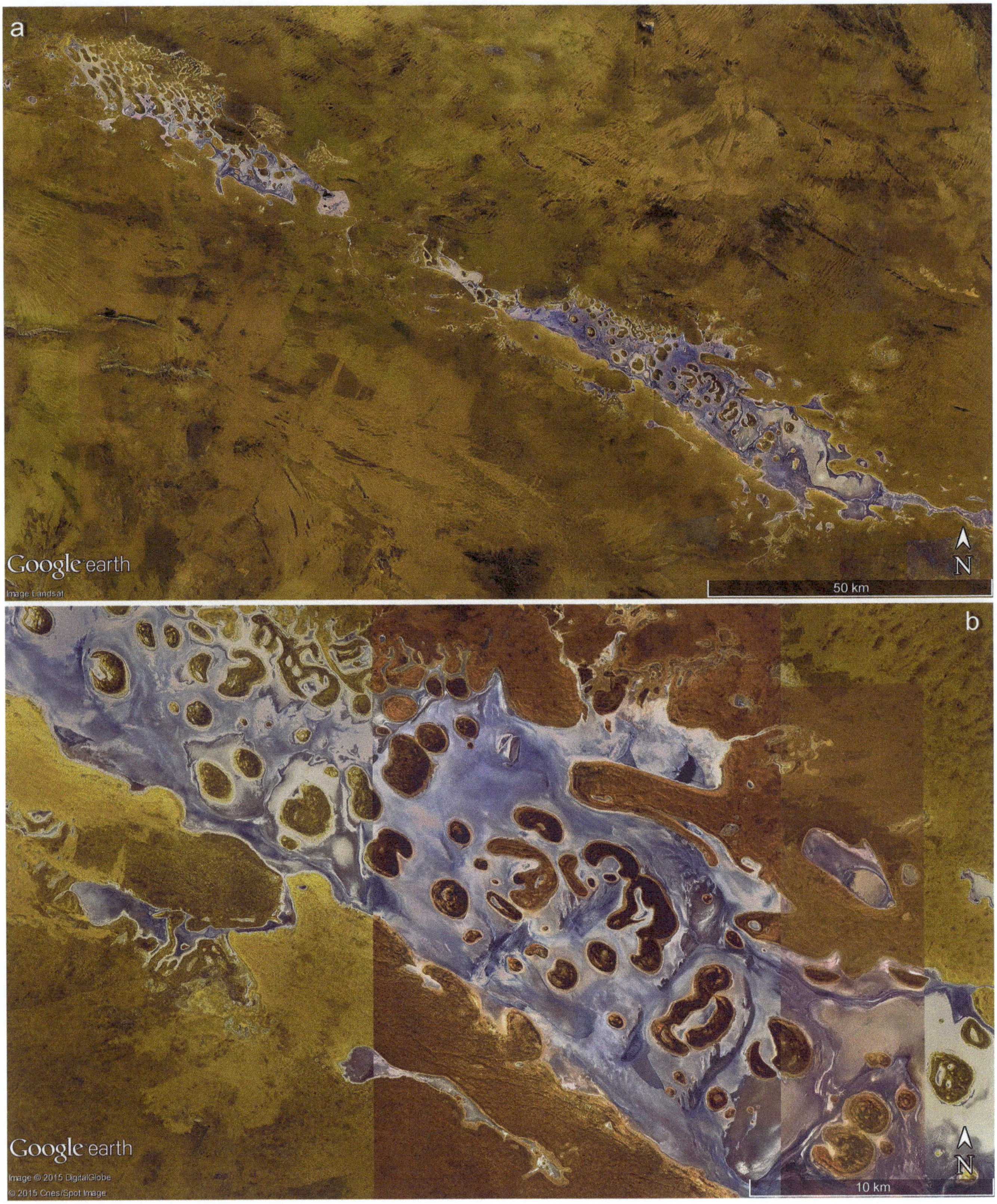

Fig. 4.59 **(a)** The 190 km long Lake Amadeus in the Northern Territory, Australia (approximately *24°35'S, 130°39'E)* is part of a palaeo-river system (Image credit: ©Google earth 2013). **(b)** A 47 km section of (a) in the southern section of the Northern Territory, Australia (approximately *24°45'S, 130°53'E).* Lake Amadeus is a mixture of saltpans and remnant dunes where as a lake may expand outward 190 × 30 km covering areas around 1000 km^2, and although fresh water from discharge and rain feeds the many springs throughout the area, most evaporates (Images credits: ©Google earth 2015)

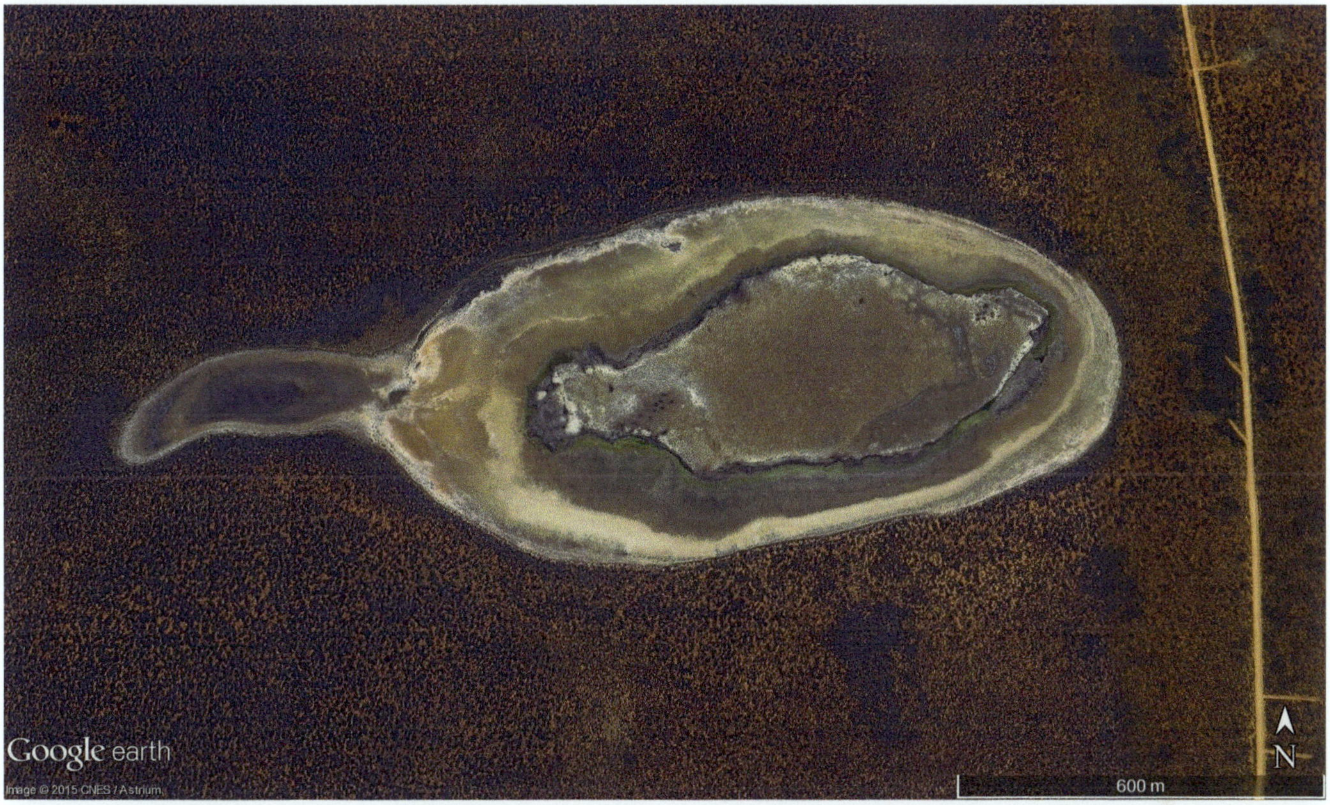

Fig. 4.60 The 1.8 km colourful saltpan is a deflation depression in the Shark Bay region of Western Australia *(25°52'52.41"S, 113°32'25.69"E)*. The colour arrives from a long lasting weathering of rocks, soil formation and mobilization of the element iron, which stains all landscapes with different intensities of red (Image credit: ©Google earth 2015)

Fig. 4.61 A saltpan of Lag de Paraguaya, east of the large Laguna de Epecuén in the Buenos Aires Province of Argentina *(37°06'S, 62°47'W)*. Scene is 5.4 km wide (Image credit: ©Google earth 2015). Interestingly, Laguna de Epecuén is the lake that drowned a tourist village where a long-term weather event swelled the waters and submerged the 50-year-old resort town to 10 m below the lake level. The wet weather reversed in 2009 and has now exposed the ruins of Villa Epecuén (the new tourist ghost-village)

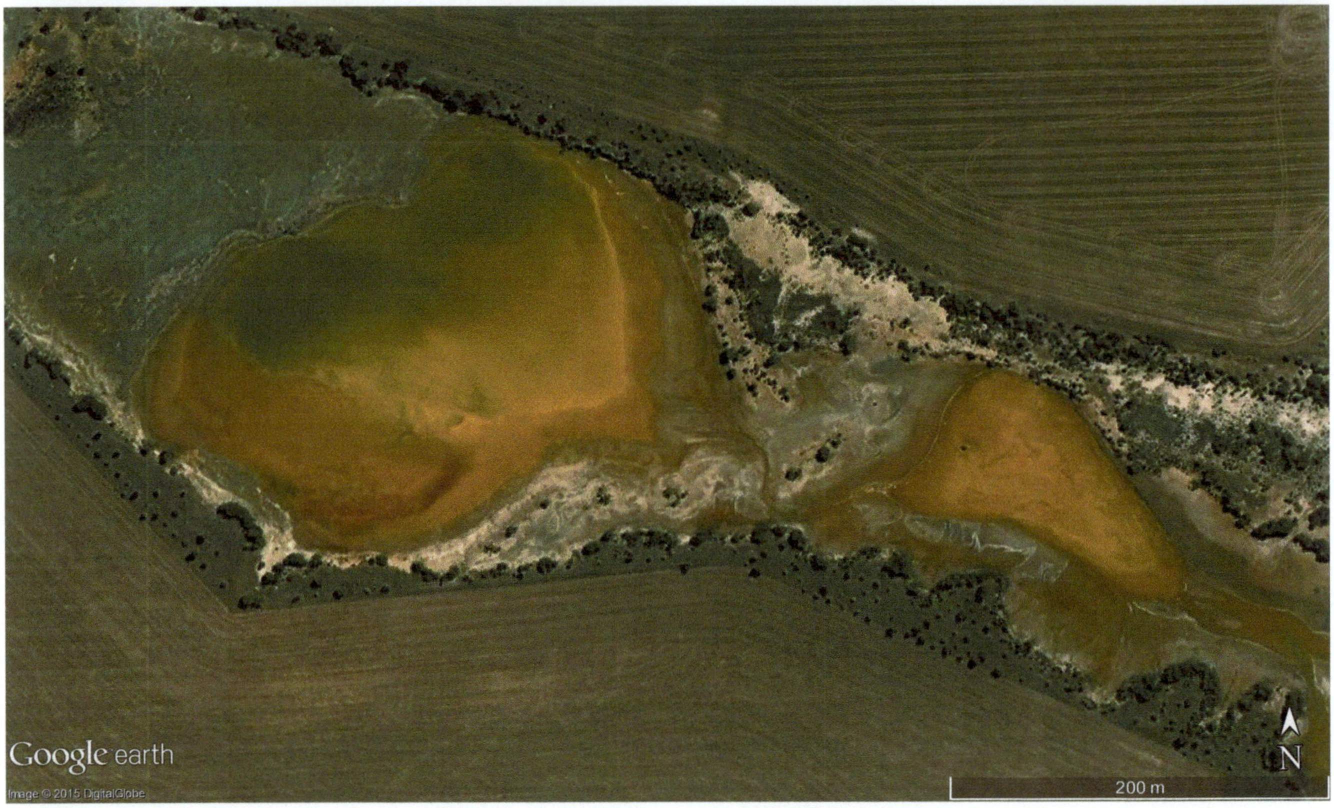

Fig. 4.62 A 660 m long section of a depression in South Australia *(34°12'S, 135°32'E)* exhibits salt evaporation and is partly coloured by bacteria, east of Lake Malata (Image credit: ©Google earth 2015)

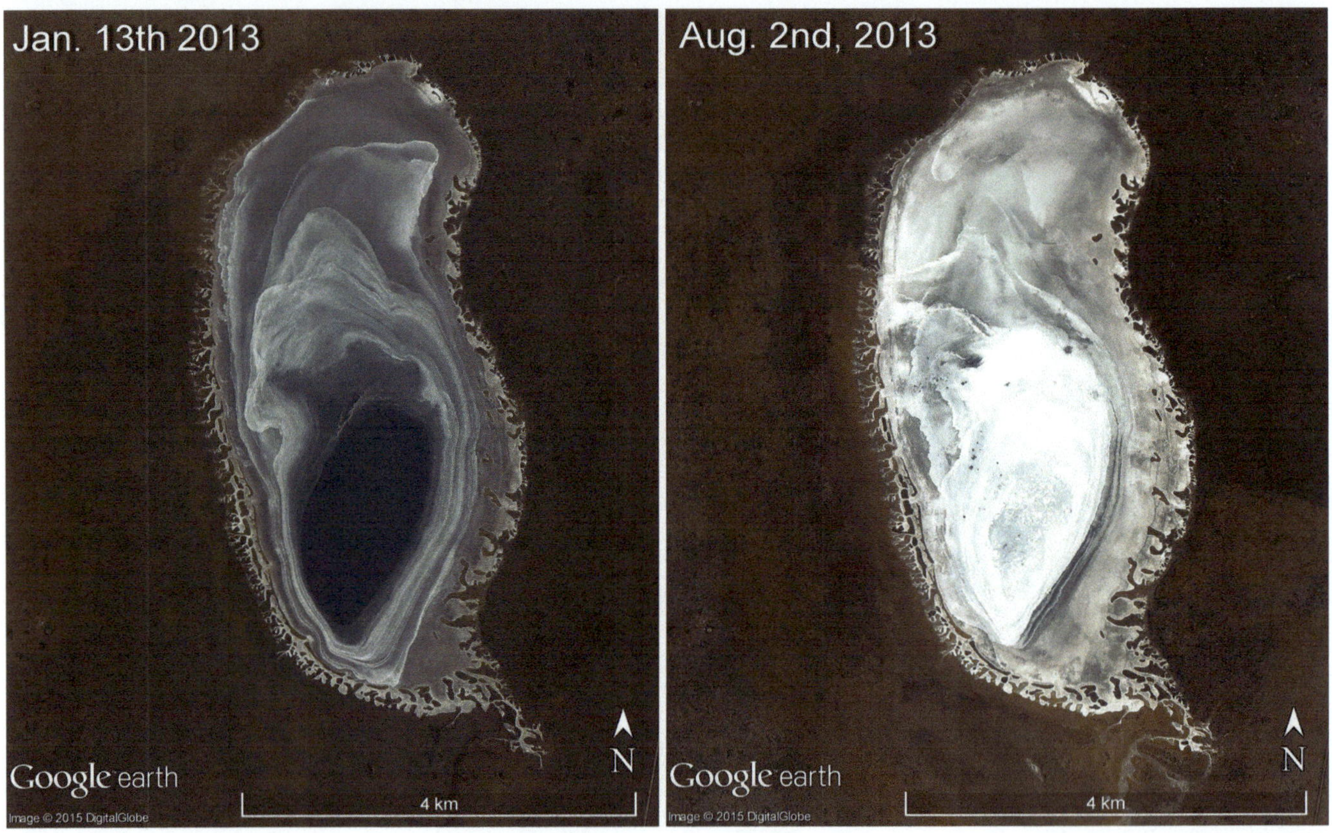

Fig. 4.63 Saltpans in various stages (7 months apart) of filling, in the Etosha region *(18°32'21.42"S, 15°37'19.02"E* at 1089 m asl). Scene 12.5 km wide (Images credits: ©Google earth 2015)

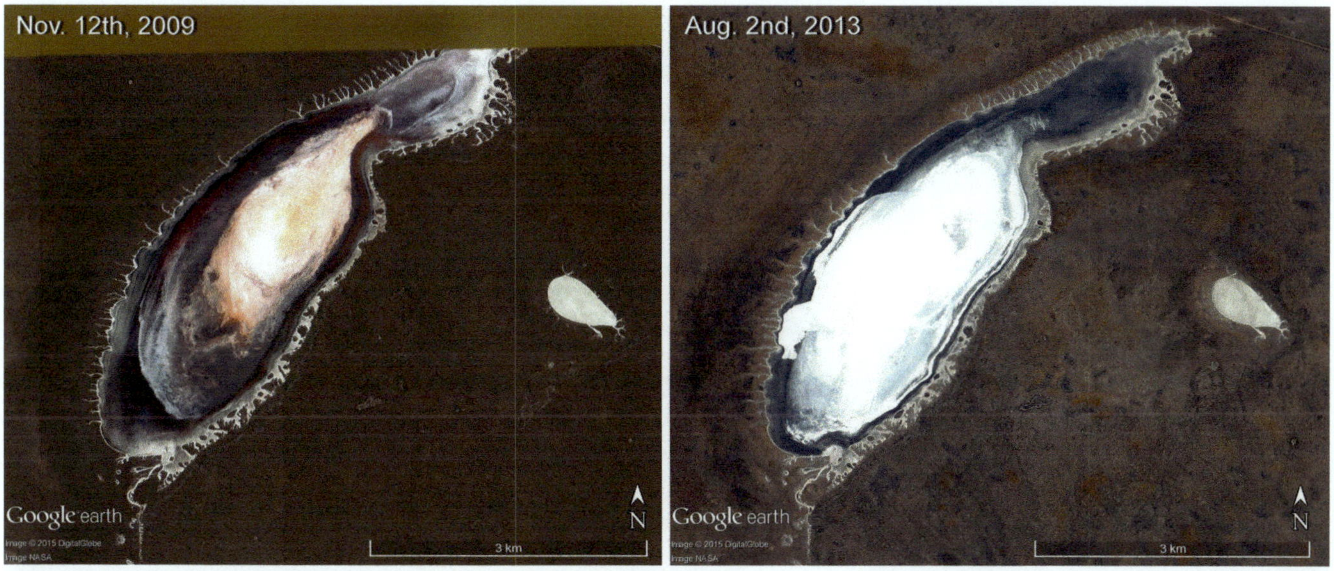

Fig. 4.64 A saltpan achieves distinctive coloration through various stages of evaporation after the last water infill of distant rains in the Etosha region *(18°39'54.70"S, 15°28'11.05"E* at 1095 m asl) Scene is 6.5 km wide (Images credits: ©Google earth 2015)

Fig. 4.65 A saltpan shown in three different time periods presenting distinctive colours and various water levels in Victoria Australia *(38°04'23.50"S, 143°39'31.65"E)* Scene is 1 km wide (Images credits: ©Google earth 2015)

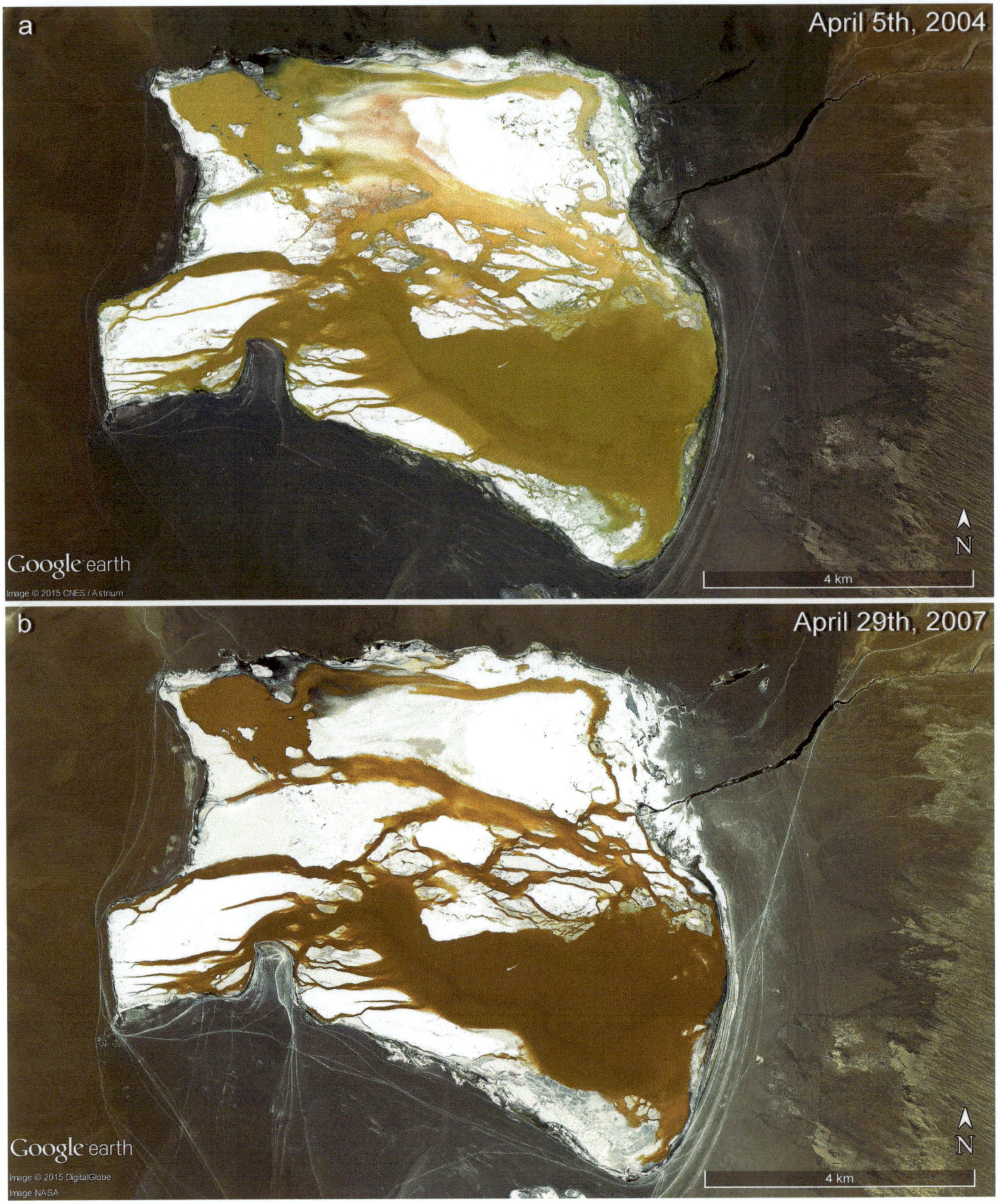

Fig. 4.66 (**a**) Laguna Colorada in southwest Bolivia *(22°12'12.69"S, 67°47'07.01"W* at 4293 m asl and max. 10 km wide). (**b**) Laguna Colorada (max. 10 km wide) in southwest Bolivia (refer to (a) after 3 years. (**c**) Laguna Colorada (max. 10 km wide) in southwest Bolivia (refer to (b) after 6 years (Images credits: ©Google earth 2015)

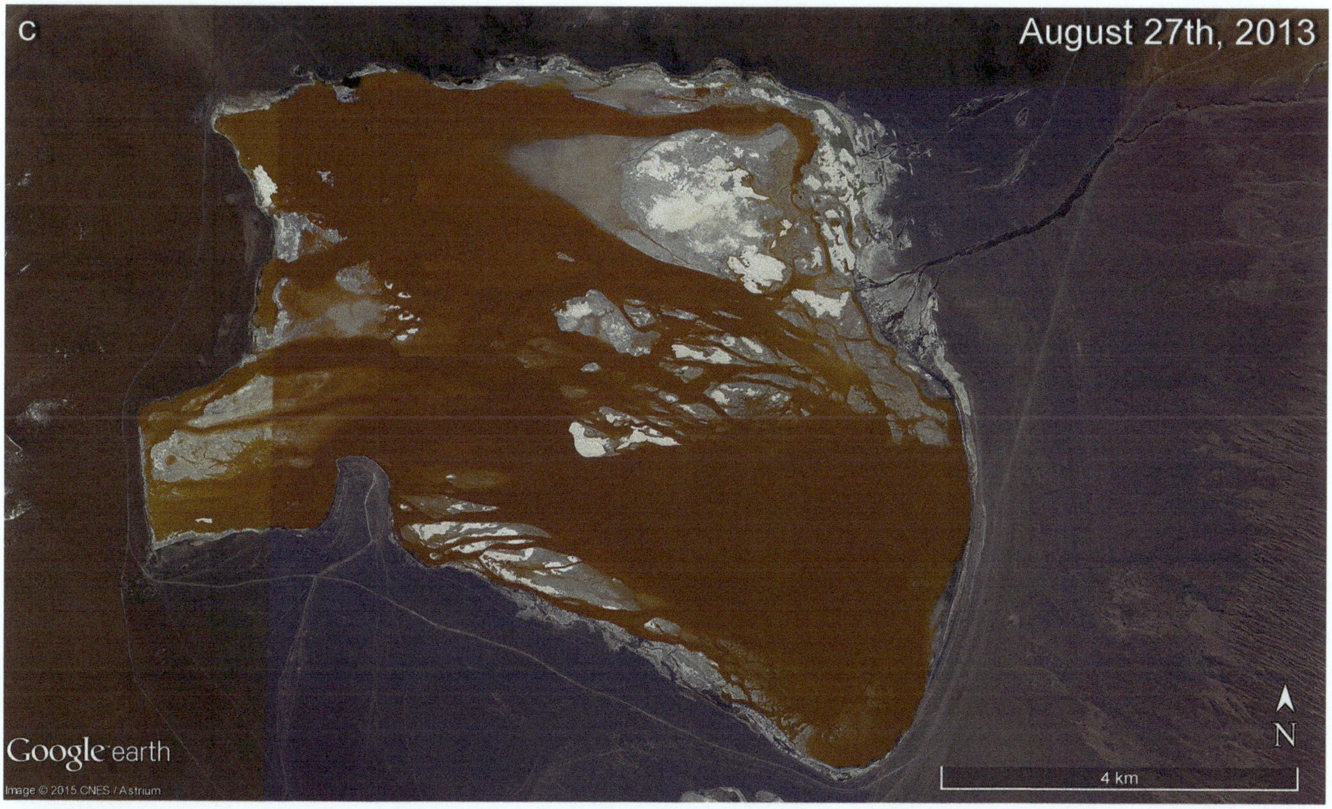

Fig. 4.66 (continued)

Fig. 4.67 An aged riverbed and surrounding dunes create an interesting saltpan formation east of Lake Gairdner in the South Australia State *(31°58'S, 136°23'E)*. Scene is 3 km wide (Image credit: ©Google earth 2015)

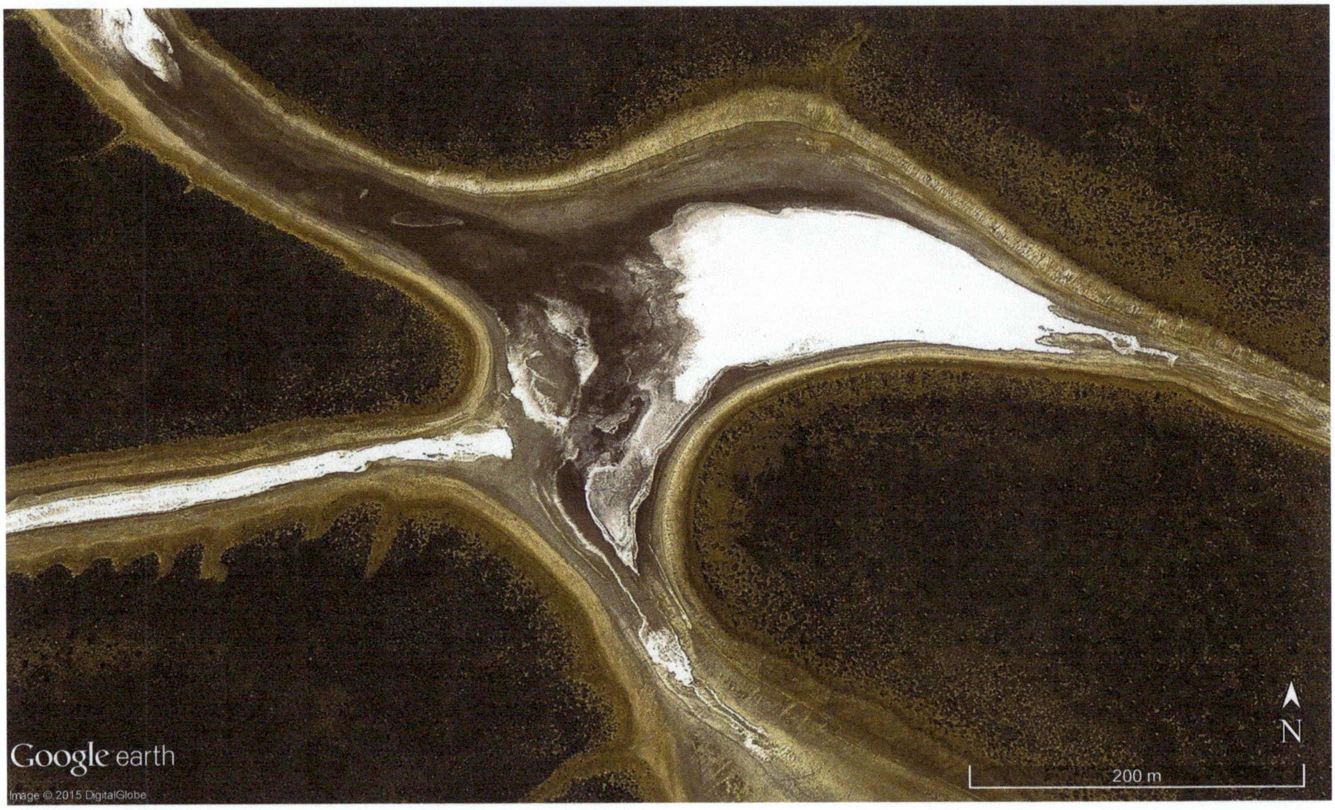

Fig. 4.68 Tidal creeks often create saltpans in the east of Onslow, Western Australia *(21°40'S, 115°14'E)*. Scene is 1 km wide (Image credit: ©Google earth 2015)

Fig. 4.69 Salt crusts border a southern section of Algodones Lagoon near the Gulf of Mexico in west Mexico *(24°20'N, 97°49'W)*. Scene is 2.6 km wide (Image credit: ©Google earth 2015)

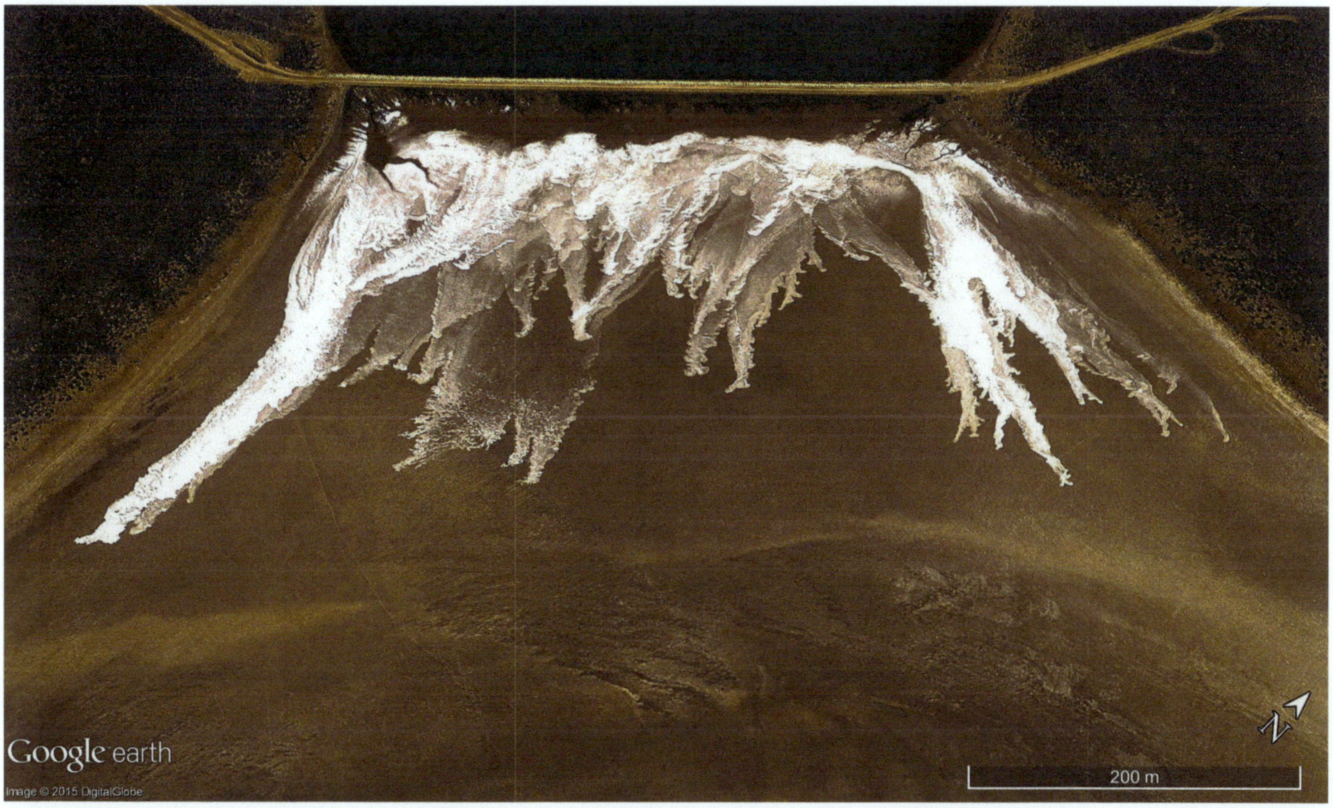

Fig. 4.70 The salt cover on the red soil deposits demonstrate evaporated wind-driven over-wash of marine saltwater near Onslow, Western Australia *(21°39'10.53"S, 115°14'37.81"E)*. Scene is 800 m wide (Image credit: ©Google earth 2015)

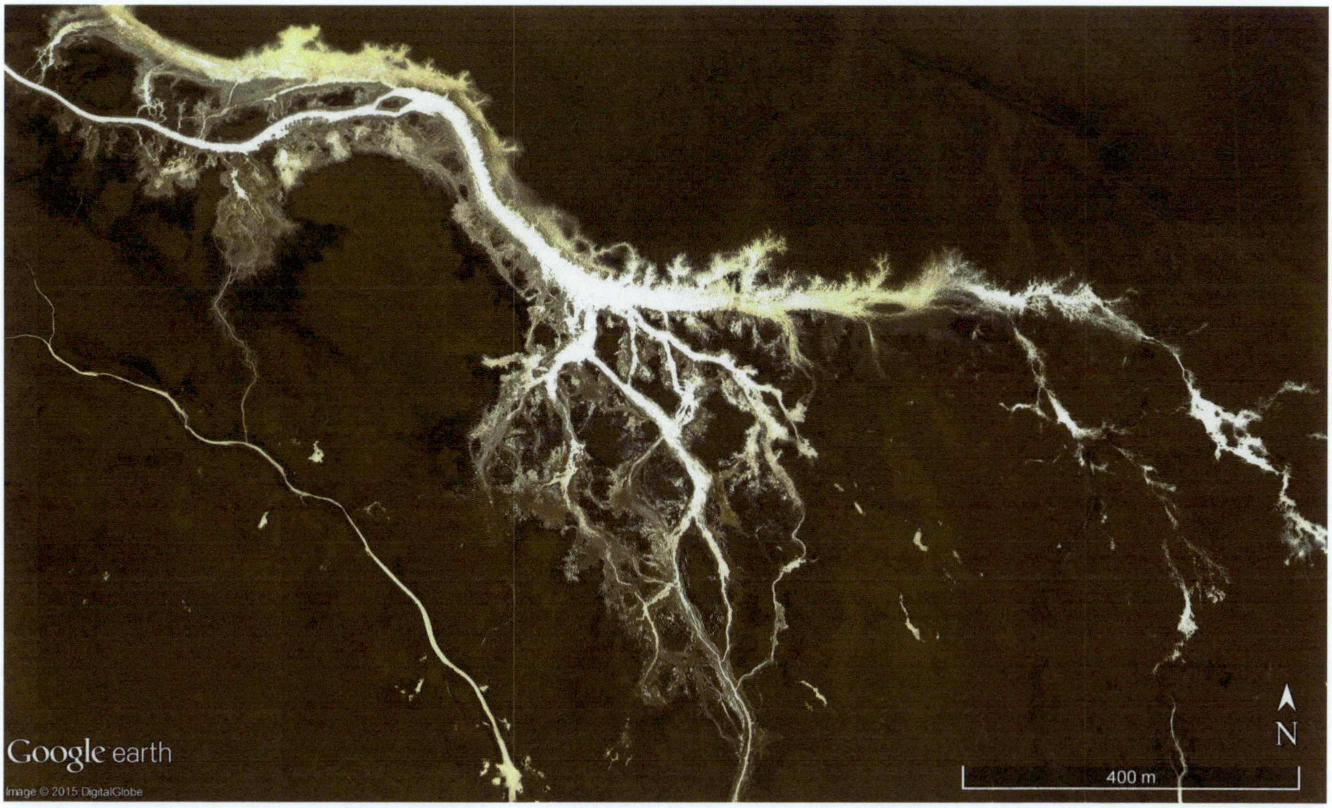

Fig. 4.71 Meltwater creeks evaporate into pure salt in the eastern Chilean Andes *(23°20'S, 68°16'W)*. Scene is 1.5 km wide (Image credit: ©Google earth 2015)

Fig. 4.72 A saltpan is partly surrounded by branching arms of salt (shallow gullies) from the surrounding slopes in Kazakhstan *(49°17'26.52"N, 48°38'07.51"E)*. Scene is 7.6 km wide (Image credit: ©Google earth 2015)

Fig. 4.73 A 2.8 km long saltpan surrounded with interesting forms resembling insect antennas in the Etosha region, Namibia *(18°36'33.11"S, 15°32'44.55"E)* (Image credit: ©Google earth 2015)

Fig. 4.74 A colourful centre of a saltpan caused by beta-carotene producing micro-organisms surrounded on the periphery with tiny gullies of antenna-like arms of salt from the Etosha region, Namibia *(18°36'24.52"S, 15°21'36.60"E* at 1085 m asl). A 2.7 km wide scene (Image credit: ©Google earth 2015)

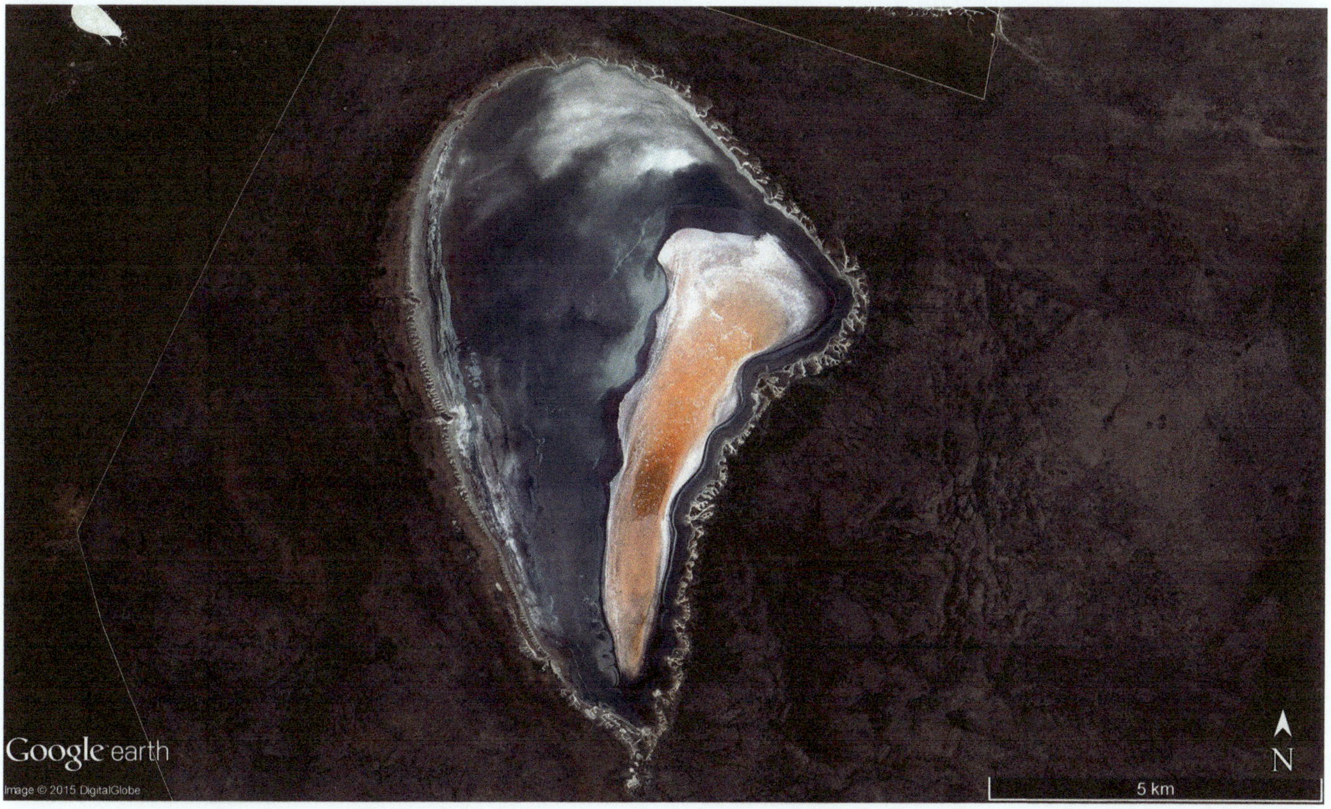

Fig. 4.75 A narrow saltpan coloured by bacteria lies within a larger saltpan (10 km long) and represents a lake of more humid climates in the Etosha region of northern Namibia, southwest Africa *(18°44'S, 15°35'E)* (Image credit: ©Google earth 2015)

Fig. 4.76 A 1 km wide southern section of a saltpan from Fig. 4.75, locally called Otjiwahmda (Image credit: ©Google earth 2015)

Fig. 4.77 A natural "painting" of salt illustrates strange insects shapes from a saltpan in southern Kazakhstan *(49°19'57.25"N, 46°01'57.63"E)*. Scene is 30 km wide (Image credit: ©Google earth 2015)

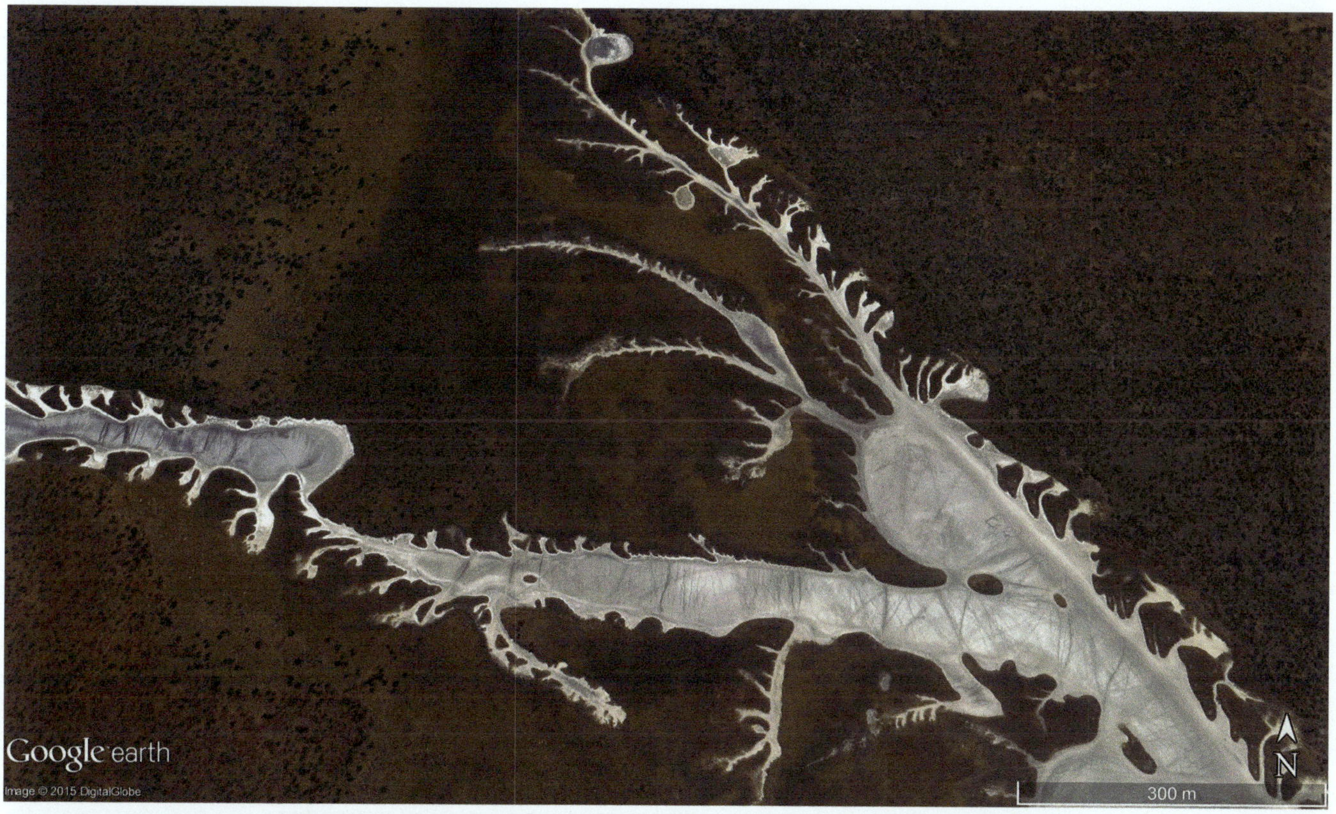

Fig. 4.78 The detail of "insect-antenna-shaped" fringes of saltpans includes tracks of wild beasts etched into the salt crust (grey lines) in the Etosha region of northern Namibia *(18°28'S, 16°02'E)*. Scene is 1.5 km wide (Image credit: ©Google earth 2015)

Fig. 4.79 (**a**) Laguna Kara with a diameter of 4.2 km shows a pattern formed by wind drift of floating salt crusts in southwest Bolivia (*21°54'S, 67°52'W* at 4530 m asl), (Image credit: ©Google earth 2015, July 27th, 2012). (**b**) Detail of Laguna Kara illustrating the central salt crust pattern. Scene is 2 km wide (Image credit: ©Google earth 2015). (c) Detail of the central section of Laguna Kara. Scene is 500 m wide (Image credit: ©Google earth 2015)

Fig. 4.79 (continued)

Fig. 4.80 A colourful algal community flourishes along a section of the northeast flank of the massive Potosi saltpan covering over an area of 20,000 km^2 in south-western Bolivia *(19°46'S, 67°17'W* at 3666 m asl). Scene is 2.3 km wide (Image credit: ©Google earth 2015)

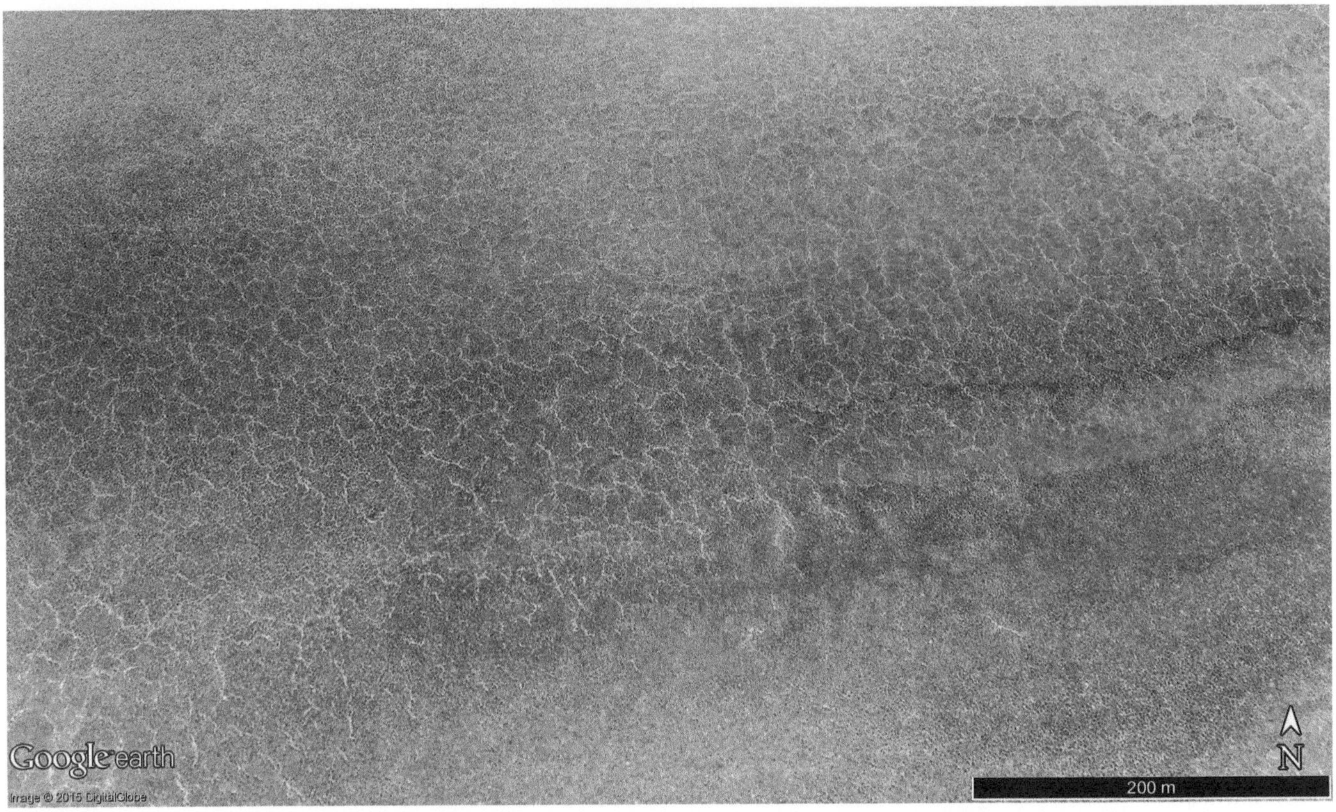

Fig. 4.81 Delicate polygonal patterning of distinctive algae and bacteria in a saltpan from Iran *(33°14'N, 56° 54'E)*. Scene is 860 m wide (Image credit: ©Google earth 2015)

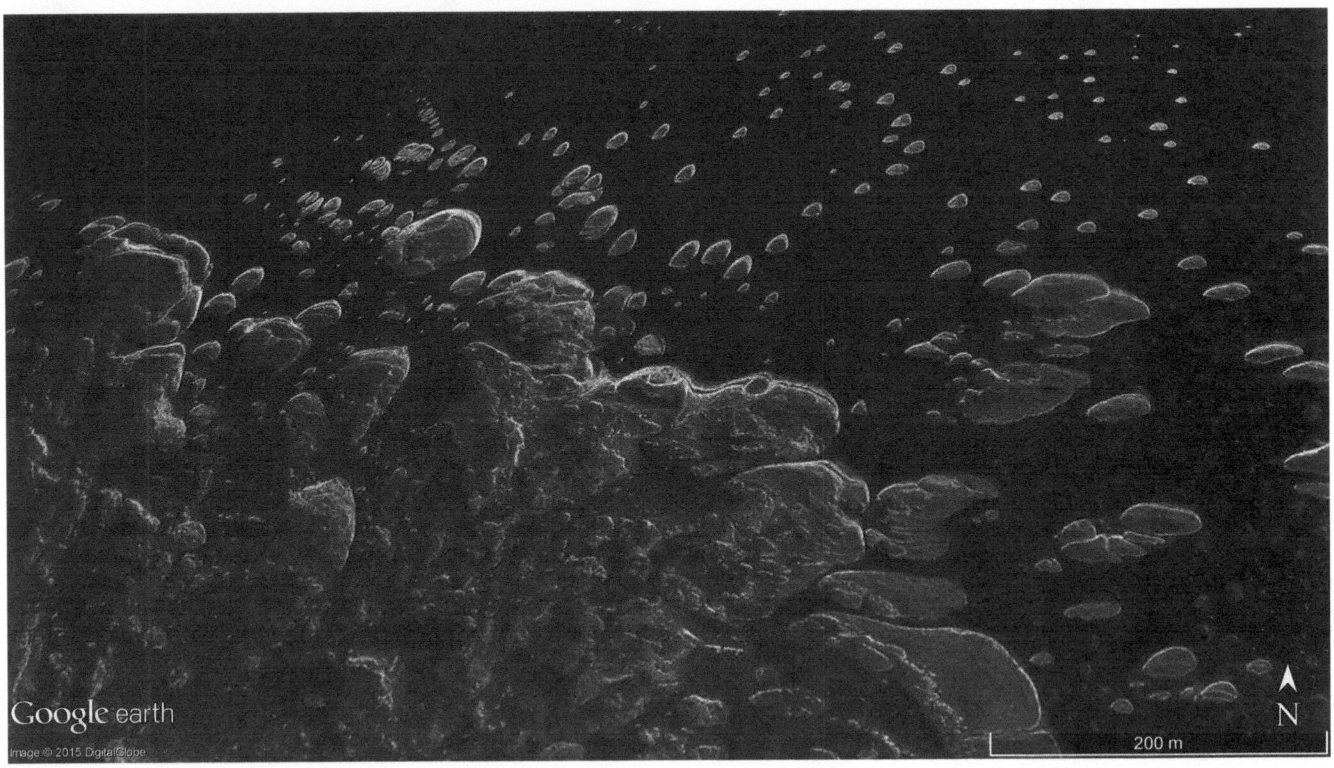

Fig. 4.82 Detail of a salt island complex in Lake Natron, eastern Africa (approximately *2°25'S und 36°01'E)*. The pattern is the result of constant easterly winds. Scene is 1 km wide, Center of the image *2°25'S, 36°01'E* (Image credit: ©Google earth 2015, February 2nd, 2008)

Fig. 4.83 Lake Natron in east Africa (Tanzania, *2°15'S, 35°58'E)* is a soda lake (high concentrations of carbonate salts, typically sodium) with an area of 1040 km^2 and being only 2 deep, is partly covered by a floating salt crust that breaks and collides by wind and current processes. Scene is 900 m wide (Image credit: ©Google earth 2015)

Fig. 4.84 A saltpan illustrates polygon patterning of cracks caused by evaporative processes in northern Tibet. Width of scene is about 600 m at *37°07'09.42"N, 86°48'12.18"E* at 4190 m asl (Image credit: ©Google earth 2015)

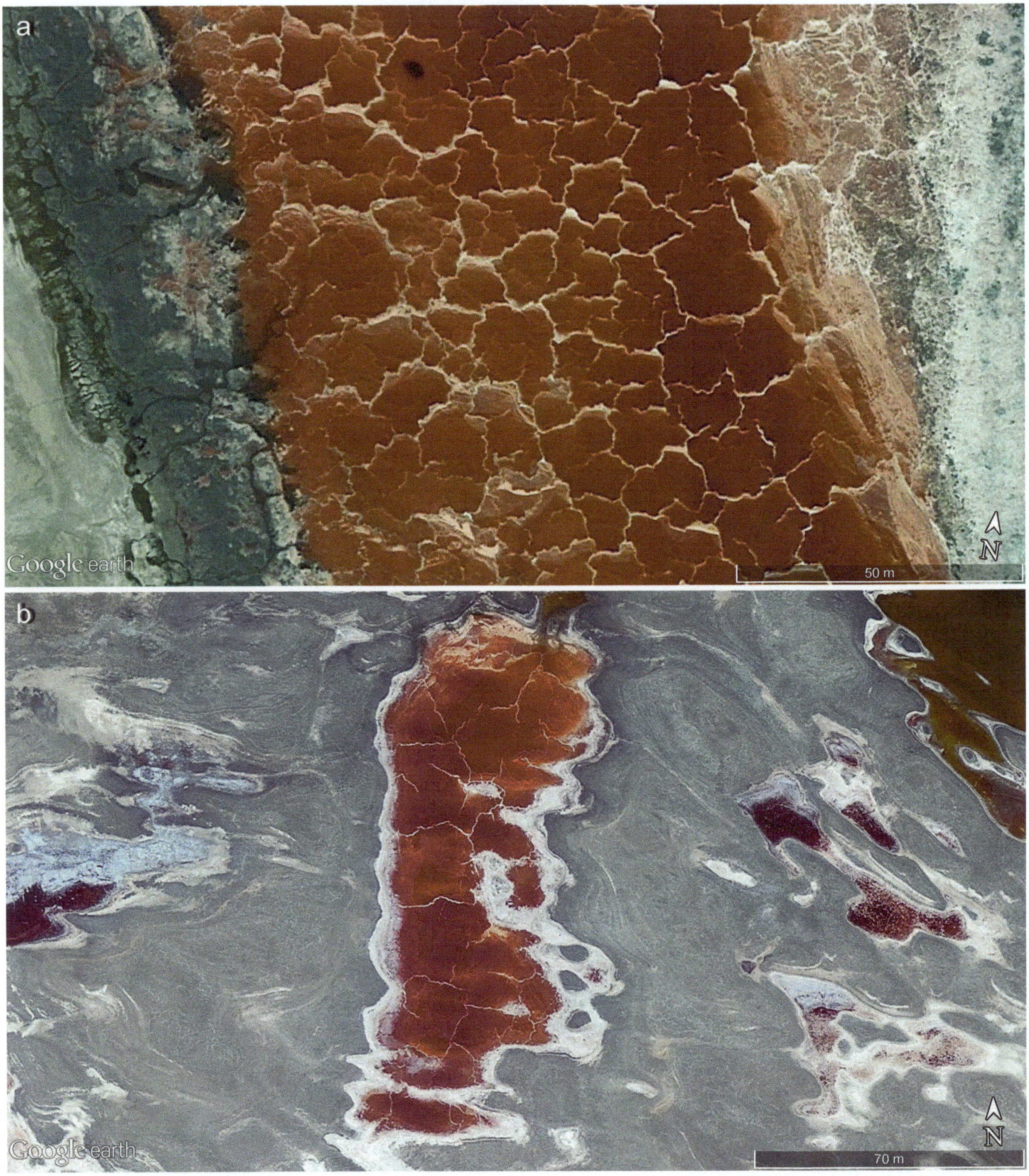

Fig. 4.85 (**a**) The dried out Owens Lake displaying a polygonal salt crust coloured with algae in eastern California *(36°25'32.95"N, 118°00'10.66"W)*. Scene is 200 m wide. (**b**) An alternative salt crust site *(36°30'09.48"N, 117°56'16.37"W)* in the former Owens Lake with large polygons exceeding 25 m in width. Scene is 240 m wide. (**c**) A salt crust pattern on Lake Natron in northernmost Tanzania, east Africa (approximately *2°17'S, 35°59'E)*. Scene is 480 m wide (Images credits: ©Google earth 2015)

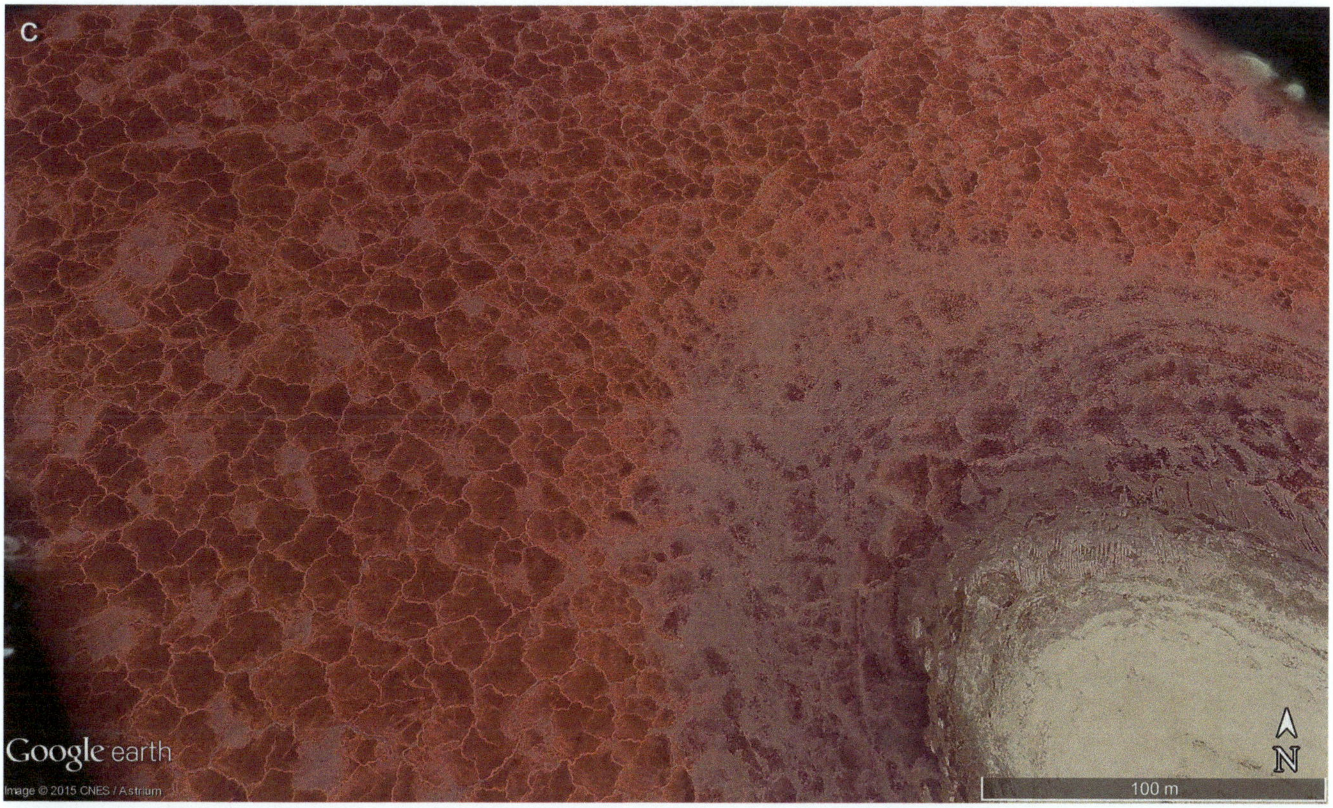

Fig. 4.85 (continued)

References

Alsharhan AS, Kendall CGSC (2003) Holocene coastal evaporites of the southern Arabian Gulf and their ancient analogues. Earth Sci Rev 61:191–243

Arnow T, Stephens D (1990) Hydrologic characteristics of the Great Salt Lake, Utah: 1847–1986. USGS water-supply paper, p 2332

Baker PA, Rigsby CA, Seltzer GO et al (2001) Tropical climate changes at millennial and orbital timescales on the Bolivian Altiplano. Nature 409(6821):698–701

Eugster HP, Hardie LA (1978) Saline lakes. In: Lerman A (ed) Lakes – chemistry, geology, physics. Springer, New York, pp 237–293

Evans MS (1999) Paleolimnological studies of saline lakes. J Paleolimnol 8:97–101

Hammer UT (1986) Saline lakes ecosystems of the world, Monographiae Biologicae 59. Springer, Dordrecht

Kendall CGStG, Alsharhan AS (eds) (2011) Quaternary carbonate and evaporite sedimentary facies and their ancient analogues: a tribute to douglas James Shearman. IAS special publication 43. Wiley, Oxford

Lashab MI, Daluob HS, Saqer NH (2004) Geological and geochemical studies on recent sabkha of Karkurah, northeastern Libya. In: 7th international conference on the geology of the Arab world, Cairo University, Egypt, February 2014, pp 1–10

Last WM (2002) Geolimnology of salt lakes. Geosci J 6(4):347–369

Lokier SW, Steuber T (2007) Seasonal dynamics of a modern sabkha surface. Geophys Res Abstr 9: 01874

Lowenstein T, Hardie LA (1985) Criteria for the recognition of salt-pan evaporates. Sedimentology 32(5):627–644

Melack JM, Jellisson R, Herbse DB (2001) Publications from the 7th international conference on salt lakes, held in Death Valley National Park, California, USA, September 1999. Springer, New York

Meris SM, Compton JS (2004) Origin and evolution of major salts in the Darling pans, Western Cape, South Africa. Appl Geochem 19(5):645–664

Nield DA, Simmons CT, Kuznetsov AV et al (2008) On the evolution of salt lakes: episodic convection beneath an evaporating salt lake. Water Resour Res 44. doi:10.1029/2007WR006161

Nielsen DL, Brock MA, Rees GN et al (2003) Effects of increasing salinity on freshwater ecosystems in Australia. Aust J Bot 51(6):655–665

Oren A, Nafiz DL, Palacios P, et al (eds) (2009) Saline lakes around the world: unique systems with unique values. Nat Resour Environ Issues 15(1). University of Utah, Salt Lake City

Renaut R, Last WM (eds) (1994) Sedimentology and geochemistry of modern and ancient saline lakes. SEPM special publication 50. Tulsa, Oklahoma

Savvaitova KA, Petr T (1992) Lake Issyk-kul Kirgizia. Int J Salt Lake Res 1(2):21–46

Singh G, Joshi RD, Singh AB (1972) Stratigraphic and radiocarbon evidence for the age and development of three salt lake deposits in Rajasthan, India. Quat Res 2(4):496–505

Tyler SW, Munoz JF, Wood WW (2006) The response of playa and sabkha hydraulics and mineralogy to climate forcing. Ground Water 44:329–338

West IM, Lashhab MI, Muhan IM (2000) North African sabkhas and lagoons compared to those of the Arabian Gulf. In: Proceedings of the sixth mediterranean petroleum conference and exhibition, November 23–25th, 1999, Tripoli, Libya, pp 512–530

Williams WD (1986) Limnology, the study of inland waters: a comment on perceptions of studies of salt lakes, past and present. In: De Deckker P, Williams WD (eds) Limnology in Australia. CSIRO Australia, Melbourne, pp 471–486

Williams WD (2002) Environmental threats to salt lakes and the likely status of inland saline ecosystems in 2025. Environ Conserv 29(2):154–167

Yechieli Y, Wood WW (2002) Hydrogeologic processes in saline systems: playas, sabkhas, and saline lakes. Earth Sci Rev 58:343–365

Lakes Adapted to Landscapes and Climate

5

Abstract

Lakes may be considered from a geomorphological aspect investigating the processes which have formed the depressions and basins, from a hydrological perspective examining water balance, and from an ecological point of view understanding flora and fauna interrelationships (and their uses for fishing, energy supply, irrigation and much more). The existence of lakes in certain landscapes, in particular, lakes formed by volcanoes, those in high mountain ranges and in the inner tropics, can also be analyzed by their distribution, number, size and climate dependencies to name a few. This chapter presents examples of lakes (and ponds, swamps/wetlands or pans) associated to desert dune landscapes and regions of permafrost. Both are extraordinary regions; the first is permeable (sand) and arid (lack of water), the second landscape has an impermeable underground where even low precipitation results in surface water accumulation due to reduced evaporative processes in cold climates. As there is a lack of scientific research around these special features with only exceptional examples of lakes in these regions and latitudes being studied in detail (in particular the lakes in deserts), this chapter focuses on several examples based on satellite image interpretation.

5.1 Lakes (Perennial, Periodic or Episodic) in Dune Landscapes and Deserts

Wind processes shape the landscape (even in rock) into deflation hollows characterized by wide and shallow depressions, or leave behind a landscape of sand hills (dunes) with starkly different forms. A multitude of closed morphological depressions occur in between these dunes, but as in karst landscapes, dune landscapes are not very suitable for lakes, ponds or swamps. Because dune landscapes appear in dry climates (deserts and semi-deserts) as well as along many coastlines, the existing sandy substratum is permeable and incapable of containing open water—the exception being where groundwater is close to the surface (again like karst landscapes) and small lakes may occur—more often, however, pans (saltpans) result due to high evaporation rates. Active dune fields are in constant movement and the depressions lying in between may shift in the direction of the dominating winds, carrying the lakes and pans along as well. The relationship of dune to basin (pans or lakes) morphology allows an evaluation to their relative age: are the basins older and later partly filled by shifting dunes, or is basin formation and filling formed simultaneously (Figs. 5.1, 5.2, 5.3, 5.4, 5.5, 5.6, 5.7, 5.8, 5.9, 5.10, 5.11, 5.12, 5.13, 5.14, 5.15, 5.16, 5.17, 5.18, 5.19 and 5.20).

5.1.1 Ancient and Modern Lakes in Deserts of Africa (Figs. 5.21 and 5.22a–c)

For many decades, rock art of mainly petroglyphs carved into desert varnish (a thin red to black coating on exposed rock surfaces) are well known throughout numerous deserts of the world—such as the Sahara Desert—depicting beasts including gnus, zebras, elephants, giraffes and buffaloes that can only exist in an environment with grass (e.g. savannah) and with sufficient water in surrounding rivers or lakes. Ancient river formations can easily be detected by their

A.M. Scheffers, D.H. Kelletat, *Lakes of the World with Google Earth*, Coastal Research Library 16,
DOI 10.1007/978-3-319-29617-3_5

Fig. 5.1 An inactive old dune-field exhibits many saltpans in basins that originally developed as blow-outs (now mostly inactive) and are set along the peninsulas of Shark Bay in Western Australia *(26°10'S, 113°52'E)*. The last dune movement belongs to the Younger Pleistocene time, older than 10,000 years. Scene is 40 km wide (Image credit: ©Google earth 2015)

modern appearance as wadis (dry valleys), and occasionally as remnants left in the form of carbonate deposits (shell) and old shorelines. The original flows of these ancient rivers have fed what is today's groundwater and document the past as a much wetter climate. Anthropologists and archaeologists have documented ancient sites of human occupancy by unearthing fire places, stone tools, shell and bone middens (even jewellery made of ostrich shells), and are able to map an accurate extension of the old lakes by their shorelines.

Determining the age of a lake's existence is enhanced with techniques such as radiocarbon dating (^{14}C), dating the rock art, dating rock art petroglyphs by desert varnish analysis (the chemical differences between engraved and non-engraved varnish present micro-laminations of manganese dioxide (MnO^2) rich layers, induced by climatic change of increased humid conditions), and importantly, sometimes to simply count the fine silt and mud layers in the basins where a yearly rhythm (warves) can be interpreted depending on precipitation from trade wind systems or monsoon. Worldwide, throughout the Last Glacial Maximum (14,000–26,000 years ago) the size of lakes were much larger than today, when climate was cooler and evaporation lower. Evidently during the last millennia, the climate changed into dryer conditions and surface water evaporated causing lakes and rivers to disappear creating modern deserts where groundwater can now only be found in wells and occasionally emerge in depressions as oases. As well as dating these periods, much research is needed in identifying past open-water expanses and in reconstructing moist climate conditions.

5.1.2 Lakes, Swamps and Pans in the Great Artesian Basin of Australia (Figs. 5.23, 5.24, 5.25, 5.26, 5.27, 5.28, 5.29, 5.30, 5.31, 5.32, 5.33, 5.34, 5.35, 5.36 and 5.37)

The largest uninterrupted basin on Earth holding artesian water exists in the eastern part of Australia and covers an astounding 22 % of the continent (1.7 Mio km^2). According to particular geological pre-conditions, artesian water and artesian springs require (as in Australia) a wide basin structure, layered by a sequence of sedimentary rocks that fill the base to form an extremely wide syncline (a trough). If the bedrock is/sedimentary deposits are impermeable (siltstone and mudstone prevent water percolating into the ground) and the intermediate stratum consists of water-bearing permeable

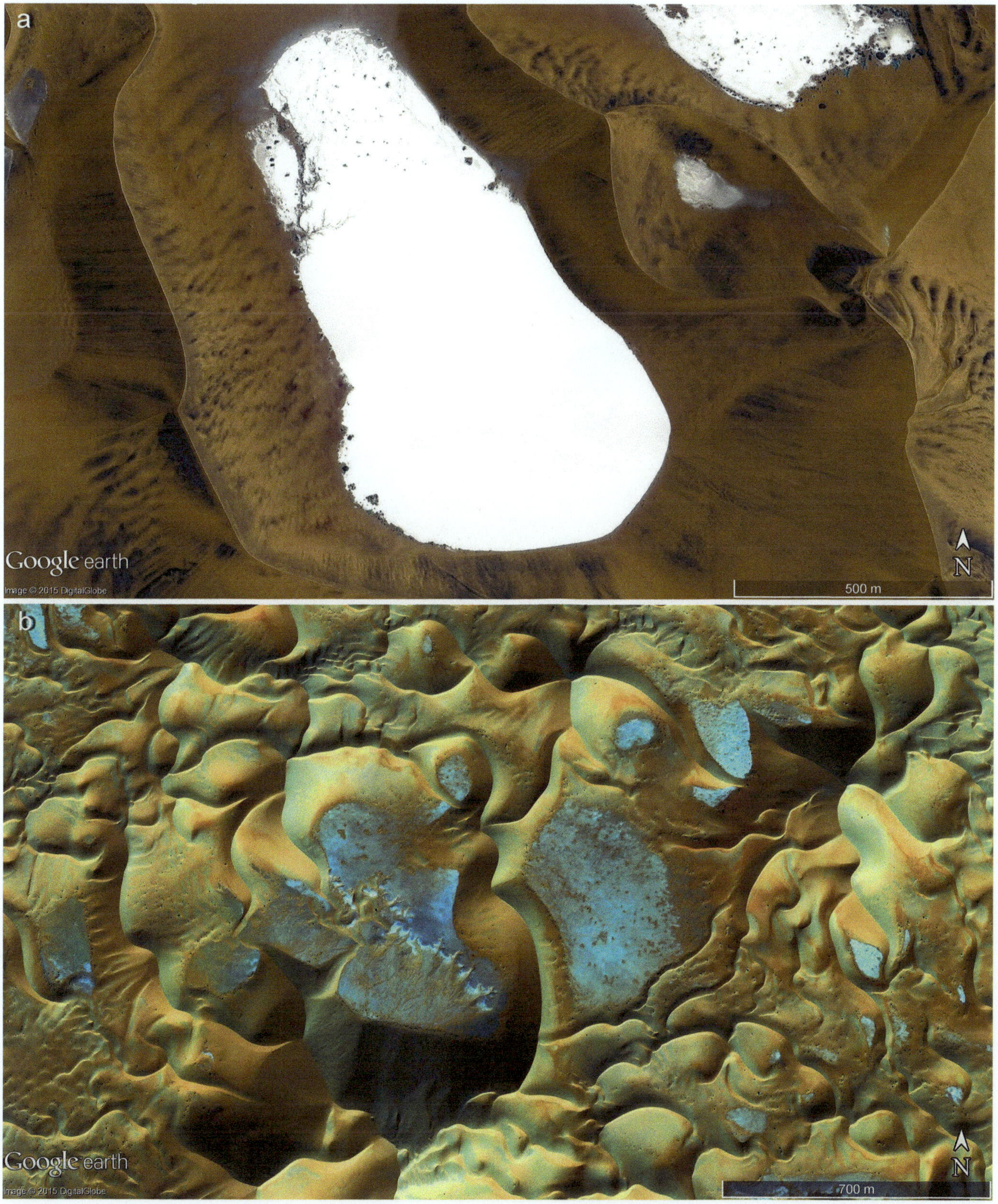

Fig. 5.2 (**a**) Saltpans are set amongst giant sand dunes (occasionally appearing as episodic ponds) of Sossusvlei in Namibia *(24°46'S, 15°18'E)*. The dunes are approximately 100 m high and the large saltpan is 1.3 km long. (**b**) Huge sand dunes with heights reaching 140 m edge over a flat salt crust in western Algeria (approximately *29°29'N, 1°26'W*). Scene is 2.5 km wide (Images credits: ©Google earth 2015)

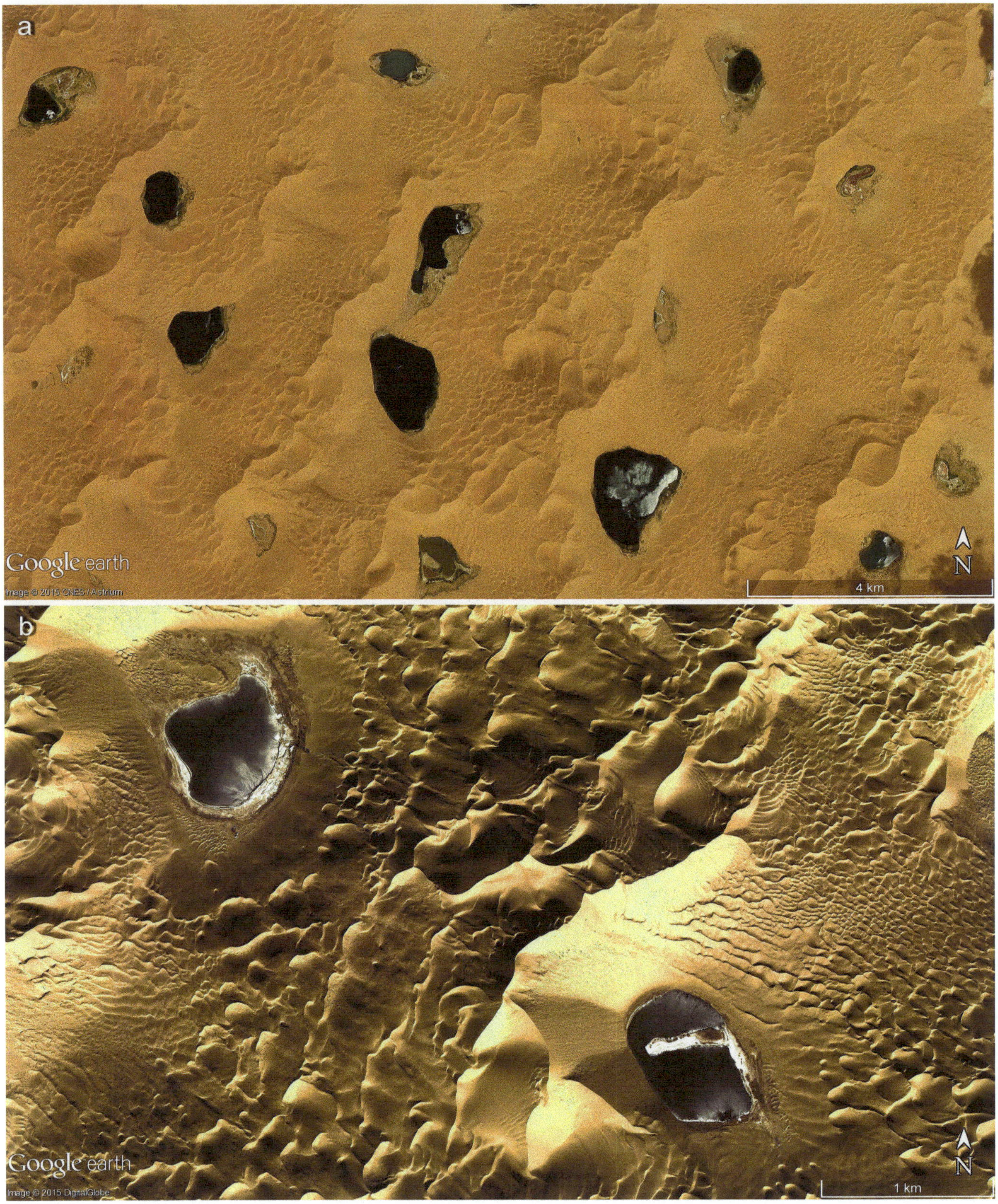

Fig. 5.3 (**a**) Groundwater ponds lie in active dune fields of the Gobi Desert in China *(39°48'N, 102°27'E)*. Scene is 17 km wide. (**b**) Saline ponds are set in an active dune field of the Gobi Desert in China *(40°01'40.05"N, 102°12"07.44"E)*. Scene is 6 km wide (Images credits: ©Google earth 2015)

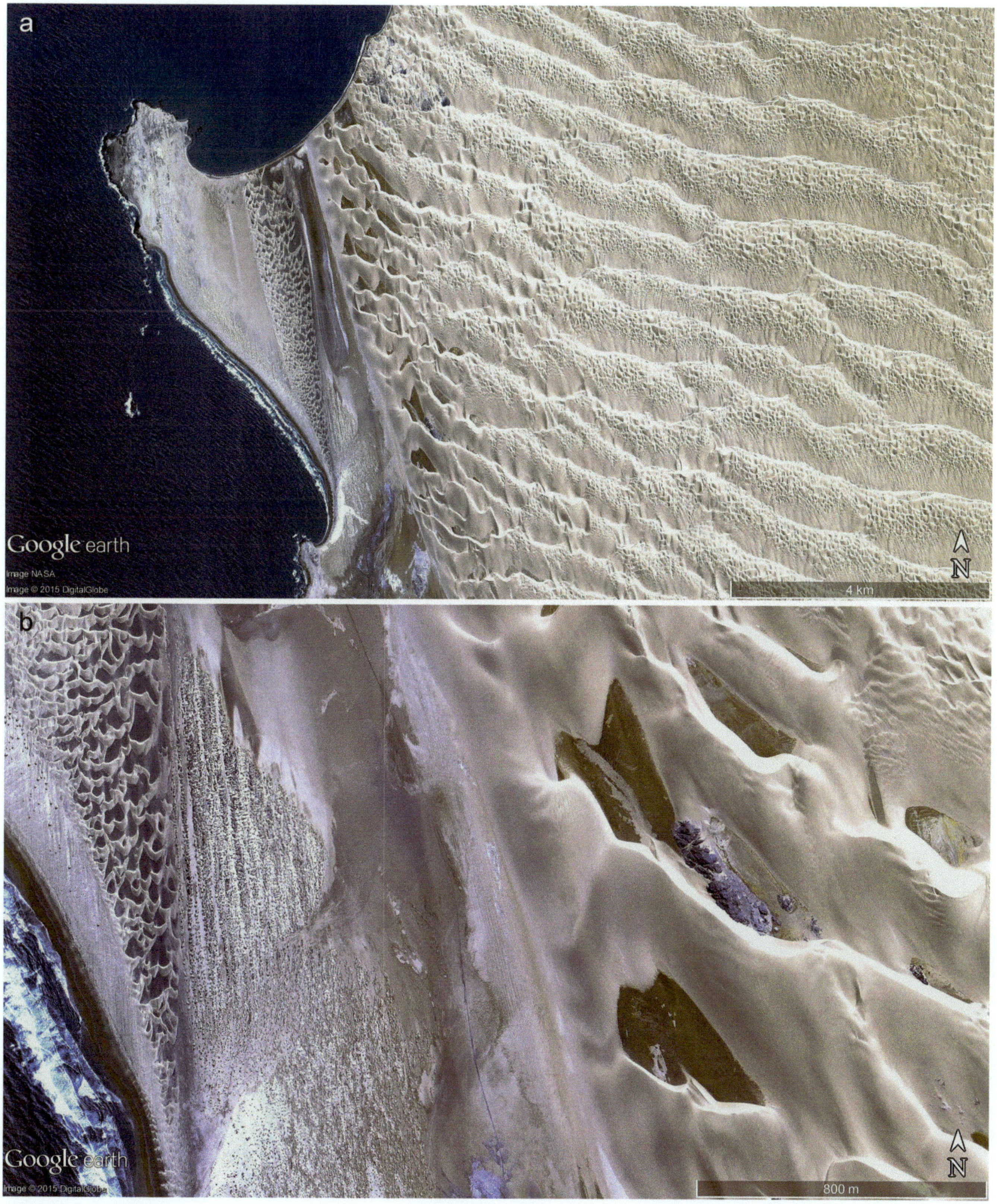

Fig. 5.4 (**a**) A field of crescent-shaped shifting sand dunes (barchans) lie amid groundwater ponds near Hottentotspunt in coastal Namibia *(26°10'S, 14°59'E)*. Scene is 15.5 km wide (Image credit: ©Google earth 2015). (**b**) Magnified detail of (a). Width is 2.8 km (Images credits: ©Google earth 2015)

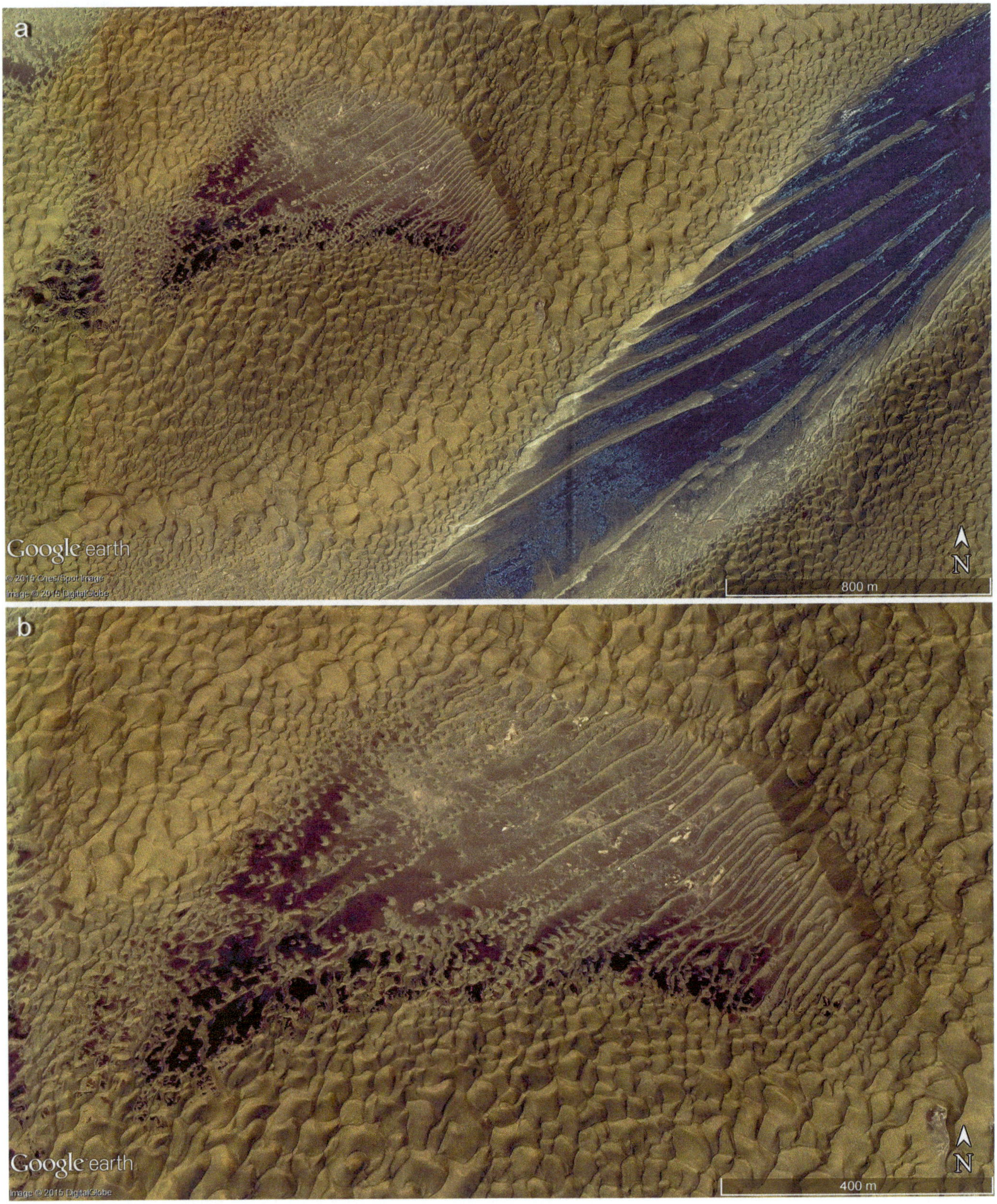

Fig. 5.5 (**a**) Groundwater emerges amid systems of linear and barchanoid dune fields in the Taklamakan Desert of China (approximately *40°37'N, 86°11'E)*. Scene is 3 km wide. (**b**) Detail of (a) demonstrates this depression was dry for a period of time where a different type of dune (linear and tiny barchans) has formed in contrast to the higher outer frame of barchanoids. Scene here is 1.5 km wide (Images credits: ©Google earth 2015)

Fig. 5.6 Several decades ago Lake Chad presented a periphery of drowned barchanoid dune systems located in the southern Sahara between Mali and Nigeria (approximately *13°24'N, 14° 7'E*). Today these areas are predominantly dry (compare Introduction (Part I) for more information on Lake Chad). The 110 km wide scene is part of the eastern section (Image credit: ©Google earth 2015)

Fig. 5.7 Ephemeral lakes lie within numerous longitudinal dunes in the southwest section of Australia's Northern Territory *(20°53'S, 129°28'E)*. Scene is 18.7 km wide (Image credit: ©Google earth 2015)

Fig. 5.8 Lakes occur in a field of barchans in southern China (approximately *34°27'N, 98°49'E)*. The braided river in the northwest section of the image is evidence of groundwater reaching the surface. Scene is 8.4 km wide (Image credit: ©Google earth 2015)

Fig. 5.9 Impressive giant ripple marks and lakes resulting from the outburst of Lake Missoula's late-glacial meltwater 14,000 years before, is located in the landscape known as the Channeled Scablands of Washington State, USA *(47°03'N, 119°20'W)*. Although resembling windblown dunes, the dry patterns are ripples from the flood unleashed when the lake drained, now partly inundated by groundwater from a river to the west. Scene is 2.3 km wide (Image credit: ©Google earth 2013)

Fig. 5.10 An ancient linear dune field lies within the western Wolga Delta, south Russia *(46°11'N, 47°31'E)*. Scene is 20 km wide (Image credit: ©Google earth 2015)

rock (as sandstone), an aquifer develops because of the many pores wide enough to yield and transmit water. This body is layered again by an impermeable siltstone and mudstone (to prevent evaporation to the surface) to become a contained, underground repository where ground water is stored for long periods. The ability for aquifer recharge relies on the layer of porous sandstone that surfaces along the fringes of the basin and refills directly from rainfall, or from rivers crossing these outcrops.

In the case of Australia's Great Artesian Basin (GAB), sedimentary rocks are extremely old and deposited in a closed basin over a long time span (240–100 Mio years ago). Over time, denudation and erosion have cut into the fringes of the syncline structure and opened the aquifer to water input—and because the Great Dividing Range (GDR) of the states Queensland and New South Wales has rather high precipitation, run off into the basin recharges the aquifer. The aquifer centre (GAB) exceeds 2000 m in thickness and diminishes to 100 m along its outer margins. Under the force of gravity recharge water slowly creeps along at only 10 mm per year where the margins and recharge areas (GDR) are elevated higher than the main section of the sandstone aquifer, and the impermeable rock layer situated above places water under pressure. Instabilities (cracks, faults) in the overlying silt and mudstone allow water under pressure to reach the surface (upwelling) to construct water fountains commonly known as "mound springs". The highest concentrations of these springs are located in the deepest part of the basin, occurring in clusters of dozens to hundreds (though often very small), organized into groups and super-groups. Today more than 4500 man-made bores are active, the deepest to 2000 m, drilled for the purpose of irrigation and cattle management. The more artificial wells are drilled the more of the pressure is lost, where the uppermost aquifer water level decreases and pumping is necessary to gain more groundwater. The residence time of artesian water can reach 2 Mio years, with increases in salt content from the dissolution of minerals in rock.

Some examples are presented from three different sites in the Great Artesian Basin of Australia.

5.2 Lakes and Ponds in Permafrost Environments

Mean annual temperatures below 0 °C characterize the cold latitudes and permafrost is defined as perennially frozen soil that occurs if the average temperature is at or below

Fig. 5.11 (**a**) A field of coastal barchans and barchanoid ridges are surrounded by swamps with groundwater lakes lying in between, located south of the Amazon mouth in Brazil *(2°26'S, 43°14'W)*. Scene is 9 km wide. (**b**) Detail of a barchanoid coastal dune field in Brazil *(2°45'27.41"S, 42°22'26.30"W)*. The pattern of parallel contours exists at sites that had been exposed as dunes, which indicate seasonal breaks in wet periods of the year, and again indicate movement of metres to decametres (distance between contours) in the dry periods of the year. The number of contours determines the number of years of movement, and also calculates the amount of sand transported by local trade winds (east to west). Scene is 1.8 km wide. (**c**) In coastal Brazil (approximately *2°46'S, 42°21'W)* large fields of barchans are evidently pushed by northeast trade winds across a periodic swampy base. Moisture fixes the windward band of sand and retains the contours of the same shape as the barchans continue to move. The number of parallel marked contours represents periods the dunes were motionless, and documents dune movement over a period of 40–50 years. Scene is 3.5 km wide (Images credits: ©Google earth 2015)

Fig. 5.11 (continued)

Fig. 5.12 Hundreds of migrating steps of barchanoid dunes have been preserved by high groundwater swamps, resulting as crescent-shaped forms in northeast Brazil *(2°44'S, 42°23'W)*. Scene 3.5 km wide (Image credit: ©Google earth 2015)

Fig. 5.13 (**a**) A field of chevron-shaped dunes with lakes set in between located in southern Venezuela (approximately *6°34'N, 69°17'W)*. Scene is 56 km wide. (**b**) A 18 km wide detail from (a) of a chevron-shaped-dune field demonstrating the dunes are actively migrating into the southern swampy region (Images credits: ©Google earth 2015)

Fig. 5.14 The contours of a chain of small lakes (called the Archibald Lakes), are influenced by shifting sand dunes and barrier formation of longshore drift by strong westerly winds, and are located to the south of the large Athabasca Lake in central Canada *(59°02'N, 108°34'W)*. Scene is 21 km wide (Image credit: ©Google earth 2015)

−2 °C. Permafrost covers around 26 Mio km^2 in the northern hemisphere alone, where in the coldest (pole-ward) sections of these latitudes continuous permafrost exists and in summer only thaws the surface down to less than a metre (active-layer). The waterproof frozen ground causes the surface to become muddy and with any available depressions the water is stored as small or large lakes, ponds (small and very shallow), or swamps and bogs (if vegetation grows in the ponds) (Figs. 5.38, 5.39 and 5.40). Although in warmer latitudes the permafrost is discontinuous and its active-layer summer melt is thicker, the soil processes of freezing and thawing still produce the same forms as the continuous permafrost zone (Figs. 5.41, 5.42 and 5.43).

Three processes are responsible for the millions of ponds and lakes occurring in permafrost landscapes:

- former glaciations have left behind a cover of ground moraine with numerous kettle lakes in different forms (mostly circular in shape). However during the Ice Ages, glaciation had not covered all the cold latitudes due to the extremely cold conditions, in particular at pole-ward regions and in low lying elevations where insufficient (snow) precipitation accumulated to nourish the glaciers—here permafrost reaches its greatest depths in excess of 1 km.
- ice-wedge formation is the process responsible for most of the ponds and occurs when the surface is extremely cold (below −20 °C). Ice-wedges demonstrate a "patterned ground" represented as netting of elevated polygonal rims with shallow ponds enclosed (Figs. 5.44, 5.45, 5.46, 5.47,5.48,5.49, 5.50, 5.51, 5.52, 5.53, 5.54, 5.55, 5.56, 5.57, 5.58, 5.59). Contraction fissures expand in drying mud (Fig. 5.59a) and expand when meltwater from thawing on the surface or from snow, fills the fissures in the extremely cold ground instantly freezing it to ice. Water expands during freezing and forces the fissures to widen, often repeating the process in summer. Muddy surface-soil and debris is thrust aside from all the wedges to develop two parallel ridges along the ice wedge network, which results in enclosed honeycombs surrounded by low dams (Fig. 5.59b) that fill with shallow water from ground frost melting. Additionally, freezing of these filled ponds further push material aside and helps to elevate and stabilize the dams (Fig. 5.59c). In several images the network of ridges and dams exhibit central dark lines at the site of the ice wedges. In time, most ponds transform into

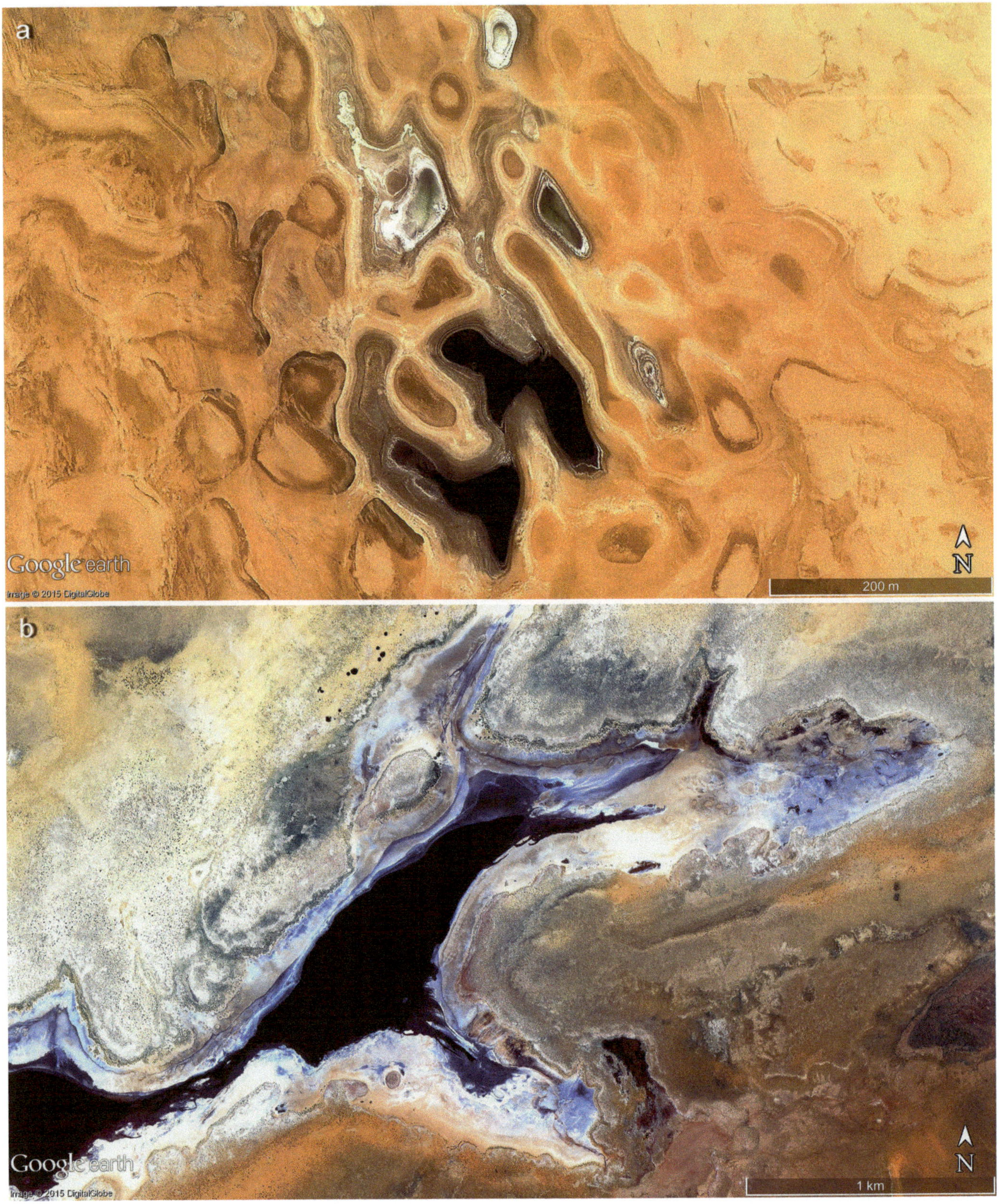

Fig. 5.15 (**a**) A sandstone landscape is sculpted by aeolian processes into deflation basins that can fill during rare rainfall events and evaporate rapidly culminating as saltpans near the west coast of Mauretania *(20°18'37.85"N, 16°16'19.96"W)*. Scene is 1 km wide (Image credit: ©Google earth 2015). (**b**) A rare rainfall event has filled a wadi (Oued) in a sandstone landscape along the northern fringe of the Sahara Desert in western Algeria *(33°23'N, 0°08'W)*. Scene is 4 km wide (Image credit: ©Google earth 2015, June 14th, 2011). (**c**) The same site and scene as in (b) after 1 month with only several saltpans remaining (Image credit: ©Google earth 2015, July 22nd,2011)

Fig. 5.15 (continued)

Fig. 5.16 One of the central Toshka Lakes (length 20 km) in southern Egypt *(23°10'N, 30°43'E)*. The dune systems presented here are partly drowned by water from the lake. The easternmost lake appeared in November 1998 and by late 1999, three additional lakes formed sequentially westward (the lake shown is the westernmost), following, the north-western most lake began developing sometimes between September 2000 and March 2001. In total, the Toshka Lakes (the lakes are not named individually) cover approximately 1300 km^2 at capacity whereby the water derives from overflow of the Aswan High Dam, and explains the young ages to these endorheic lakes. Although the fringe of swamps around the lake's periphery expands rapidly, the water increases salinity from year to year due to high evaporation rates (Image credit: ©Google earth 2015)

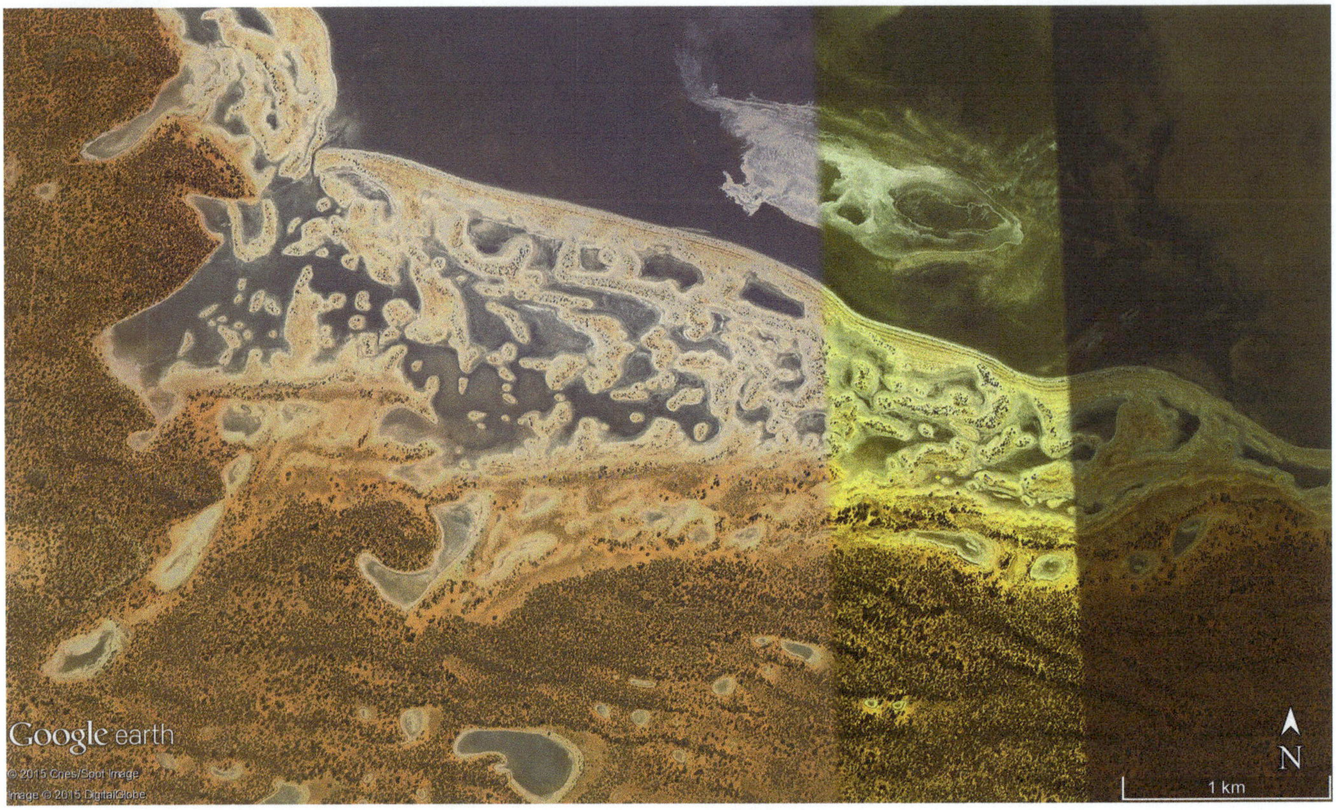

Fig. 5.17 A field of dunes and salt flats in South Australia *(32°01'S, 136°39'E)*. Although the dunes have formed in an existing basin (part of a riverbed deepened by deflation) the vegetation covering the dunes demonstrates inactiveness. Scene is 7 km wide (Image credit: ©Google earth 2015)

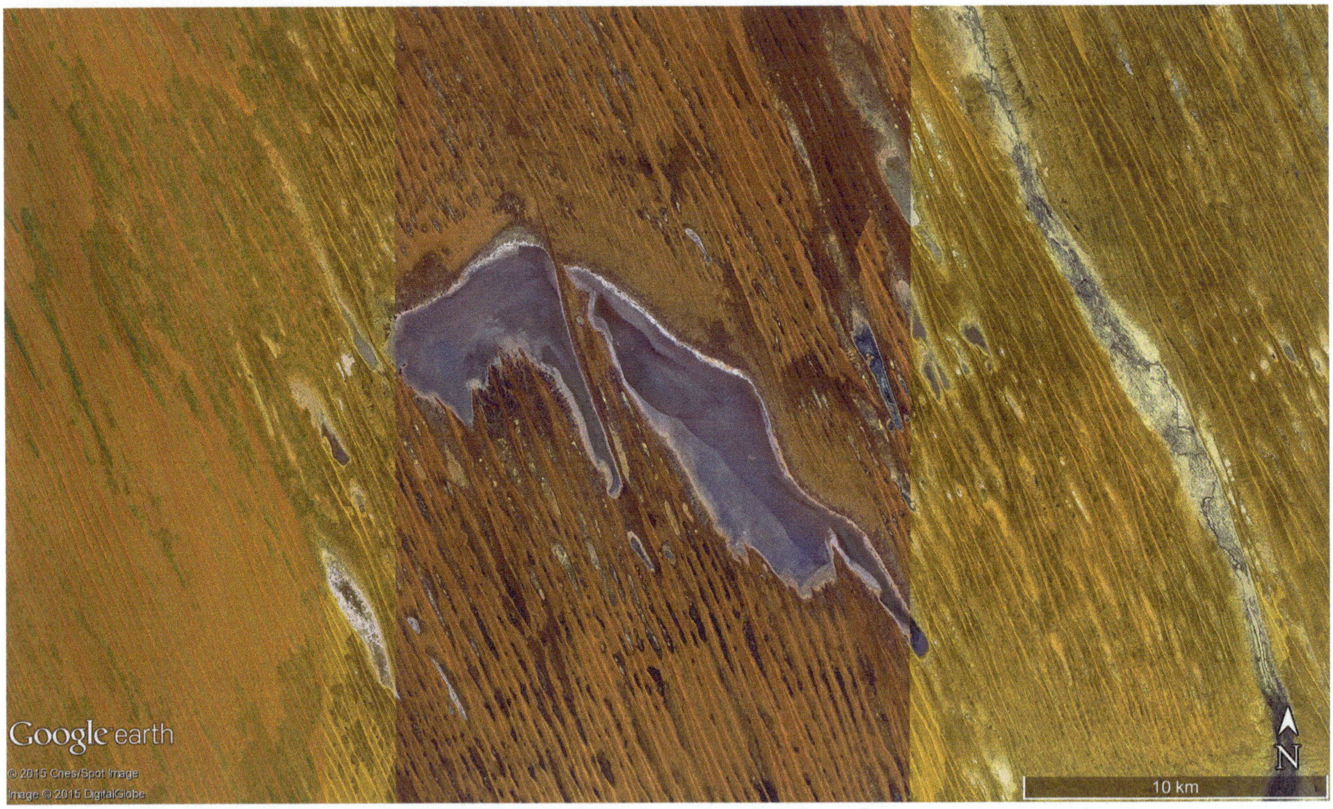

Fig. 5.18 An ephemeral salt lake and elongated saltpans amid a linear dune field in western Queensland, Australia (approximately *24°09'S, 138°45'E)*. The saltpans have developed during dune formation and are restricted from areas due to deposition of moving sands. Scene is 43 km wide (Image credit: ©Google earth 2015)

Fig. 5.19 Three white saltpans lie within linear dunes and occupy spaces left behind from an active dune system, located in the southern part of Australia's Northern Territory *(24°14'S, 136°30'E)*. Scene is 7 km wide (Image credit: ©Google earth 2015)

Fig. 5.20 An older saltpan is traversed by younger dune fields in southwestern Algeria *(27°14'N, 6°49'W)*. Scene is 7 km wide (Image credit: ©Google earth 2015)

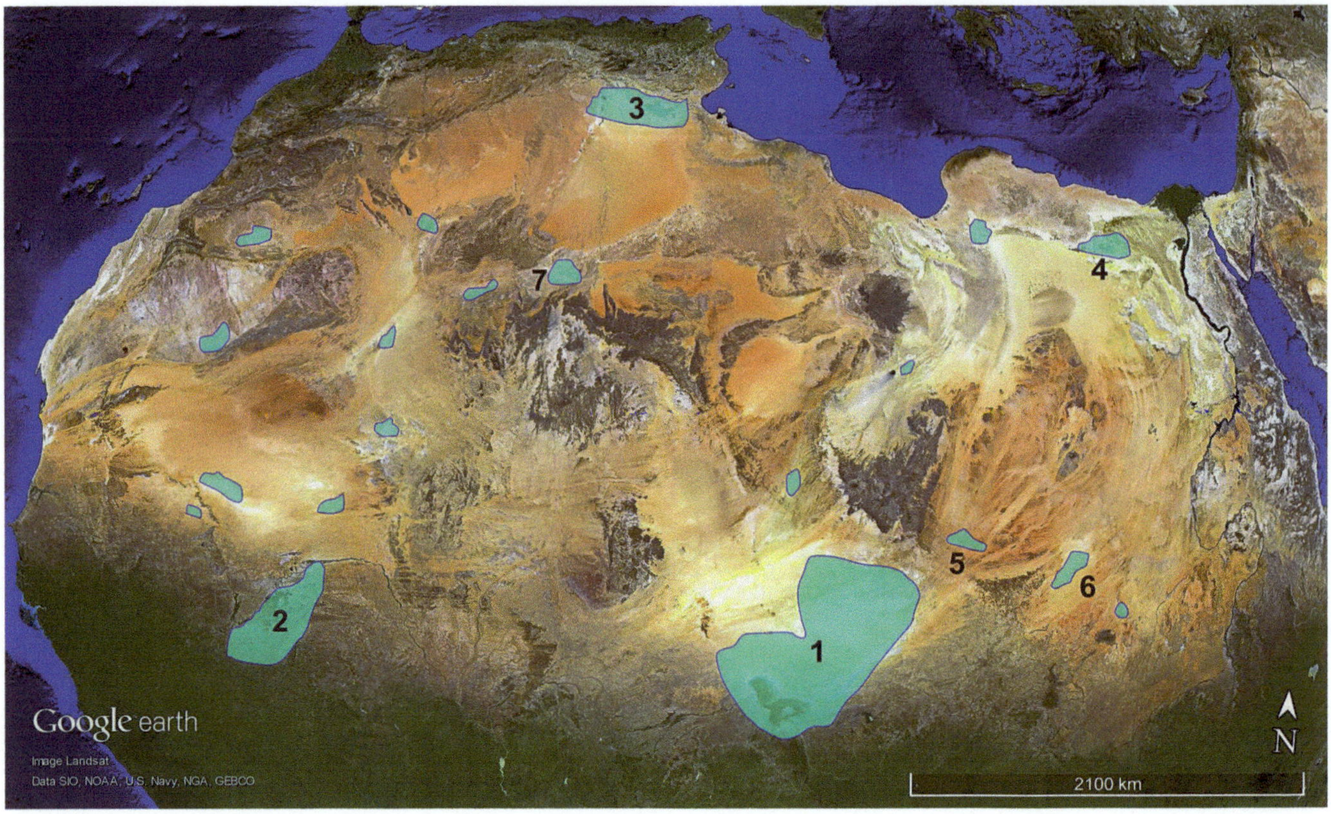

Fig. 5.21 Large former lakes in the Sahara Desert of northern Africa (their approximate positions and forms are according to particularly with wave action, are coastal forms known to open , names from modern locations:1-Lake Mega-Chad; 2-Lake along old Niger course; 3-Chott el Jerid/El M'Ghair; 4- Qattara depression; 5-Ounianga; 6-Darfour; 7-Figuig). The sites 1–5 show remnants of open water found within depressions. The age determination of the ancient lakes is not clear for all sites—however are relatively young in a geological sense with several ancient lakes existing into the Holocene period up to 5000 years ago—this indicates the contemporary Sahara Desert is rather young and has expanded during the existence of modern humans (*Homo sapiens sapiens*) (Image credit: ©Google earth 2015, modified)

Fig. 5.22 (**a**) The Tibesti Mountains lie to the west with the Ennedi Plateau to the east in Chad of North Africa (see Fig. 5.19b). The two tiny black regions in the centre *(18°55'N, 20°52'E)* are the western and eastern Ounianga lakes. Scene is 800 km wide (Image credit: ©Google earth 2015). (**b**) The Ounianga lakes in Chad *(18°55'N, 20°52'E)* occur at sites where groundwater from the sandstone aquifer is exposed by active erosion and deflation. Scene is 52 km wide and the distance between the two groups of lakes is 40 km. (Image credit: ©Google earth 2015). (**c**) The eastern Ounianga Lakes in Chad *(18°55'53.32"N, 20°52'20.36"E)*. The sizes of all lakes are around 20 km^2 (the largest is

Fig. 5.22 (continued) Lake Elimé) and represent the largest lakes in the Sahara (prior to the artificial filling of the Toshka Lakes from Lake Nasser). All except one lake contain fresh water, provided by a groundwater aquifer filled in more humid times. A sediment core extracted from Lake Yoan (the largest of the western group) established through pollen analyses, warve chronology and radiocarbon dating that a previous much more expansive lake flourished in a savannah-like environment with a more humid climate between 14,500 and 5500 years ago. Most sources understand the specific form of the lakes to be the result of the accumulation of elongated dunes from trade winds, but more importantly, the sandstone base sculpted by trade winds lies more than 20 m below the lake surface. This erosive wind sculpturing is evident in the elongated sandstone hills located alongside and between lakes (called yardangs). Floating reeds reduce evaporation and cover the smaller lakes, where small settlements, agricultural crops and cattle can be found along the southern shores of these lakes. Scene is 11.4 km wide (Image credit: ©Google earth 2015)

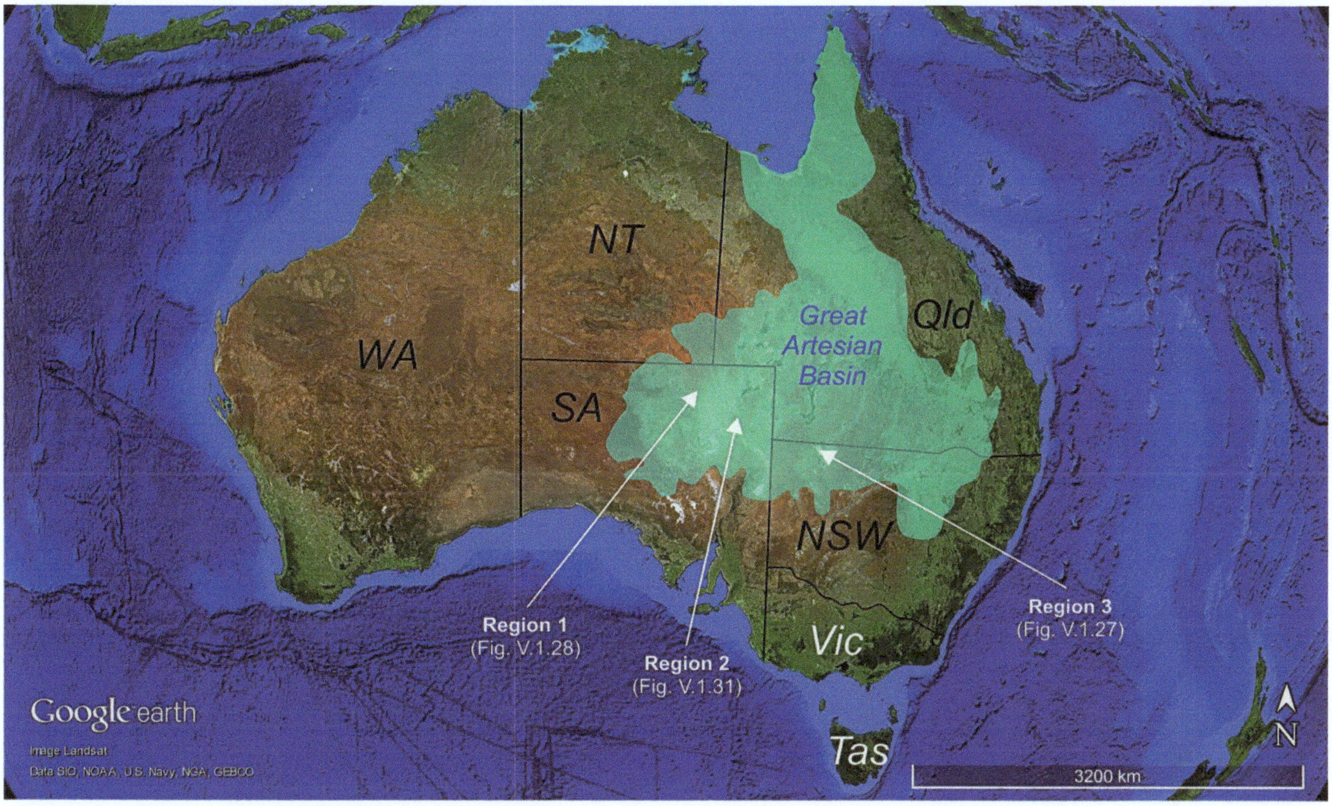

Fig. 5.23 The Great Artesian Basin of eastern Australia covers vast regions of South Australia, New South Wales, Queensland, and the Northern Territory and is situated between ancient structures over one billion years old (to the west-northwest). Younger structures emerge to the south and represent the youngest tectonic element of continental Australia, with the Great Dividing Range to the east (Image credit: ©Google earth 2015, Graphics: Anne Hager)

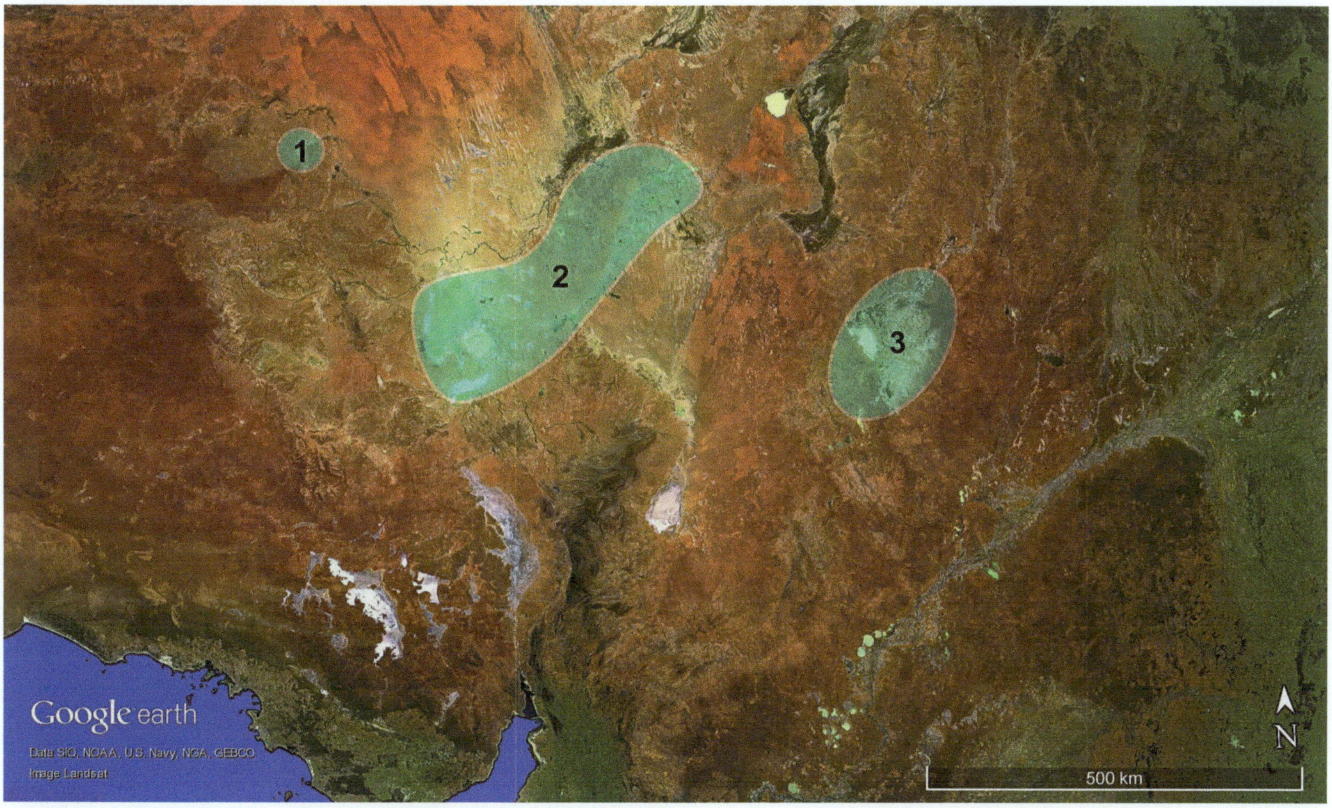

Fig. 5.24 Artesian fresh water ponds and swamps are located around the Witjira-Dalhousie springs (a super-group that contains over 60 springs extending over an area of 500 km^2) in the Witjira National Park, South Australia (Regions 1–3, for a general view compare Fig. 5.23, for the details Figs. 5.28, 5.29, 5.30 and 5.31). Centre of image at about 30°S and 140°E, width 1600 km (Image credit: ©Google earth 2015, modified)

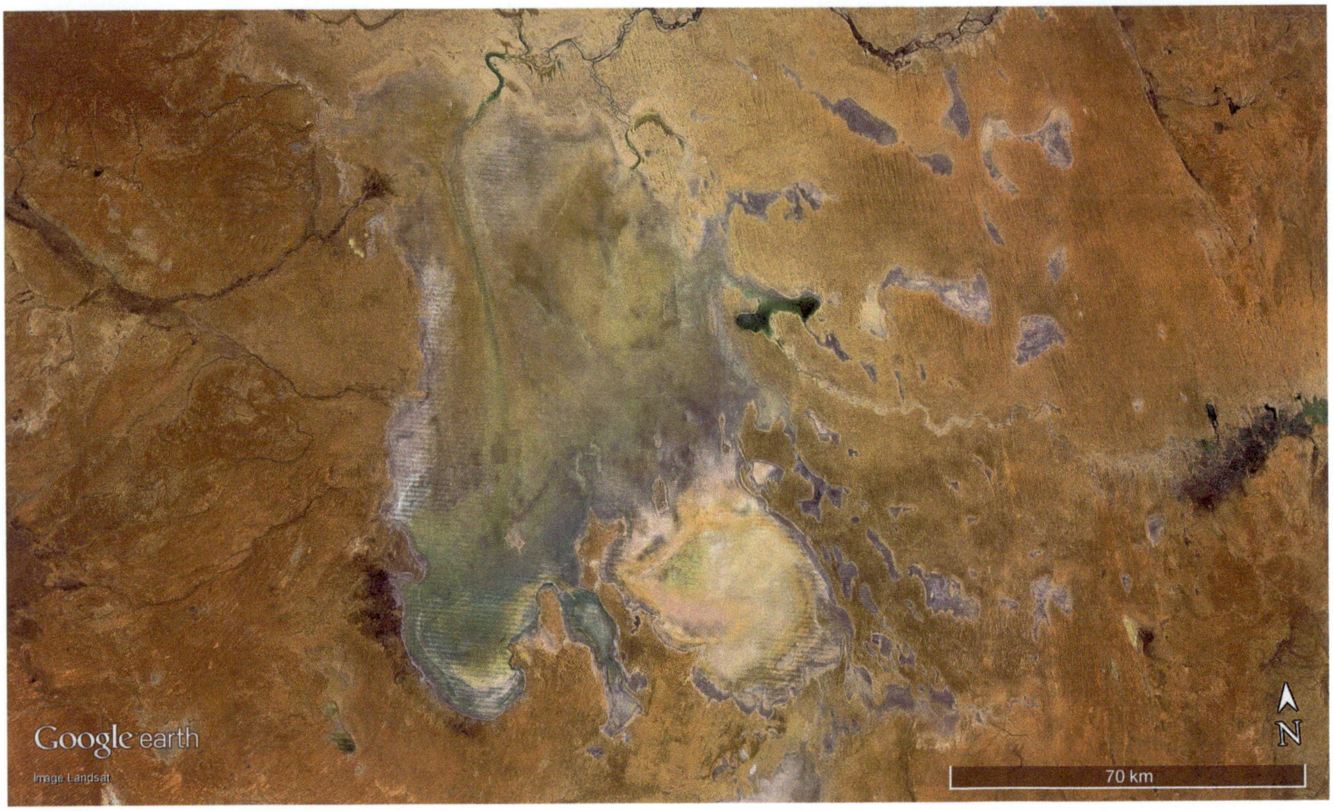

Fig. 5.25 The basin of Lake Eyre *(28°33'S, 13°32'E)* is the lowest point of the Australian continent (17 m below sea level) and consists of many diverse sections, demonstrating various elevations and time spans containing water, and time spans since total evaporation. When the main basin is filled (several times a century resulting from strong monsoonal rainfall influenced by El Nino events and/or unusually strong rains from the Great Dividing Range) the lake area exceeds 9000 km^2 (from a water catchment region covering 1.1 Mio km^2) with a maximum depth of 4 m. Figure 5.26 presents saltpans and a small remnant of open water in the north-east (16 km across) (Image credit: ©Google earth 2015)

Fig. 5.26 A remnant of open water (approximately *28°20'S, 137°42'E)* from the Lake Eyre basin to the northeast. Scene is 18 km wide (Image credit: ©Google earth 2015)

Fig. 5.27 A view of salt lakes and pans (Region 3 in Fig. 5.24, *29°22'S, 142°21'E*). Scene is 360 km wide (Image credit: ©Google earth 2015)

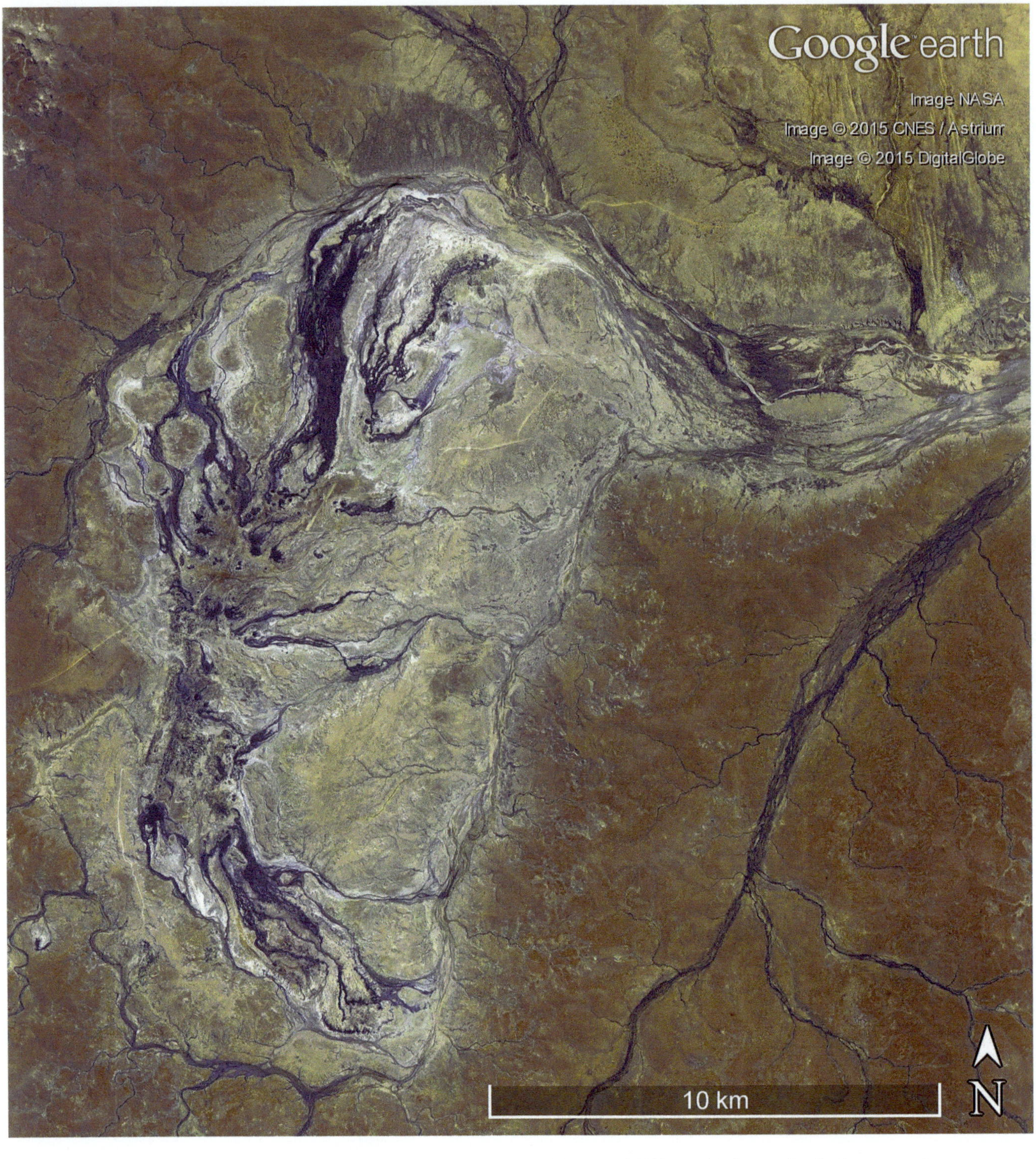

Fig. 5.28 The Witjira National Park near Dalhousie (Region 1 in Fig. 5.24, *26°27'S, 135°30'E)* with the artesian fresh water spring super-group containing ponds and swamps. Scene is 40 km wide (Image credit: ©Google earth 2015)

Fig. 5.29 A freshwater artesian swamp is located near Dalhousie *(26°23'07.14"S, 135°30'40.14"E)*, with rivers leading north fed by springs from the swampy region. Scene is 4 km wide (Image credit: ©Google earth 2015)

Fig. 5.30 Artesian springs flow into three ponds near Dalhousie *(26°25'09.47"S, 135°29'45.09"E)*. Scene is 2 km wide (Image credit: ©Google earth 2015)

Fig. 5.31 A view of several saline lakes (Region 2 in Fig. 5.24, *28°10'S, 138°42'E)*. Lake Eyre lies to the west (compare Figs. 5.24 and 5.25). Scene is 560 km wide (Image credit: ©Google earth 2015)

Fig. 5.32 Lake Lady Blanche to the left contains a strong artesian spring and Lake Sir Richard lies to the right where both lakes feed a river to the north (the north-easternmost part of Fig. 5.28 at *27°01'S, 140°22'E)*. Both lakes are 4 km across (Image credit: ©Google earth 2015)

Fig. 5.33 Four pans/lakes *(27°10'S, 140°11'E)* present from left to right—Lake Apachirie, Coongie Lake, Lake Marroocoolcannie, and Lake Marroocutchanie—and are all fed by an artesian spring via a small river in the south. Scene is 17 km wide (Image credit: ©Google earth 2013)

Fig. 5.34 A series of lakes *(27°22'S, 139°46'E)* present from left to right—Lake Mckinlay, double-lake Watchiewatchina and Wynalkana, and Lake Moolionburrinna to the northeast—and are all fed by small rivers to the north. The linear dune system is mostly inactive and covered by a thick layer of soil. Scene is 25 km wide (Image credit: ©Google earth 2015)

Fig. 5.35 The ephemeral Lake Warrakalanna is presently a saltpan (9 m asl and 4 km wide) located in south Australia *(28°11'58.49"S, 139°18'37.11" E)* that is fed episodically by a small creek from the north and when full covers an area of 13 km² (Image credit: ©Google earth 2015)

Fig. 5.36 A series of lakes are connected by a riverbed to Lake Warrakalanna (refer to Fig. 5.35 approximately *28°29'S, 138°55'E)* to the northeast. The meandering riverbed, lakes and saltpans to the east are all fed from the north where the swampy pan (at 5 m asl.) in the southwest flows into two small lakes. Scene is 130 km wide (Image credit: ©Google earth 2015)

Fig. 5.37 The large saltpan is Lake Killamperpunna in South Australia *(28°35'24.06"S, 138°43'24.74"E)* and feeds Lake Kopperamanna in the west and Lake Yandiya in the central north. The detail is of the larger scene in Fig. 5.33 and is 20 km wide (Image credit: ©Google earth 2015)

Fig. 5.38 Set in a scene spreading over 300 km are thousands of lakes in northwest Alaska (approximately *70°21'N, 156°10'W)*. The band of lakes is oriented from north to south and set in the coastal lowlands approximately 150 km wide (Image credit: ©Google earth 2015)

Fig. 5.39 A detailed view of the oriented lakes shown in Fig. 5.38 *(70°14'33 N, 159°06'18 W)* with distinctive colours that reflect the depth of water and of algal communities growing in the lakes. Their form is asymmetrical in the sense that they are wider in the southern section, which, additional to the direction of the long axes, point to starting from this end. Scene is 20 km wide (Image credit: ©Google earth 2015)

Fig. 5.40 Numerous elongated lakes in permafrost regions demonstrate all states of development – from shallow lakes to swamps to depressions filled with sediments – or a combination of all of these conditions. The lakes exhibit different contours of similar size and is often an expression of high morphological activity. The largest oval lake in this image from northern Alaska *(70°46'N, 153°56'W)* is 11 km long, width of the scene is 90 km (Image credit: ©Google earth 2015)

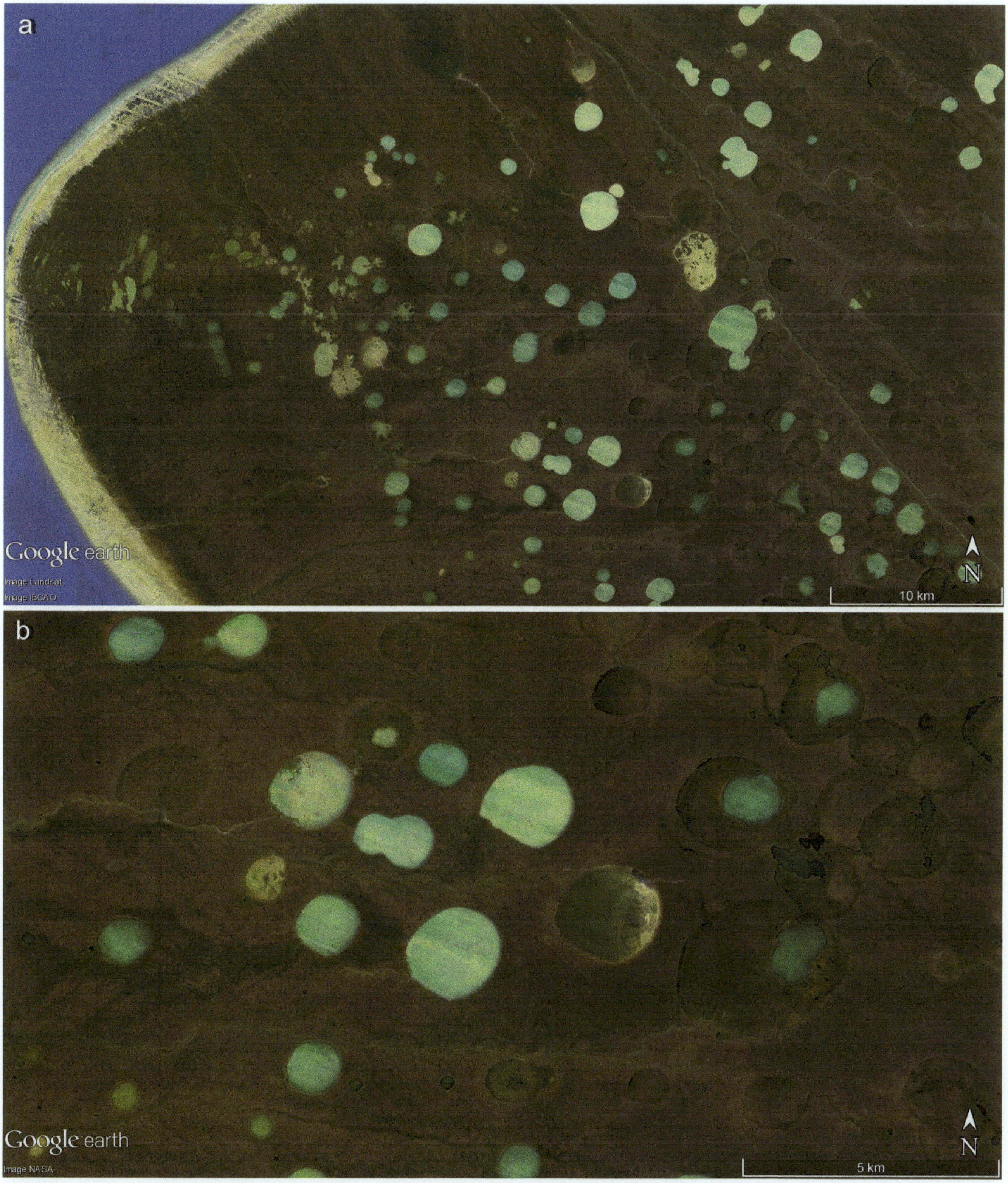

Fig. 5.41 (**a**) Due to environmental warming effects, fresh thermo-karst lakes have emerged where several have evolved into swamps lacking any open water, and are located on southwest Baffin Island in Canada *(66°07'N, 73°53'W)*. The white colours indicate receding winter ice cover. These lakes have not modified to the main wind direction, and to the west are elongated small lakes filling the swales in a beach ridge system. Scene is 60 km wide. (**b**) Detailed view of (a) *(66°26'N, 73°03'W)* presenting a group of old and fresh thermo-karst lakes. The white thermo-karst lake to the east presents winter ice cover and is 2 km across (Images credits: ©Google earth 2015)

Fig. 5.42 Inland lakes are opened by the process of thermo-abrasion in permafrost caused by warm waters (from the ocean) and supported by some mechanical abrasion by wave processes in summer. The site is in the northeast section of Mackenzie Delta in northern Canada (approximately *70°11'N, 129°37'W)*. Scene is 7 km wide (Image credit: ©Google earth 2015)

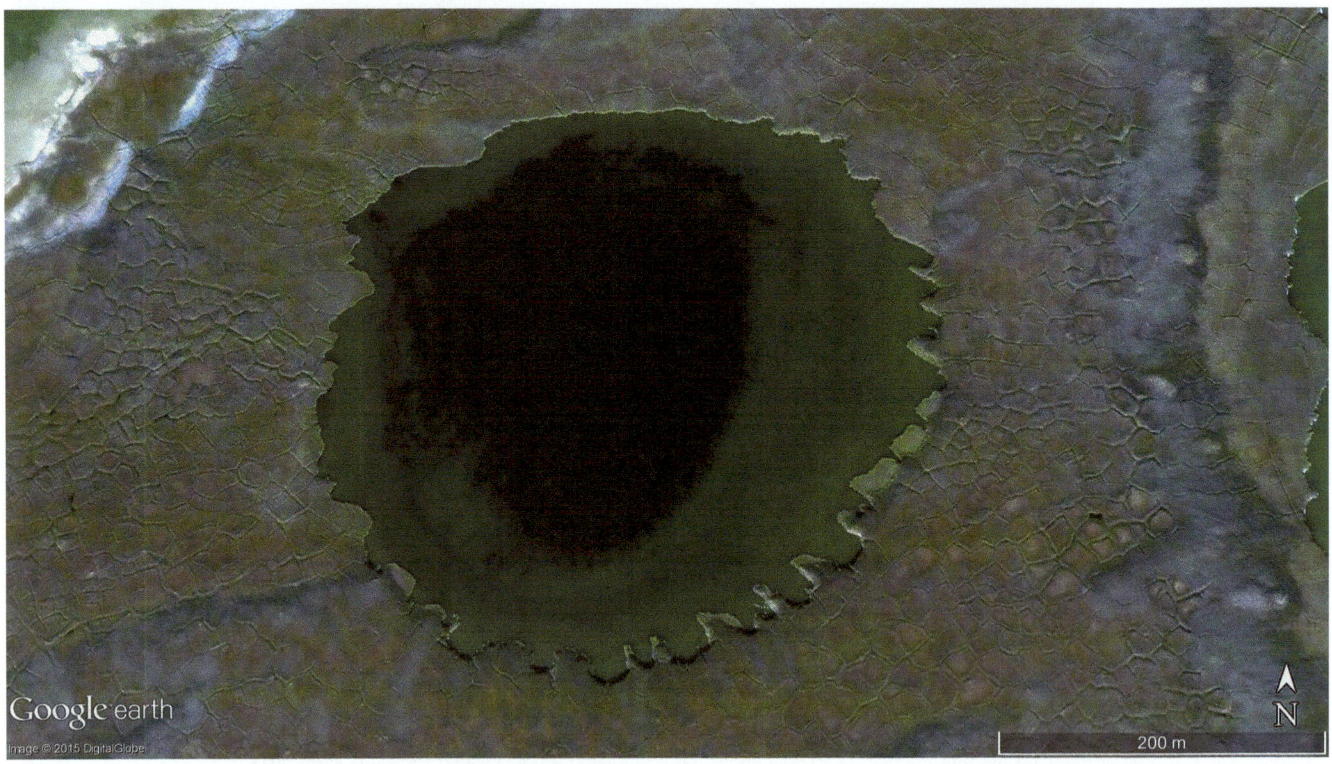

Fig. 5.43 An active thermo-karst expansion of a 420 m wide lake is set along the ice wedge patterns of its shorelines in northernmost mainland Canada *(70°08'02.29"N, 129°46'37.40"W)* (Image credit: ©Google earth 2015)

Fig. 5.44 Lakebed partly filled by polygonal structures, width of scene 1.6 km *(70°21'05.39"N, 155°11'26.11"W)* (Image credit: ©Google earth 2015)

Fig. 5.45 Lakes with irregular forms disturbing a pattern of ice wedges lay within lowlands situated in the northernmost Alaska (approximately *70°21'N, 151°W)*. Scene is 6 km wide (Image credit: ©Google earth 2015)

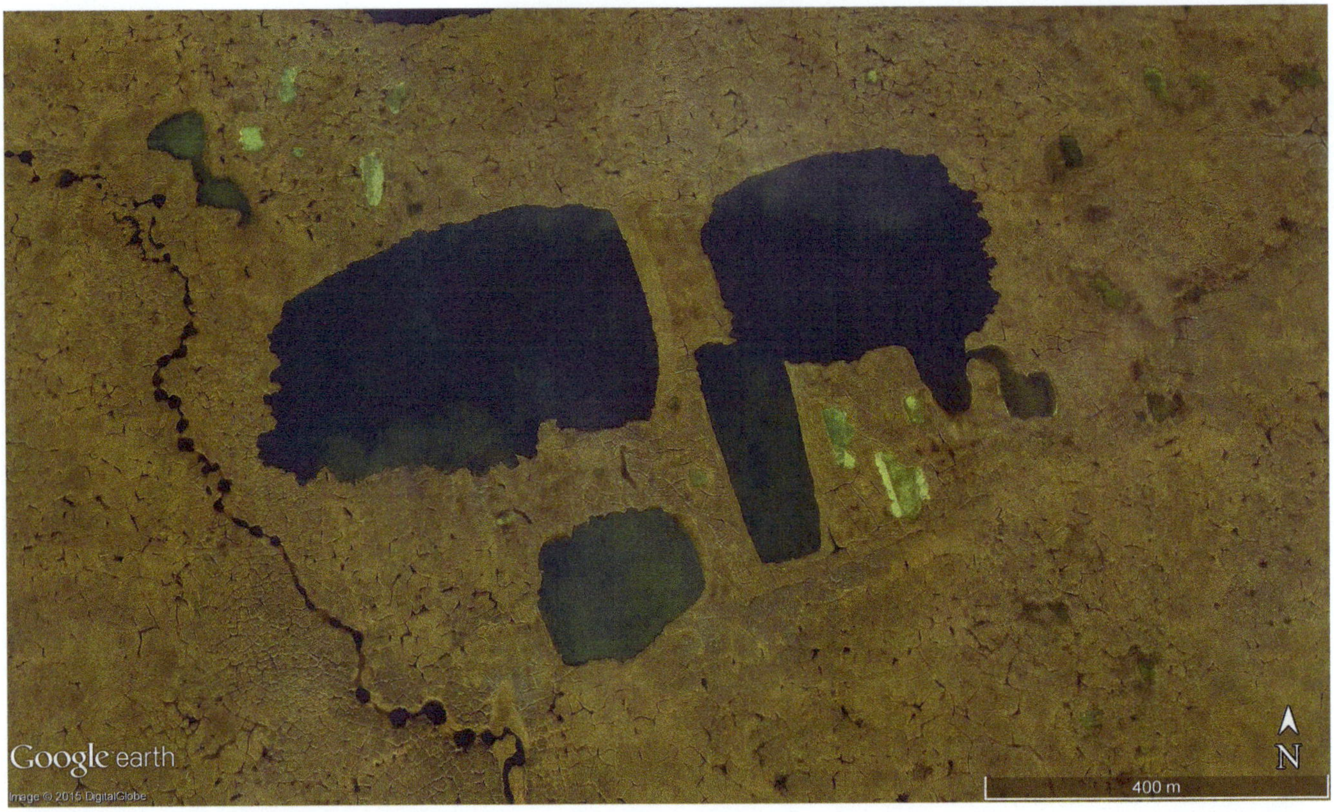

Fig. 5.46 Permafrost regions with an astonishing variety of lake forms are the results of rectangular or polygonal ice wedge patterning, the pressure of lake ice, from wind impacts and from active thermo-karst processes set in a flat landscape from Alaska *(70° 21'N, 150° 06'W)*. Scene is 1.5 km wide (Image credit: ©Google earth 2015)

Fig. 5.47 A variety of lake forms on permafrost in northern Siberia (approximately *71°20'04.76"N, 136°14'15.51"E)*. Scene is 5 km wide (Image credit: ©Google earth 2015)

Fig. 5.48 Young open-water lakes are the result of the permafrost thawing and the older lakes are evident by their contours, which have subsequently filled by polygonal ice wedge patterning and developing swampy ground in NE Alaska *(70°10'N, 151°08'W)*. Scene is 6 km wide (Image credit: ©Google earth 2015)

Fig. 5.49 A former 2.7 km long lake has breached and emptied as a result of the back-cutting processes of a small river in northern Siberia *(71°38'32.37"N, 132°21'06.68"E)* (Image credit: ©Google earth 2015)

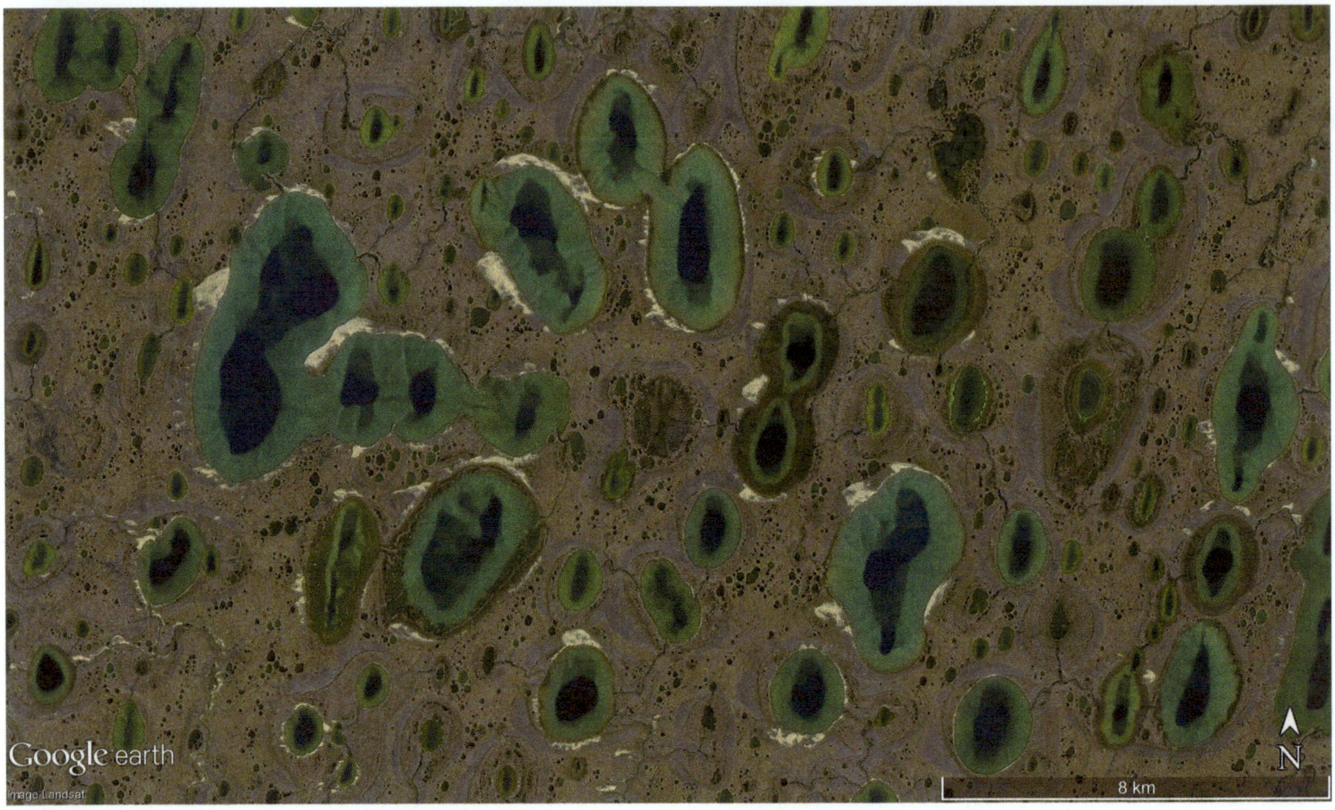

Fig. 5.50 The "cat-eye" lakes document a young phase of lake development through enlarging caused by thawing processes of the Lena Delta in northwest Siberia *(73°19'N, 124°24'E)*. The characteristic aspects of cat-eye lakes result from active expansion by water temperatures climbing significantly above zero in summer. Wide shallow fringes due to permafrost decay now surround the lakes. Scene is 26 km wide (Image credit: ©Google earth 2015)

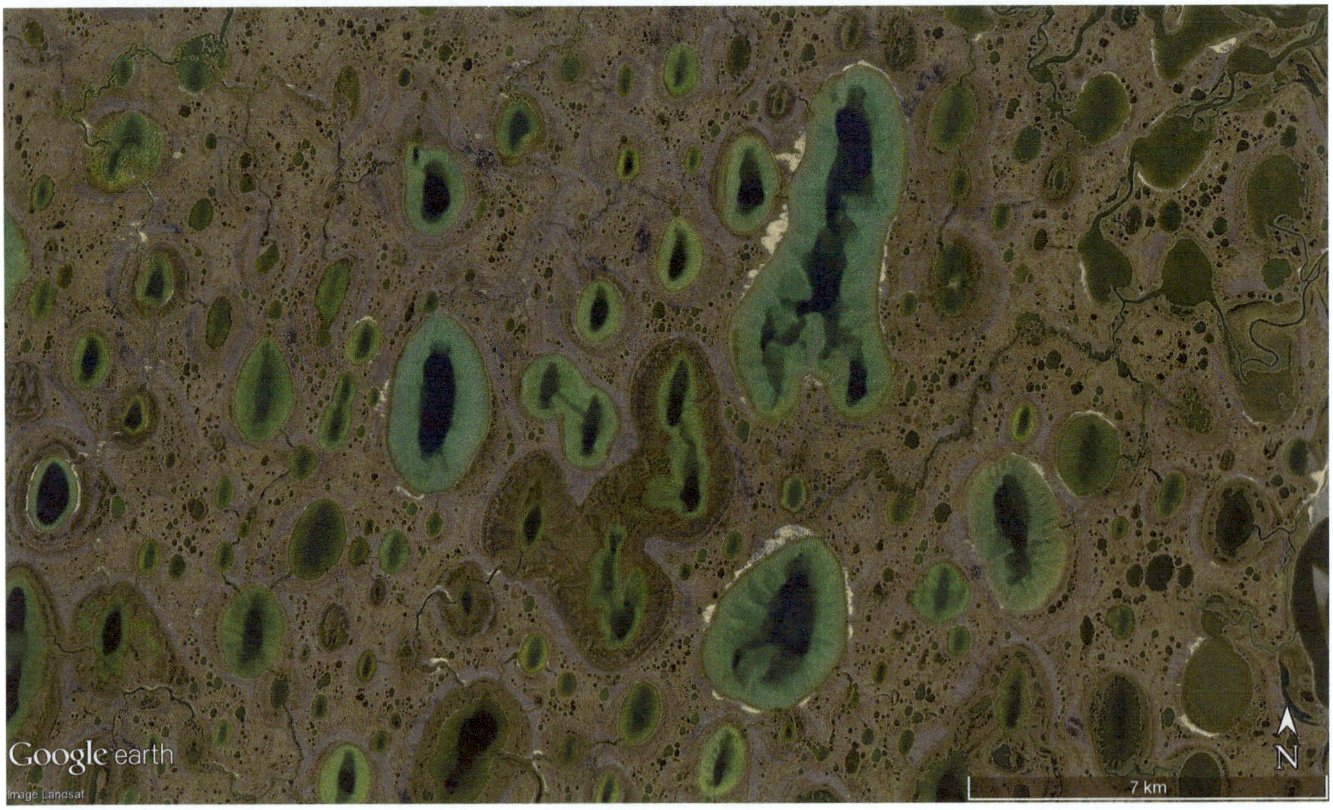

Fig. 5.51 A large group of cat-eye lakes from the Lena Delta in northern Siberia *(73° 19'N, 125° 52'E)*. Scene is 25 km wide (Image credit: ©Google earth 2015)

Fig. 5.52 (**a**) A series of string bogs in northwest Russia (*63°44'12.19"N and 37°51'31.09"E*). The tiny lakes are remnants of a former open-water lake and persist between the pressure ridges of peat formed by water expansion processes during winter freezing. Scene is 5 km wide. (**b**) A detailed view of a Siberian string bog (approximately *50°10'N, 80°W*) in a 2 km wide scene. (**c**) The string bog is a permafrost indicator in Tierra del Fuego, Argentina, (*54°45'29.50"S, 65°26'42.05"W*). Scene is 1.1 km wide (Images credits: ©Google earth 2015)

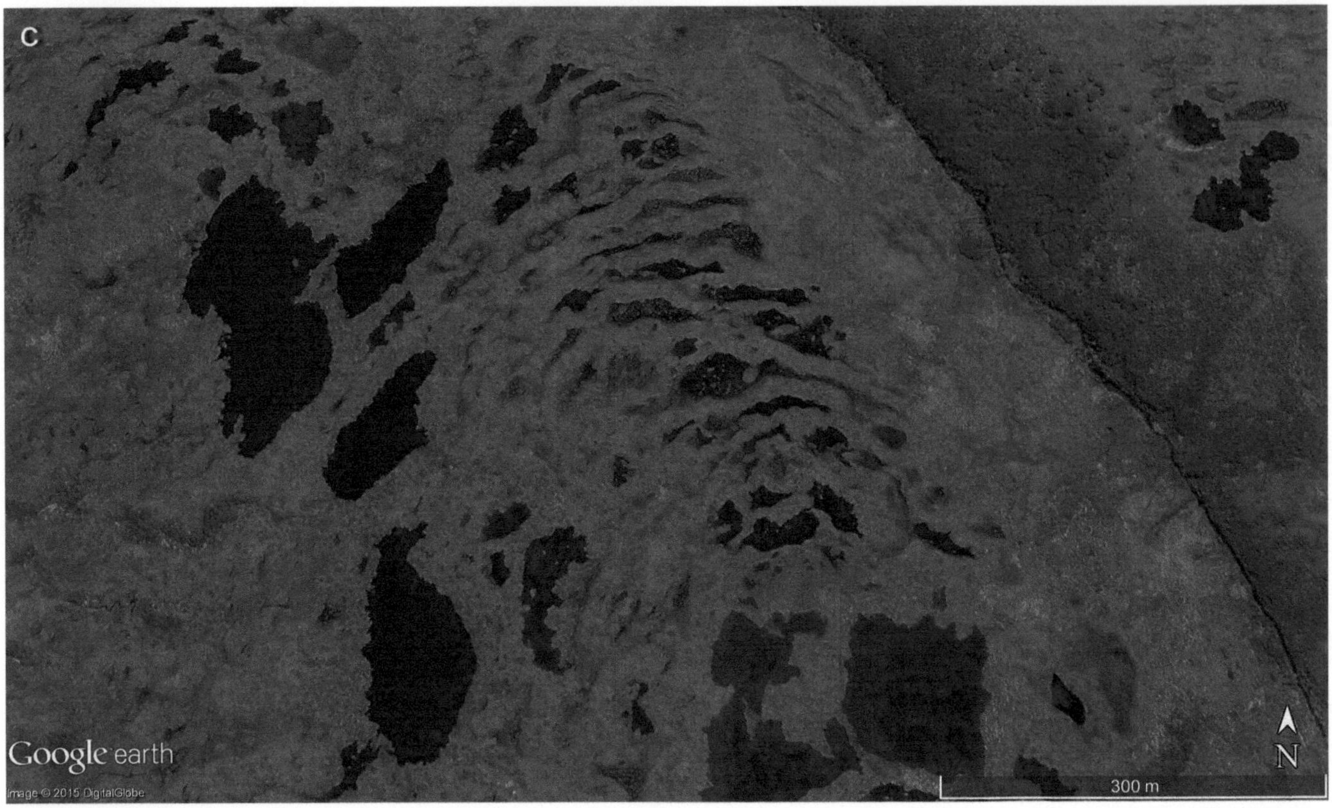

Fig. 5.52 (continued)

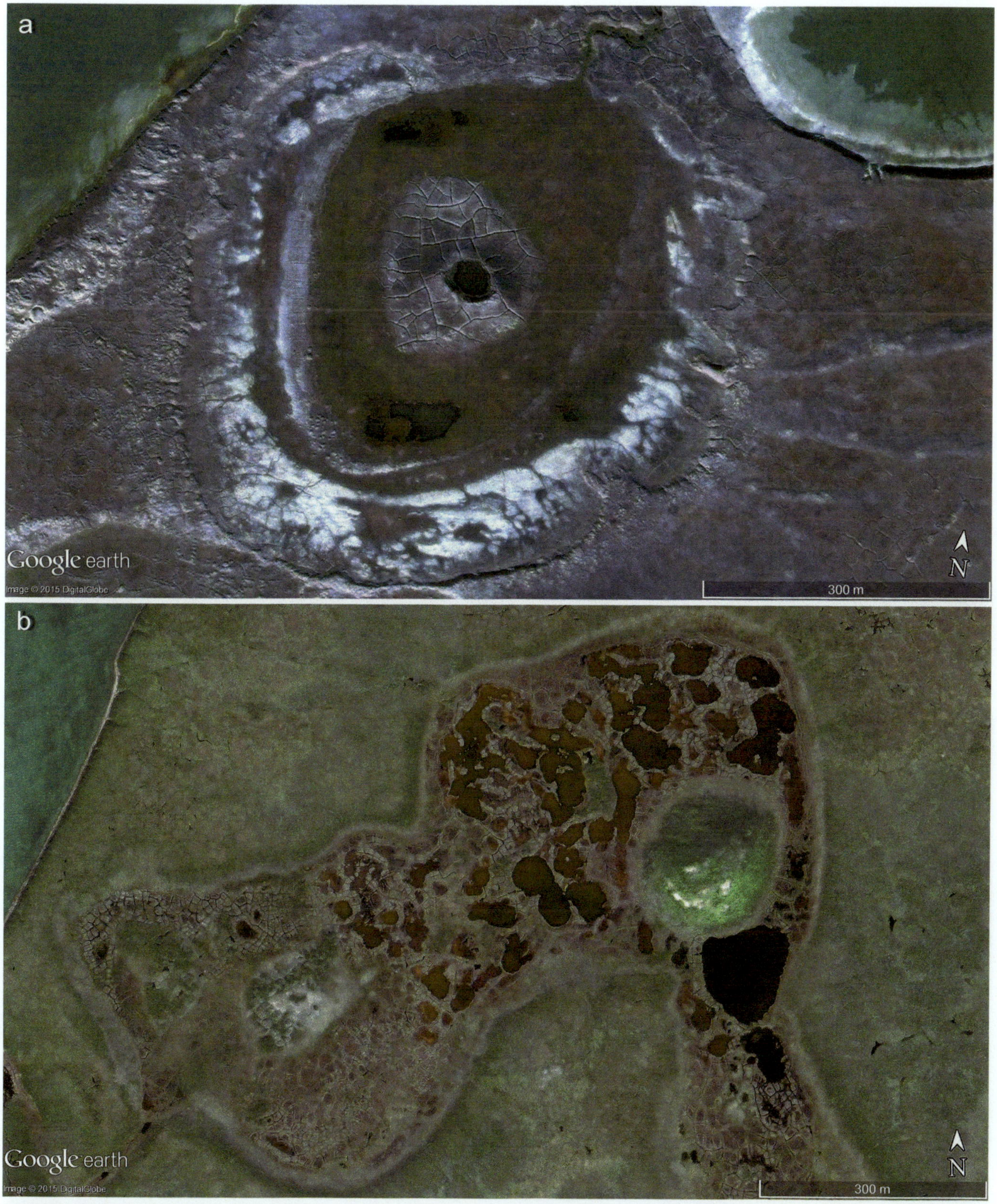

Fig. 5.53 (**a**) A dome-shaped mound consists of earth covered ice ("pingo" being 80 m across) within the Mackenzie River Delta *(70°10'51.20"N, 129°40'52.75"W)*. Pingos attract water through hygroscopic ice processes by pulling moisture from the environment to form central ice domes covered by soil and debris. Ring-like ditch then form around the pingo and are subsequently filled by ponds and polygonal patterns. (**b**) A pingo is in a state of decay and lies in the northern section of the Mackenzie Delta *(69°36'50.55"N, 132°09'51"W)*. Width of the scene is 1.3 km (Images credits: ©Google earth 2015)

Fig. 5.54 A detailed view of patterned ground exhibits minute dark lines in the network, fringed by two low ridges with dams and are ice wedges located in northernmost Alaska *(70°22'22.48"N, 151°01'41.51"W)*. Young thermo-karst lakes disturb the polygons. Scene is 1.5 km wide (Image credit: ©Google earth 2015, 8th December, 2012)

Fig. 5.55 The variance in vegetation cover of polygonal patterns including the various sizes of the honeycomb shapes is triggered by small differences in elevation of this swampy ground, set in northernmost Alaska *(70° 24'N, 151° 01'W)*. Scene is 850 m wide (Image credit: ©Google earth 2015)

Fig. 5.56 Recently formed thermo-karst lakes and relic riverbeds are disturbed by ice wedges in northern Siberia *(71°18'24.83"N, 136°22'51.04"E)*. Scene is 5 km wide (Image credit: ©Google earth 2015)

Fig. 5.57 Small ponds lie in patterned ground of ice wedges and are oriented along old river-bend sediment deposition in northern Siberia *(71°23'N, 136°25'E)*. Scene is 5 km wide (Image credit: ©Google earth 2015)

Fig. 5.58 In ice wedge polygons, pentagonal and hexagonal forms dominate and rectangular ones are rare. These rectangular polygons lie in a river bend of northern Russia *(71°24'N, 135°20'E)*. Scene is 2 km wide (Image credit: ©Google earth 2015)

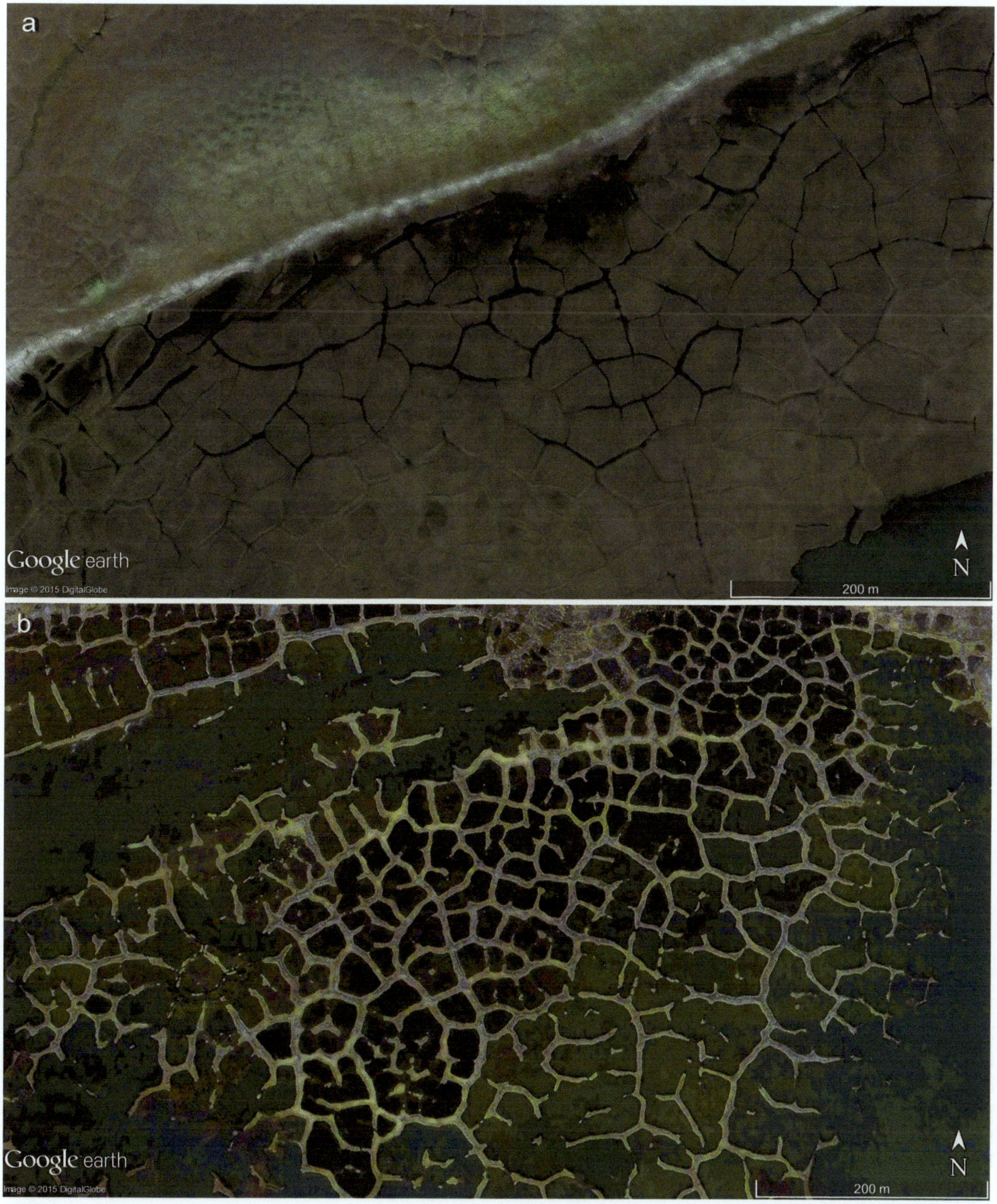

Fig. 5.59 (**a**) Initial phase in the development of ice wedge polygons0 presenting fresh frost cracks in a young fluvial sediment from northern Russia *(73°04'13.04"N, 120°02'10.82"E)*. Single polygons are more than 50 m across and the scene is 750 m wide. (**b**) Second phase in the development of an ice wedge polygon patterned ground presents ponds of irregular forms (rectangular to hexagonal) fringed by the pressure ridges of ice wedges in the permafrost region of northern Siberia, Russia Site *(71°28'40.68"N, 135°58'24.09"E)*. Scene is 1 km wide. (**c**) Third phase in the development of an ice wedge polygon landscape presents ridges on both sides of the central fractures, and the progression of water-containing ponds into swamps, from the coastal lowlands of northernmost Alaska *(70°23'00.16"N, 151°01'33.39"W)*. The remarkable difference in polygon size is the result of different pore sizes and the pore water found within the fluvial sediments. Scene is 1.1 km wide (Images credits: ©Google earth 2013)

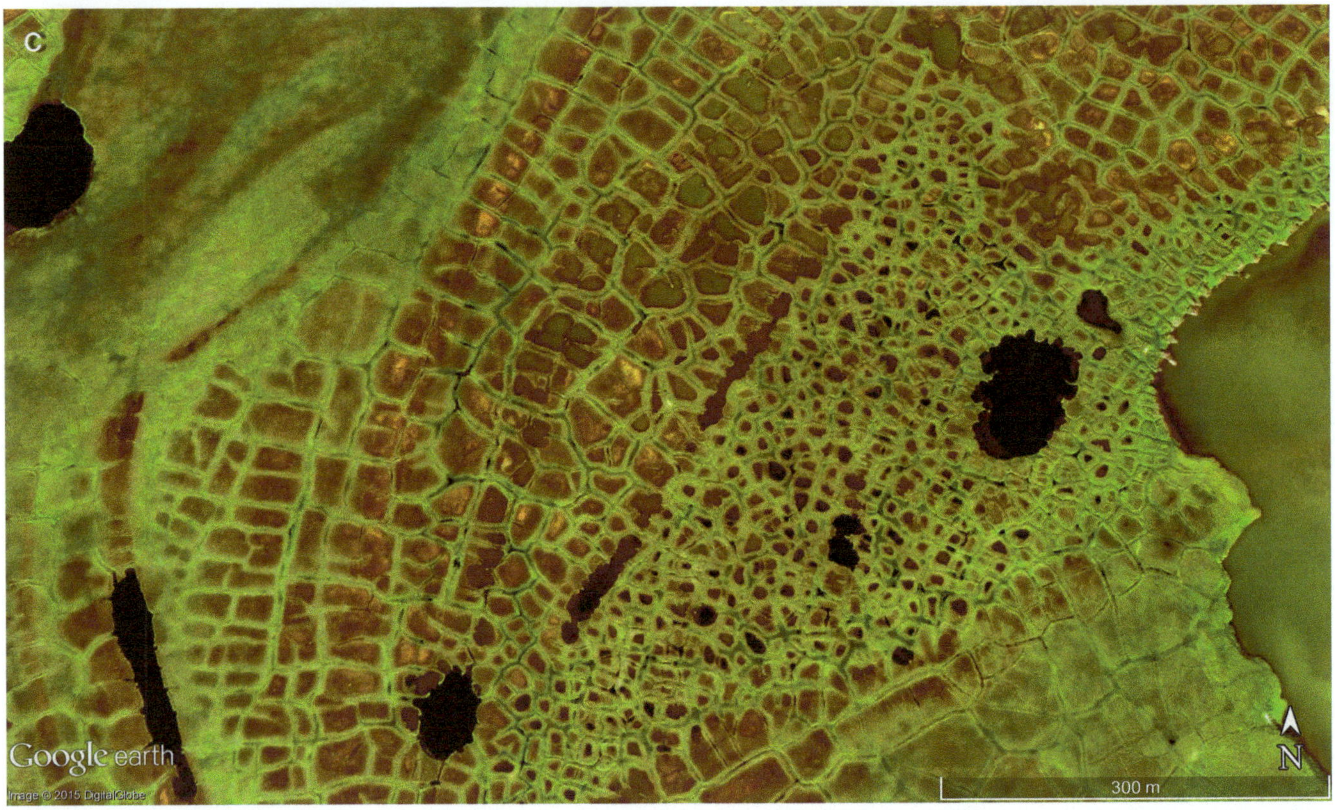

Fig. 5.59 (continued)

swamps when algae flourish during the long daylight of arctic summers.

- the third important process is permafrost decay defined as thermo-karst (Figs. 5.42, 5.43, 5.44, 5.45, 5.46 and 5.47) where open lakes of irregular size are occasionally modified to ice wedge polygons. Climate change and in particular the exceptional warming of high arctic latitudes has accelerated melting of the permafrost resulting in a deeper thawing where the number, size and depth of cryogenic lakes and ponds have increased significantly (although only very small sections are mapped thus far).

Generally, landscapes with patterned ground exhibit fascinating aspects not present in other parts of the planet and include forms such as: string bogs (Fig. 5.52a–c); ice hills covered by soil, vegetation or peat ("pingos", Fig. 5.53a, b); lobes of creeping soils ("solifluction lobes"); and stone networks and stone stripes. This group of patterned surfaces belong to the "periglacial" environment—being the periphery of a glacier, or cold climatic regions—the larger periglacial forms can be seen by modern satellite technique in lakes, ponds or swamps and are presented here in a series of image.

References

Allenby RJ (1989) Clustered, rectangular lakes of the Canadian Old Crow Basin. Tectonophysics 170(1–2):43–56

Arp CD, Jones BM (2008) Geography of Alaska lake districts. U.S. Geological Survey Scientific Investigations Report, 2008–5215:1–40

Burn CR (2002) Tundra lakes and permafrost, Richards Island, western Arctic coast, Canada. Can J Earth Sci 39:1281–1298

Burn CR (2005) Lake-bottom thermal regimes, western Arctic coast, Canada. Permafr Periglac Proc 16:355–367

Castiglia PJ, Fawcett PJ (2005) Large Holocene lakes and climate change in the Chihuahuan Desert. Geology 34(2):113–116

Chen XY, Bowler JM, Magee JW (1991) Aeolian landscapes in central Australia: gypsiferous and quartz dune environments from Lake Amadeus. Sedimentology 38:519–538

Chen J, Wang CY, Tan H et al (2012) New lakes in the Taklamakan desert. Geophys Res Lett. doi:10.1029/2012GL053985

Costelloe JF, Shields A, Grayson RB et al (2007) Determining loss characteristics of arid zone river waterbodies. River Res Appl 23(7):715–731

Costelloe JF, Irvine EC, Western AW et al (2009) Groundwater recharge and discharge dynamics in an arid-zone ephemeral lake system. Australian Limn Oceanogr 54(1):86–100

Dietze E, Wünnemann B, Hartmann K et al (2013) Early to mid-Holocene lake high-stand sediments at Lake Donggi Cona, north-eastern Tibetan Plateau, China. Quat Res 79(3):325–336

Drake NA, Blench RM, Armitage SJ et al (2010) Ancient watercourses and biogeography of the Sahara explain the peopling. Proc Nat Acad Sci Am 108(2):458–462

Duguay CR, Lafleur PM (2003) Determining depth and ice thickness of shallow sub-Arctic lakes using space-borne optical and SAR data. Int J Remote Sens 24(3):475–489

Felauer T, Schlütz F, Murad W et al (2012) Late Quaternary climate and landscape evolution in arid Central Asia: a multidisciplinary research of lake archive Bayan Tohomin Nuur, Gobi desert, southern Mongolia. J Asian Earth Sci 48:125–135

Frohn RC, Hinkel KM, Eisner WR (2005) Satellite remote sensing classification of thaw lakes and drained thaw lake basins on the North slope of Alaska. Remote Sens Envir 97:116–126

Gibbs L (2006) Valuing water: variability and the Lake Eyre Basin, central Australia. Austral Geogr 37(1):73–85

Grosswald MG, Hughes TJ, Lasca NP (1999) Oriented lake-and-ridge assemblages of the Arctic coastal plains: glacial landforms modified by thermokarst and solifluction. Polar Rec 35(194):215–230

Grunert J, Stolz C, Hempelmann N et al (2009) The evolution of small lake basins in the Gobi desert in Mongolia. Quat Sci 29:678–686

Habermehl MA (1982) Springs in the Great Artesian Basin, Australia – their origin and nature, Report no. 235. Bureau of Mineral Resources, Geology and Geophysics, Australia

Harris C (1989) Dalhousie Springs – an introduction. In: Zeidler W, Ponder WF (eds) Natural history of Dalhousie springs. South Australian Museum, Adelaide, pp 1–4

Henriksen M, Mangerud J, Matiouchkov A, Paus A, Svendsen JI (2003) Lake stratigraphy implies an 80,000 yr delayed melting of buried dead ice in northern Russia. J Quat Sci 18:663–679

Herr A, Smith T, Brake L (2009) Regional profile of the Lake Eyre Basin catchments. In: Measham TG, Brake L (eds) People, communities and economies of the Lake Eyre Basin, DKCRC research report 45. Desert Knowledge Cooperative Research Centre, Alice Springs, pp 41–88

Hinkel KM, Frohn RC, Nelson FE et al (2005) Morphometric and spatial analysis of thaw lakes and drained thaw lake basins in the western Arctic Coastal Plain, Alaska. Permafr Periglac Process 16:327–341

Jones B, Arp C, Hinkel K et al (2009) Arctic lake physical processes and regimes with implications for winter water availability and management in the National Petroleum Reserve Alaska. Environ Manag 43(6):1071–1084

Labrecque S, Lacelle D, Duguay CR et al (2009) Contemporary (1951–2001) evolution of lakes in the Old Crow Basin, northern Yukon, Canada: remote sensing, numerical modeling, and stable isotope analysis. Arctic 62(2):225–238

Lauriol B, Lacelle D, Labrecque S et al (2009) Holocene evolution of lakes in the Bluefish Basin, northern Yukon, Canada. Arctic 62(2):212–224

Loope WL, Fisher TG, Jol HM et al (2004) A Holocene history of dune-mediated landscape change along the southeastern shore of Lake Superior. Geomorphology 61:303–322

Mahowald NM, Bryant RG, Del Corral J et al (2003) Ephemeral lakes and desert dust sources. Geophys Res Lett 30(2):1–4

Marsh P, Russell M, Pohl S et al (2009) Changes in thaw lake drainage in the Western Canadian Arctic from 1950 to 2000. Hydrol Process 23:145–158

McMahon TA, Murphy RE, Peel MC et al (2008a) Understanding the surface hydrology of the Lake Eyre Basin: part 1 – rainfall. J Arid Envir 72:1853–1868

McMahon TA, Murphy RE, Peel MC et al (2008b) Understanding the surface hydrology of the Lake Eyre Basin: Part 2 – streamflow. J Arid Envir 72(10):1869–1886

Mischke S, Demske D, Wünnemann B et al (2005) Ground-water discharge to a Gobi desert lake during mid and late Holocene dry periods. Palaeogeo Palaeoclimatol Palaeoecol 225(1–4):157–172

Ponder WF (1986) Mound springs of the Great Artesian Basin. In: DeDeckker P, Williams WD (eds) Limnology in Australia. Springer, The Hague, pp 403–420

Reheis MC, Bright J, Lund SP et al (2012) A half-million-year record of paleoclimate from the Lake Manix Core, Mojave Desert, California. Palaeogeo Palaeoclimatol Palaeoecol 365–366:11–37

Sannel ABK, Brown IA (2010) High-resolution remote sensing identification of thermokarst lake dynamics in a subarctic peat plateau complex. Can J Remote Sens 36(1):S26–S40

Schwamborn G, Andreev AA, Rachold V et al (2002) Evolution of Lake Nikolay, Arga Island, western Lena River delta, during late Pleistocene and Holocene time. Polarforschung 70:69–82

Smith PC (1989) Hydrology. In: Zeitler W, Ponder WF (eds) Natural history of Dalhousie springs. South Australian Museum, Adelaide, pp 27–40

Smith LC, Sheng Y, MacDonald GM, Hinzman LD (2005) Disappearing Arctic lakes. Science 308:1429

Vincent WF, Hobbie JE (2000) Ecology of Arctic lakes and rivers. In: Nuttal M, Callaghan TV (eds) The Arctic: environment, people, and policy. Overseas Publishers Association, Amsterdam, pp 197–274

Zhang HC, Wünnemann B, Ma YZ et al (2002) Lake level and climate change between 40,000 and 18,000 14C years BP in Tengger Desert, NW China. Quat Res 58:62–72

Zhang HC, Peng JL, Ma YZ et al (2004) Late quaternary palaeolake levels in Tengger Desert, NW China. Palaeogeo Palaeoclimatol Palaeoecol 211(1–2):45–58

Epilogue: History and Future of Lakes in Times of Climate Change and Rising World Population

We hope this book has highlighted the dynamic and complex history of lake development. Existing freshwater lakes may diminish to saline lakes and finally become saltpans. Enormous late-glacial lakes dammed by Ice Age glaciers once belonged to the largest freshwater bodies that ever existed on Earth had disappeared over 10,000 years ago. In contrast, today thousands of new lakes occur where valley glaciers now recede as a result of climate change, nonetheless are small and will disappear with the glaciers they now feed. Along the margins of arctic oceans with the high capacity to hold warmth, thermo-abrasion unlocks freshwater lakes to flood wide coastal lowlands with salt water. With the present increasing temperatures, rising evaporation diminishes open-waters in arid zones and converts freshwater into saline lakes. In terms of glacial episodes, periods of increasing numbers of freshwater lakes toward the end of each Ice Age caused by exposing wide landscapes from former ice cover, has come to an end. The numbers and sizes of freshwater lakes are constantly reducing.

As discussed, by far the largest number of freshwater lakes exist in the northern high-latitudes—from western Alaska to eastern Canada, and from Finland to eastern Siberia—in these regions the population density is very low, industry poorly developed, and the flat relief prevents the use of water for energy. On the other hand, in regions with the highest concentrations of populations and highest rates of population growth—such as China, India, Pakistan, Bangladesh and Southeast Asia—many of the large basins contain saline waters and the best source for fresh water in these regions is from the snow and ice from the Himalayas, Hindukush and Karakorum mountains. The number and size of lakes, however is small, where most of the water arrives as meltwater directly from the snow and ice fields. As we have seen in several examples glaciers are receding worldwide, either by a natural process of warming, or accelerated through human-induced emissions of greenhouse gases. This results in the shift of snowlines to higher mountain belts, the reduction in size of existing large glaciers and the complete melting of our smaller glaciers. The consequence is the constant diminishing of water reservoirs for more than two billion people, who will increase to at least three billion by the year 2050.

Shortage of fresh water is now a problem in many other regions of the world including the less populated regions such as the desert belt of Africa, Australia and the Near East. The water budgets to most of the longer permanent rivers are over-used, and conflicts among nations sharing the same river systems have become more severe. Exploring new fresh water resources with better technology for drilling deep wells is also coming to an end, because most of the groundwater reservoirs were fed in previous more humid climate periods. Groundwater reservoirs are exhausted and cannot regenerate again in the short time span of human generations. Quantitative groundwater data demonstrates new problems developing with fresh water contamination; by the infiltration of saline waters from the ocean and evaporate deposits of the surface, and from the uncontrolled disposal of bacterial-waste water from human use, including agricultural poisons.

Every day we learn from media all of these issues are imminent and any plan to solve them is lacking. The consequence is, that mankind will face more conflicts over fresh water shortages in the foreseeable future.

A.M. Scheffers, D.H. Kelletat, *Lakes of the World with Google Earth*, Coastal Research Library 16,
DOI 10.1007/978-3-319-29617-3

Index

A.M. Scheffers, D.H. Kelletat, *Lakes of the World with Google Earth*, Coastal Research Library 16,
DOI 10.1007/978-3-319-29617-3

FSC
www.fsc.org
MIX
Papier | Fördert gute Waldnutzung
FSC® C083411

Kolibri 360 GmbH
Ferdinand-Jühlke-Straße 7
99095 Erfurt, Deutschland
produktsicherheit@kolibri360.de